Integrating Insurance and Risk Management for Hazardous Wastes

Sponsored by THE GENEVA ASSOCIATION

Integrating Insurance and Risk Management for Hazardous Wastes

edited by
Howard Kunreuther
M. V. Rajeev Gowda

Kluwer Academic Publishers
Boston/Dordrecht/London

Distributors for North America:
Kluwer Academic Publishers
101 Philip Drive
Assinippi Park
Norwell, Massachusetts 02061 USA

Distributors for all other countries:
Kluwer Academic Publishers Group
Distribution Centre
Post Office Box 322
3300 AH Dordrecht, THE NETHERLANDS

Library of Congress Cataloging-in-Publication Data
Integrating insurance and risk management for hazardous wastes/
[edited by] Howard C. Kunreuther and M. V. Rajeev Gowda.
    p. cm.
Papers presented at a conference held at the Wharton School.
Includes index.
    ISBN-13: 978-94-010-7475-9   e-ISBN-13: 978-94-009-2177-1
    DOI: 10.1007/978-94-009-2177-1
    1. Risk (Insurance) — Congresses. 2. Risk assessment — Congresses. 3. Hazardous waste management — Congresses. I. Kunreuther, Howard. II. Rajeev Gowda, M. V. (Mothakapalli Venkatappa)
H88054.5.I58 1989
628.4 — dc19   89-2668
                                        CIP

Copyright © 1990 by Kluwer Academic Publishers
Softcover reprint of the hardcover 1st edition 1990

All rights reserved. No part of this publication may be reproduced,
stored in a retrieval system or transmitted in any form or by any
means, mechanical, photocopying, recording, or otherwise, without
the prior written permission of the publisher, Kluwer Academic
Publishers, 101 Philip Drive, Assinippi Park, Norwell, Massachusetts
02061.

Contents

Conference Participants	ix
Preface	xv
Introduction Howard Kunreuther and Rajeev Gowda	1
I Cleaning Up Inactive Waste Sites	13
1 Risk Assessment Issues Associated with Cleaning Up Inactive Hazardous Waste Sites Elizabeth Anderson, Paul Chrostowski, and Judy Vreeland	15
Discussants: Ruth Patrick and James Wilson	
2 Insurability Issues Associated with Cleaning Up Inactive Hazardous Waste Sites Leslie Cheek, III	41
Discussants: Kenneth Abraham and Cornelius Smith	
3 Risk Management Issues Associated with Cleaning Up Inactive Hazardous Waste Sites James Seif and Thomas Voltaggio	83
Discussants: Jack Schramm and Paul Portney	
II Managing Existing Waste Sites	103
4 Risk Assessment Issues Associated with Managing Existing Hazardous Waste Sites Peter Schroeder	105
Discussants: John Amore and Martin Katzman	

vi CONTENTS

5
Insurability Issues Associated with Managing Existing
Hazardous Waste Sites 131
Baruch Berliner and Juerg Spuehler

Discussant: Dennis Connolly

6
Risk Management Issues Associated with Managing Existing
Hazardous Waste Sites 169
Richard Fortuna

Discussants: Nicholas Ashford and Marcia Williams

III
The Small Group Discussions 191

7
A Risk Communication Perspective on an Integrated Waste
Management Strategy 195
Paul Slovic

Summary: Paul Slovic
Chairperson's Remarks: Frederick Allen

8
An Environmental Perspective on an Integrated Waste
Management Strategy 217
Sheldon Novick

Summary: Sheldon Novick
Chairperson's Remarks: Marcia Williams

9
An Industrial Perspective on an Integrated Waste
Management Strategy 245
Isadore Rosenthal and Lynn Johnson

Summary: Lynn Johnson
Chairperson's Remarks: Paul Arbesman

10
An Insurance Perspective on an Integrated Waste
Management Strategy 271
Neil Doherty, Paul Kleindorfer, and Howard Kunreuther

Summary: Neil Doherty
Chairperson's Remarks: John Morrison

CONTENTS   vii

11
A Legal Perspective on an Integrated Waste Management Strategy   303
Michael Baram

Summary: Michael Baram
Chairperson's Remarks: Dennis Connolly

12
A Legislative Perspective on an Integrated Waste
Management Strategy   317
Jack Clough

Summary: Lee Fuller (representing Jack Clough)
Chairperson's Remarks: Lee Fuller

IV
Toward an Integrated Waste Management Strategy   331

13
A Synthesis of the Conference   333
F. Henry Habicht II

14
Conclusion   343
Howard Kunreuther and Rajeev Gowda

Index   351

Conference Participants

Kenneth Abraham
School of Law
University of Virginia
Charlottesville, VA 22901

Frederick Allen
U. S. Environmental Protection Agency
401 M Street, SW
PM 221
Washington, DC 20460

John Amore
American Home Assurance Co.
70 Pine Street
New York, NY 10270

G. Anandalingam
Systems Department
University of Pennsylvania
Philadelphia, PA 19104

Elizabeth Anderson
ICF Clement
9300 Lee Highway
Fairfax, VA 22031

Paul Arbesman
Allied-Signal
P. O. Box 3000R
Morristown, NJ 07960

Nicholas Ashford
Massachusetts Institute of Technology
Center for Technology, Policy, and
    Industrial Development
Building E 40239
Cambridge, MA 02139

Warren Azano
Aetna Insurance Company
151 Farmington Avenue
Hartford, CT 06156

Michael Baram
Bracken & Baram
33 Mount Vernon Street
Boston, MA 02108

Baruch Berliner
Swiss Reinsurance Company
Mythenquai 50/60
CH-8022 Zurich
Switzerland

CONFERENCE PARTICIPANTS

William Black
Law Engineering
1000 Abernathy
Atlanta, GA 30356

Edward Bowman
Reginald H. Jones Center
2041 SH-DH/6370
University of Pennsylvania
Philadelphia, PA 19104

Gene Capaldi
ARCO
3801 W. Chester Pike
Newtown Square, PA 19073

Leslie Cheek, III
Crum & Forster Insurance
1025 Connecticut Avenue, NW
Washington, DC 20036

Paul Chrostowski
ICF Clement
9300 Lee Highway
Fairfax, VA 22031

Joseph Chu
Environmental Activities Staff
GM Technical Center
Warren, MI 48090-9015

Thomas Clarke
Marsh & McLennan
1221 Avenue of the Americas
New York, NY 10020

John Clough
Committee on Energy and Commerce
U. S. House of Representatives
2125 Rayburn House Office Building
Washington, DC 20515

Dennis Connolly
Johnson & Higgins
125 Broad Street
New York, NY 10004

Patricia Danzon
Department of Health Care Systems
208 CPC/6218
University of Pennsylvania
Philadelphia, PA 19104

Bruce Diamond
U.S. Environmental Protection Agency
401 M Street, SW
EN 336
Washington, DC 20460

Neil Doherty
Insurance Department
305 CPC/6218
University of Pennsylvania
Philadelphia, PA 19104

William N. Edwards
American Reinsurance
1 Liberty Plaza
91 Liberty Street
New York, NY 10006

Harold Elkin
Sun Oil
100 Matsonford Road
Radnor, PA 19087

Linda Erdreich
Clement Associates, Inc.
Metro Park III
399 Thornall Street
Edison, NJ 08837

Gerald Faulhaber
Public Policy and Management
  Department
3020 SH-DH/6372
University of Pennsylvania
Philadelphia, PA 19104

Robert Field
CIGNA
8 Home Office
1600 Arch Street
Philadelphia, PA 19103

Richard Fortuna
Hazardous Waste Treatment Council
1440 New York Avenue, NW
Suite 310
Washington, DC 20005

CONFERENCE PARTICIPANTS

xi

Hansjorg Fricker
Swiss Reinsurance Company
Mythenquai 50/60
CH-8022 Zurich
Switzerland

Lee Fuller
Charles E. Walker Associates
1730 Pennsylvania Avenue, NW
Washington, DC 20006

Orio Giarini
Geneva Association
18, chemin Rieu
CH-1208 Geneve
Switzerland

Rajeev Gowda
Risk and Decision Processes Center
The Wharton School
University of Pennsylvania
Philadelphia, PA 19104

Thomas Graves
National Paint & Coatings Assoc.
1500 Rhode Island Avenue, NW
Washington, DC 20005

F. Henry Habicht, II
Environmental Protection Agency
401 M Street, SW
Washington, DC 20460

David Havanich
National Corporate Cash Management
   Associaton
52 Churchill Hill Road
Newton, CT 06470

George Henderson
CIGNA
1600 Arch Street
Philadelphia, PA 19103

Elizabeth Hoffman
Department of Economics
University of Wyoming
Laramie, WY 82071

Arthur K. Ingberman
9704 Mossy Stone Court
Vienna, VA 22180

Daniel E. Ingberman
Public Policy & Management Department
3027 SH-DH/6372
University of Pennsylvania
Philadelphia, PA 19104

James Janis
ICF
9300 Lee Highway
Fairfax, VA 22031

Lynn Johnson
Rohm & Haas
Independence Mall West
Philadelphia, Pa 19105

Martin Katzman†
Energy and Economic Analysis Section
Oak Ridge National Laboratory
P. O. Box 2008
Oak Ridge, TN 37831-6205

James L. Kimble
American Insurance Association
1130 Connecticut Avenue, NW
Washington, DC 20036

Paul Kleindorfer
Decision Sciences Department
1150 SH-DH/6366
University of Pennsylvania
Philadelphia, PA 19104

William Kronenberg
Environmental Compliance Services, Inc.
721 East Lancaster Avenue
Dowingtown, PA 19335

Howard Kunreuther
Risk and Decision Processes Center
The Wharton School
University of Pennsylvania
Philadelphia, PA 19104

CONFERENCE PARTICIPANTS

Lester Lave
Graduate School of Industrial
  Administration
Schenley Park
Carnegie Mellon University
Pittsburgh, PA 15213

Peter Linneman
Finance Department
2258 SH-DH/6367
University of Pennsylvania
Philadelphia, PA 19104

John Lower
American Risk Management
Spectrum Analysis
7724 Eagle Lane
Spring, TX 77379

Randy Lyon
General Accounting Office
401 M Street, SW
Washington, DC 20460

Jeff Malkoeski
CIGNA
1600 Arch Street
Philadelphia, PA 19103

Ernest Merklein
Alexander and Alexander
1700 Buckner Square
Shrevesport, LA 71101

Troy Jeannine Meyer
Occidental Petroleum
10889 Wilshire Drive
Los Angeles, CA 90024

Sharon Moran
Massachusetts Institute of Technology
Center for Technology, Policy, and
  Industrial Development
Cambridge, MA 02139

John Morrison
CIGNA
1600 Arch Street
Philadelphia, PA 19103

John Mulroney
Rohm & Haas
Independence Mall West
Philadelphia, PA 19105

Sheldon Novick
S. Stratford, VT 05070

Timothy O'Leary
Chemical Manufacturers Association
2501 M Street, NW
Washington, DC 20037

Robert Owens
Rohm & Haas
Independence Mall West
Philadelphia, PA 19105

Ruth Patrick
Academy of Natural Sciences
19th and B F Parkway
Philadelphia, PA 19103

Mark Pauly
Leonard Davis Institute
210 CPC/6218
University of Pennsylvania
Philadelphia, PA 19104

Robert Pollak
520 McNeil/6297
University of Pennsylvania
Philadelphia, PA 19104

Paul Portney
Resources for the Future
1616 P Street, NW
Washington, DC 20036

Jack Pulley
Dow Chemical Corporation
Administrative Law CO1242
Midland, MI 48686-0994

Bernard Reilly
Du Pont - Legal Dept.
1007 Market Street
Wilmington, DE 19801

CONFERENCE PARTICIPANTS

xiii

Isadore Rosenthal
Rohm & Haas
Safety, Health & Environmental Affairs
Box 584
Bristol, PA 19007

Don Schaefer
Aetna Insurance Company
151 Farmington Avenue
Hartford, CT 06156

Margaret Schneider
U.S. Environmental Protection Agency
401 M Street, SW
WH-563
Washington, DC 20460

Jack Schramm
Waste Management, Inc.
Suite 800
1155 Connecticut Avenue, NW
Washington, DC 20036

Peter Schroeder
Zurich Insurance
Mythenquai 10
8002 Zurich
Switzerland

James Seif
Dechert, Price and Rhoads
3450 Center Square W
1500 Market Street
Philadelphia, PA 19102

Judith Selvidge
Hudson Strategy Group
1211 Avenue of the Americas
43rd Floor
New York, NY 10020

Paul Slovic
Decision Research
1201 Oak Street
Eugene, OR 97401

Cornelius Smith
Union Carbide
1100 15th Street, NW
Washington, DC 20005

Sherry Sterling
U.S. Environmental Protection Agency
401 M Street, SW
WH-527
Washington, DC 20460

Paul Stolley
Epidemiology Department
229 L NEB/6095
University of Pennsylvania
Philadelphia, PA 19104

Robert Stone
Massachusetts Institute of Technology
Center for Technology, Policy, and
    Industrial Development
Cambridge, MA 02139

Robert Toth
Monsanto Company
800 N. Lindbergh Blvd.
St. Louis, MO 63167

Juerg Spuehler
Swiss Reinsurance Company
Mythenquai 50/60
CH-8022 Zurich
Switzerland

Thomas Voltaggio
U. S. Environmental Protection Agency
401 M Street, SW
Washington, DC 20460

Judy Vreeland
ICF Clement
9300 Lee Highway
Fairfax, VA 22031

William Wallace
CH2M Hill
P.O. Box 22508
Denver, CO 80222

Marcia Williams
Browning Ferris Industries
1150 Connecticut Avenue, NW
Suite 500
Washington, DC 20036

xiv CONFERENCE PARTICIPANTS

Michael Willingham
Institute of International Education
1400 K Street, NW
Suite 650
Washington, DC 20005

James Wilson
Environmental Policy Staff
Monsanto Company
800 N. Lindbergh Boulevard
St. Louis, MO 63167

Dennis Yao
Public Policy & Management Department
3018 SH-DH/6372
University of Pennsylvania
Philadelphia, PA 19104

Student Rapporteurs

Jan Ambrose
Gilles Bernier
James Boyd
Ann Butler
Rebecca Coles
Nancy Dong
Douglas Easterling
Rajeev Gowda
Mark Hanna
Ajai Rastogi
Kenneth Richards
Kim Staking

PREFACE

A challenge facing society today is how to develop a meaningful strategy for integrated hazardous waste management. Meeting this challenge was the principal motivation for the conference on "Risk Assessment and Risk Management Strategies for Hazardous Waste Storage and Disposal Problems," held at the Wharton School of the University of Pennsylvania on May 18–19, 1988. The conference brought together representatives from the major interested parties — environmentalists, government, insurance, law, manufacturing, and the university community — who have been concerned with the waste management process.

The conference was the third cosponsored by the Wharton Center for Risk and Decision Processes addressing the knotty problem of hazardous waste. The first, held at the International Institute for Applied Systems Analysis in 1985, examined the transportation, storage, and disposal of hazardous materials. It suggested steps that industry, insurers, and government agencies could take to improve the safety and efficiency with which hazardous materials are produced and controlled in industrialized societies. Specifically, it focused on the risk-management tools of insurance, compensation, and regulation.

xv

xvi   PREFACE

The second conference, held at the Wharton School, University of Pennsylvania in 1986, concentrated on the role of insurance and compensation in environmental pollution problems. It characterized a set of problems related to the environmental pollution liability insurance crisis as presented by key interested parties and proposed a set of research needs for providing a sound basis for constructing socially appropriate measures to deal with the problem.

These two meetings put key issues and points of difference on the table. This conference was a search for solutions. It sought the elements of a waste management strategy which makes clear the costs and benefits of different actions to society as a whole, yet which facilitates some compromises among the interested parties.

In this book we bring together the papers prepared for and presented at the conference, along with discussant comments and reports of the small group discussions. The first day of the conference was devoted to the issues of Risk Assessment, Insurability, and Risk Management, for the cleanup of inactive hazardous waste sites, and for managing existing facilities. In the evening participants representing various interested parties met in small groups, each exploring one of the following perspectives: Communications, Environmental, Legal, Legislative, Industrial, and Insurance. The second morning was devoted to summaries and an open discussion of key points from these small group sessions. The structure of the conference thus emphasized the multifaceted nature of hazardous waste management and the need for concerted collaborative action to tackle the past, present, and future.

Our special thanks go to all the participants at the conference, especially to the paper writers, discussants and chairpersons of the small group sessions. Hank Habicht deserves special mention as he was actively involved in the planning of the conference and provided a synthesis at the end of the meeting of the key points that emerged. We are grateful to Jack Mulroney, former Chairman of our Advisory Committee, for helping to plan the conference and for guiding the session on recommendations on the final morning. We greatly appreciate the efforts of Linda Schaefer, Wharton Center Administrative Assistant, in preparing and organizing the workshop. Our thanks also go to the student rapporteurs who took detailed notes of the lively interchanges at the small group sessions. We thank Seema Shrikhande for her key role in editing and preparing the manuscript. We are very grateful to Zachary Rolnik of Kluwer Academic Publishers for playing a lead role in bringing the proceedings of this conference to print. Finally we want to express our gratitude to the Geneva Association

PREFACE xvii

and the Swiss Reinsurance Company who provided financial support for the conference.

The Wharton Center for Risk and Decision Processes hopes that this book provides useful input into the development of a more rational policy for managing hazardous waste. We look forward to taking an active part in the development of an integrated waste management strategy.

Howard Kunreuther
Rajeev Gowda
August 1989

INTRODUCTION
Howard Kunreuther and Rajeev Gowda

I.1 Extent of the Problem

Our industrialized society generates an enormous amount of waste. Annually the United States alone produces about 25 tons of waste per resident, adding up to a staggering total of six billion tons.[1] Only 3% of this waste is produced directly by households. The rest is generated by industry through its manufacturing and distribution of goods for human use, ranging from tubes of toothpaste to giant nuclear missiles (Conservation Foundation, 1987).

Of these wastes, over 254 million tons (more than one ton per person) are classified as hazardous to human health and the environment. This total could be a substantial underestimate of the true extent of the hazardous wastes pervading our environment. Chemicals compose a significant portion of these wastes, but of the 65,000 commercial chemicals identified as being currently marketed, there is little or no data on their potential to cause cancer, birth defects, or chronic diseases. A recent National Academy of Sciences report concluded that sufficient information to complete a health hazard assessment is available for less than 2% of commercial chemicals. Exposure data are quite poor, and most testing for toxic effects has focused on cancer rather than reproductive and neurological disorders or on other chronic problems. Almost nothing is known about the synergistic effects of exposure to combinations of chemicals (An Environmental Agenda for the Future, Island Press, Washington D.C., 1985). To illustrate the state of our knowledge, dioxin, one of the few chemicals that has been investigated extensively, has been shown to be a carcinogen if given in large enough doses to animals, but to date has produced only skin rashes in humans. (Anderson et al., 1988).

Tackling the potential problems of hazardous wastes then, will require action in the face of uncertainty over the nature and extent of the risks involved. Sound management of these wastes calls for a two-pronged approach: (1) the introduction of technological changes to reduce the amount of waste produced (source reduction); and (2) effective management of even the reduced amount of wastes that industry generates.

Until very recently, most potentially hazardous wastes were disposed of on land. A Congressional Budget Office study states that two-thirds of its estimate of 266 million tons of industrial hazardous wastes generated in 1983 were managed on land: 25% went into deep wells, another 23% went into landfills, and 19% into surface impoundments. Only 1% of the wastes were incinerated (Conservation Foundation, 1987).

Improper land disposal is fraught with danger. Landfills often leak toxic substances into the air, earth, and groundwater, leading to severe effects on health and enormous costs of cleanup. Hence, there has been a movement away from land disposal towards other technologies, notably incineration, for dealing with hazardous wastes.

But incineration, is not a risk-free alternative. It may release dangerous metals and chemical compounds to the atmosphere. Incomplete incineration may create new (and possibly more dangerous) compounds not part of the original waste. The extent of the risks from incineration is determined by the composition of the materials being burned, the temperature of incineration, how well releases to air and land are controlled, and the care taken in operating and maintaining the facility (Conservation Foundation, 1987).

Today, the public demands that risks be reduced — that past mistakes be corrected and that wastes be handled safely in the future. Dangerous landfills must be cleaned up. Industry is therefore simultaneously trying to develop new production processes for reducing the amount of potentially hazardous waste, and to utilize incineration and other new technologies to dispose of wastes in a safer manner.

There is thus a link between an awareness of past mistakes, with their consequent enormous remedial costs, and the improvement of current

INTRODUCTION 3

production, transport, and disposal processes for dealing with hazardous wastes. Cleanup activities and waste management practices must be integrated so that human health and the environment can be safeguarded.

Below we delineate the factors that need to be considered in developing a more integrated approach to the hazardous waste program. Specific attention is given to the factors leading to environmental pollution and to the efforts taken to tackle such damage. The pitfalls and problems of those efforts are the subject of intense debate, as are the attempts to overcome them. We conclude by summarizing the key themes and questions that emerged from the one-and-a-half day conference at the Wharton School entitled *Risk Assessment and Risk Management Strategies for Hazardous Waste Storage and Disposal Problems*.

I.2 The Nature of the Environmental Pollution Problem

There are a set of key economic, social, and political issues that need to be considered in addressing the problems of pollution. The economic analysis of the situation has generally assumed that the economy is an open system: resources can be taken from the environment, transformed, and discarded, and any resulting changes in the environment can be ignored (Enthoven and Freeman, 1973). The environment has been normally used free of charge, unlike the physical inputs of a production process, (such as labor and raw materials) which are not assumed to be free. The natural environment has long been treated as a public good — a good that belongs to all and that is not diminished if used by any one person. Yet we know that when industry pollutes the air with smoke or the water with waste, it is consuming limited resources (i.e., clean air and water). This side effect is called an externality or a spillover of the industry's activity. When externalities arise, the system of free-market competition fails to provide an efficient allocation of resources, and too much pollution occurs.

The problem of externalities involves the absence of well-defined property rights. Though it is abundantly clear that we do not have unlimited quantities of clean air and water, we have failed to use the price system to apportion these scarce resources properly. This suggests an obvious instrument for policymakers. They can require the generators of (detrimental) externalities to bear the cost — make polluters pay — thereby discouraging the flow of undesirable spillovers. This approach is favored by many economists (Baumol and Oates, 1988).

But what price should be paid for this environmental degradation? There is no easy answer to this question. It depends on how many people are affected, on what the consequences are to both current and future generations residing in the affected areas, and to the environment per se.

We shall trace the development of policy responses to hazardous waste by examining the context in which this problem is tackled. Figure I-1 outlines the different elements that bear on an integrated hazardous waste management program. There are a set of exogenous factors — social, legislative, judicial — that influence the production process, transport, and disposal of wastes. At the same time, the competitive pressures that an industry faces play a role in this process. The bottom portion of the figure indicates that hazardous waste affects a number of different interested parties. Each of these groups has their own objectives, information base, and strategy for dealing with the problem. Their actions eventually feed back to the exogenous factors depicted above.

I.2.1 Nature of Exogenous Factors

The current programs for managing hazardous waste are influenced by different institutional features. Let us look at each of the features in turn.

I.2.1.1 Social. The public was first made aware that chemicals might harm the environment in Rachel Carson's 1962 book, *Silent Spring*. For millions of readers, Carson painted a picture of the sudden disappearance of living

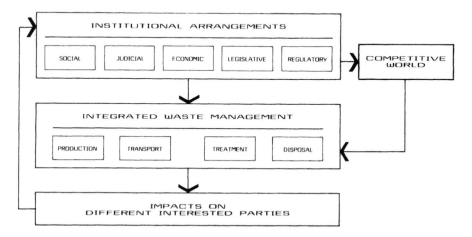

Figure I-1

INTRODUCTION 5

creatures on earth due to the poisoning of the environment through chemicals emitted into the air and water.

In the 1970s, two salient events helped trigger public concern about the potential dangers of waste to health:

(1) An explosion at a Hoffman-La Roche subsidiary in 1976 exposed the Italian town of Seveso to a toxic cloud of dioxin causing severe illnesses to animals and raising serious concerns about its effect on human lives.
(2) A leakage of 82 chemicals, 11 of them carcinogens, at a waste burial site in Love Canal, New York, led to national press coverage. Many of the families in this small community near Niagara Falls had to be relocated because some scientists claimed that exposure to the leakage might cause cancer.

These two catastrophic events and a number of smaller but similar pollution incidents have made the public fear for both personal health and the health of future generations. Despite reassurances by experts that new storage and disposal methods are fail-safe, the public has yet to be convinced. In fact, a paradoxical theme permeating many of the papers in this volume is the public's desire for zero risk and waste elimination juxtaposed with the impossibility of achieving these goals short of shutting down industries that society considers vital.

1.2.1.2 Legislation: Love Canal and Seveso spurred the burgeoning environmental movement and environmental consciousness in society. These incidents induced changes in social norms and in popular opinion that found their logical reflection in two legislative Acts:

(1) The Resource Conservation and Recovery Act of 1976 (RCRA), which dealt with the regulation of ongoing waste management activities
(2) The Comprehensive Environmental Response, Compensation and Liability Act of 1980 (CERCLA or Superfund) — an effort to tackle the problems caused by abandoned or closed waste sites

These acts were designed to make polluters pay for the damage caused by the wastes that they introduce into the environment. Liability rules and regulations administered by the Environmental Protection Agency (EPA) specify the extent of the polluters' responsibility for damages to human health and the environment. To back the rules with credible incentives, these Acts mandate a set of financial responsibility requirements.

6   INTRODUCTION

The combination of liability rules and financial responsibility requirements are expected (1) to induce waste handlers to use sound waste management practices; (2) to induce the development of technologies aimed at source reduction; and (3) to ensure that adequate finances are available for cleaning up those sites that endanger human health and the environment.

Because the features and requirements of RCRA and CERCLA form a backdrop to the discussions at this conference, it is useful to provide more detail on the key provisions of both Acts.

1.2.1.2.1 RCRA[2]   The Resource Conservation and Recovery Act of 1976, along with its amendments, the Hazardous and Solid Waste Amendments of 1984, focuses on the regulation of hazardous waste management. The main impetus is prospective, concentrating on ways to regulate waste handling and disposal operations.

The principal feature of the cradle-to-grave-policy of RCRA is the requirement that a manifest initiated by the generator and signed by the transporter accompany the waste to its final destination at a permitted treatment, storage, and disposal facility (TSDF).

RCRA requires that a waste management facility demonstrate that it has sufficient financial resources to both compensate harm to third parties caused by TSDF operations, and to conduct all necessary closure and postclosure monitoring and maintenance responsibilities. A facility can meet the financial responsibility requirements via (among other methods) trust funds, surety bonds, financial guarantee bonds, letters of credit, and insurance. The legislation was designed to encourage the utilization of new treatment technologies to reduce the volume, toxicity, or mobility of hazardous substances wherever practicable. It is clear that source reduction, is an important part of a long-range program for dealing with hazardous waste.

1.2.1.2.2 CERCLA or Superfund.   The Comprehensive Environmental Response, Compensation, and Liability Act of 1980 established a comprehensive system of notification, emergency response, enforcement, and liability for hazardous substance spills and uncontrolled hazardous waste sites. It authorizes the EPA to set up an administrative system for responding to releases of hazardous substances from currently inactive hazardous waste sites, and provides funds for both immediate cleanup and long-term corrective action. It also authorizes the federal government to recover cleanup and other costs from responsible parties.

Congress reauthorized CERCLA through the Superfund Amendments

INTRODUCTION 7

and Reauthorization Act of 1986 (SARA). This legislation invested \$8.5 billion in new funds in the Superfund program in order to clean up hazardous waste sites that are considered dangerous and ensure that permanent solutions would be applied to those sites if possible.

I.2.1.3 Judicial Rulings. Environmental legislation, emphasizing the *polluter pays principle,* has led to new developments in environmental liability law and in judicial rulings. These have had far-reaching effects on the operation of industrial firms which must dispose of their waste and those involved in cleanup operations.[3]

Two developments in statutory liability and judicial interpretation are particularly noteworthy.

I.2.1.3.1 Retroactive Strict Liability. The Superfund Act imposes strict liability on responsible parties for damages and clean-up costs associated with hazardous waste sites. This provision effectively makes parties retroactively liable for cleaning up hazardous waste sites, even though they may have employed the best available technology or appropriate procedures at the time that they disposed of the waste.

I.2.1.3.2 Joint and Several Liability. Courts, in interpreting the Superfund Act, imposed joint and several liability on generators and transporters of hazardous substances for the expense of cleaning up waste sites where more than one party deposited the wastes. Theoretically, any party that handled the waste is thus potentially liable for a substantial portion of the damage even if that party's actions played only a minor role in causing the pollution. For example, if the parties who are potentially liable cannot be identified or declare bankruptcy, then a single party whose involvement in the waste site was minimal may be forced to cover the entire loss.

I.2.2 Two Hazardous Waste-Related Scenarios

The following two examples (with fictitious names) illustrate the dilemmas that arise in dealing with the problems associated with hazardous waste.
Example 1: Cleaning up a Landfill. The EPA has identified the Leftover Landfill as potentially dangerous to the environment and to the health of nearby residents. A risk assessment of the contents of the landfill indicates the presence of certain chemicals suspected of being carcinogens. Four companies have disposed of their wastes in the landfill, but there is no easy way to determine either the relative contribution, or the nature of the

8   INTRODUCTION

waste from each company. The local citizens' action groups are demanding
that the Leftover Landfill be cleaned up in the near future. Should the
landfill be cleaned up immediately, and if so, how much should each of
the four companies and the EPA pay for cleaning it up?

Example 2: Managing an Existing Hazardous Waste Facility. The Wellbent
Chemical Company owns a waste treatment facility used to dispose of the
company's toxic waste. The facility consists of both an incinerator and a
lined landfill for incinerator ash. It is also used by other local establishments
such as dry cleaners, a small paint factory, and the local hospital. Wellbent
has been asked by EPA to show evidence of financial responsibility dem-
onstrating its ability to provide compensation to victims if there is a hazardous release from the waste treatment center. Wellbent would like to
obtain insurance to satisfy this financial responsibility requirement. Should
an insurer provide Wellbent with a policy? If so, how much coverage should
be offered and what price should be charged?

I.3 Key Questions for Consideration

Uncertainties associated with the risks created by hazardous waste make
it difficult to determine how to respond to these scenarios, even in a world
without any legal restrictions, and specific legislative enactments or government regulations.

However, the above examples raise questions about the design of hazardous waste policy. Many of these questions are addressed in the papers
and discussant comments that follow. They also provide a perspective on
what types of integrated waste management solutions need to be explored
to address economic, environmental, social, insurance, legal, and legislative issues surrounding the hazardous waste management problem.

(1) If Leftover Landfill were to be cleaned up, what residual level of
     risk would satisfy the public? Is zero risk possible, and practicable?
     How clean is clean enough?
(2) Can risk assessments clarify the extent of the risks involved in a site,
     and how credible are these assessments? Would risk assessment help
     us set priorities on the cleanup of Leftover Landfill and other such
     sites?
(3) What are the alternatives to landfills as methods of waste disposal?
     What are the alternatives to incineration for waste treatments? What
     are the risks involved? What can be recommended to Wellbent as

INTRODUCTION 9

a safer method of disposing waste than the present incinerator–landfill complex?

(4) What further resources need to be requisitioned for the cleanup of Leftover Landfill to satisfy the public about the pace and extent of cleanup activities?

(5) To what extent do inconsistencies exist between the various governmental efforts to regulate environmental pollution? For example, are there inconsistencies between the levels required under CERCLA to clean up Leftover Landfill, and the emission requirements for Wellbent's incinerator? How can these be remedied?

(6) Are existing processes adequate to ensure the participation of those affected by hazardous waste sites? Are the local communities around Leftover Landfill and Wellbent's facility aware of the hazardous nature of the operations at these facilities? How can public involvement and education be ensured?

(7) What data do insurers need in order to be able to get a better handle on environmental pollution liability claims that could arise from Wellbent's facility?

(8) Given that more than one firm is involved in disposing of their wastes at Wellbent's plant, to what extent does joint and several liability hinder the insurability of environmental pollution risks? Are there methods to insure risks in the face of joint and several liability?

(9) What alternative mechanisms for resolving disputes could take the place of tort law to settle conflicts associated with hazardous waste management? How would these mechanisms be applied to determining who is to pay for what in the cleanup of Leftover Landfill?

(10) If carcinogens in the wastes at Leftover Landfill affect the health of local residents, how can the issue of victims' compensation be adequately addressed? Would an administrative scheduled compensation system be the answer?

(11) How can public understanding of environmental risks be improved? What role can risk communication play in bringing about agreement on definitions of risk and acceptable levels of risk?

(12) To what extent are risk management approaches being used by the government as it formulates its policies towards Wellbent? What barriers impede the use of such approaches?

(13) Can the differences between large and small firms be taken into account and addressed in environmental policymaking? Do the regulations that affect Wellbent have the same impact on the other small parties utilizing the facility?

(14) Is Wellbent sharing its risk management expertise with the smaller

companies that are disposing of wastes at its plant? What are the obstacles to better information sharing between companies, both large and small? How can these be overcome? For instance, can they be overcome by setting up institutions to pass and share information on risks?

(15) What economic, legal, and regulatory incentives are necessary to enhance source reduction, thus encouraging Wellbent to change its technological processes and produce less waste?

(16) How can all parties affected by hazardous waste become involved in the policymaking process?

(17) How can we stimulate Wellbent and other companies to undertake research into new technologies to tackle the waste problem?

(18) Are the existing standards in environmental management adequate, and are enough resources being directed towards their enforcement at Wellbent and other facilities.

I.4 Objectives and Plan of the Conference

This conference intended to develop ideas for moving toward an integrated waste management program. The production, transport, and storage of hazardous waste requires a long-term strategy that balances the benefits and costs of different actions to society as a whole, while simultaneously recognizing the concerns and needs of the affected interested parties. This is a rather tall order because each stakeholder has an agenda which is top priority to that individual. The scenarios just described and the subsequent questions address different facets of the waste management problem. Our hope is that the conference will initiate a discussion among the concerned parties on formulating an integrated waste management program while also recognizing that compromises will be necessary in order to address the broader management issues.

The following themes and objectives of the conference guided the development of the papers and the discussion at the meeting itself.

(1) To understand the opportunities and limitations of risk assessment of environmental effects from cleaning up disposal facilities (e.g., landfills) and operating hazardous waste management facilities (e.g., incinerators).

(2) To gain a better understanding of the insurability issues associated with coverage for those who clean up waste facilities (e.g., potential

INTRODUCTION   11

Superfund sites) and the operation and management of hazardous waste storage and treatment facilities.

(3) To develop a set of risk management and communication strategies related to insurance, compensation, and regulation for dealing with hazardous waste treatment, storage, and disposal problems.

Based on an understanding of risk assessment, insurability, and risk management issues, a final objective of the Conference was

(4) To develop recommendations for future legislation on an integrated waste management strategy that recognizes the role of risk assessment and risk management as an integral part of the policy process.

Notes

1 This total substantially underestimates the amount of waste actually produced since it is based predominantly on dry weight measurements; i.e. this assumes that all the water has been removed from wastes that are normally produced in liquid form.

2 Report of the NAIC Advisory Committee on Environmental Liability Insurance: *Environmental Liability Insurance*, September, 1986.

3 For a more detailed description of these developments and how they affect the availability of insurance see Abraham (1988). Environmental Liability and the Limits of Insurance. Columbia Law Review 88:942.

References

Agenda Press (1985). An Environmental Agenda for the Future. Island Press, Washington, DC.

Anderson, E.L., P.C. Chrostowski and J.L. Vreeland (1989). Risk Assessment Issues Associated with Cleaning Up Inactive Hazardous Waste Sites. In H. Kunreuther and M.V.R. Gowda (Eds.) (1990). Integrating Insurance and Risk Management for Hazardous Wastes. Kluwer Academic Publishers, Norwell, MA.

Carson, R. (1962). Silent Spring. Houghton Mifflin, Boston.

The Conservation Foundation (1987). State of the Environment: A View toward the Nineties. The Conservation Foundation, Washington, DC.

Enthoven, A.C. and A.M. Freeman III (Eds.) (1973). Pollution, Resources, and the Environment. W.W. Norton & Co., New York.

Report of the NAIC Advisory Committee on Environmental Liability Insurance. Environmental Liability Insurance. September, 1986.

I CLEANING UP INACTIVE WASTE SITES

The first day of the meeting was devoted to papers that addressed the two key elements of the hazardous waste problem that have been treated separately to date: cleaning up inactive hazardous waste sites and managing existing hazardous waste facilities. In the United States, CERCLA (or Superfund) and its recent amendments addresses the first area, while RCRA deals with currently operating facilities.

Three themes guided the discussion of each of the areas: the need for risk assessments, insurability questions, and the development of risk management strategies. The papers in part I focus primarily on the United States experience of cleaning up inactive waste sites under Superfund.

Risk Assessment Issues: Elizabeth Anderson, Paul Chrostowski, and Judy Vreeland indicate that there have been significant advances in risk assessment techniques since the EPA formally adopted the use of this approach in evaluating chemicals for their potential toxicity. They stress the importance of conducting risk assessments at sites that may require remediation, in order to determine whether a site really needs to be remediated while at the same time specifying overall cleanup priorities.

Insurability Issues: Leslie Cheek is rather emphatic that CERCLA deprives the insurer of the predictability needed to offer coverage. Given the developments in environmental law, insurers are uncertain as to what their losses would be should they provide coverage to a potentially responsible party. In fact, given the number of waste sites that must be cleaned up, the insurance industry does not have enough surplus to cover cleanup costs should such a design be imposed upon them by the courts. Cheek proposes a voluntary arrangement between insurers and potentially responsible parties (PRPs) to work out settlements in cases where there have been a number of insurers who are potentially liable for cleanup losses.

Risk Management Strategies: James Seif and Thomas Voltaggio indicate that risk management strategies should be used for setting priorities on which sites must be cleaned up and determining what remedial techniques need to be utilized. With respect to prioritization, the EPA utilizes a hazard-ranking system that builds on risk assessment data and incorporates waste quantity, degree of toxicity, and persistence of the chemical as part of the ranking criteria. When a site is selected for cleanup, a remedial program is utilized based on nine different criteria. The absence of insurance for cleanup contractors and responsible parties poses a challenge for implementing the Superfund program.

1 RISK ASSESSMENT ISSUES ASSOCIATED WITH CLEANING UP INACTIVE HAZARDOUS WASTE SITES

Elizabeth Anderson, Paul Chrostowski, and Judy Vreeland.

1.1 Introduction

The broad practice of using risk assessment approaches for the evaluation of suspected human carcinogens is about 12 years old. The primary departure point was the announcement by the U.S. Environmental Protection Agency (EPA) that adopted guidelines for assessing the risk of carcinogens and a policy to regulate suspect carcinogens based essentially on a risk management approach. The scientific basis was derived from the earlier experience of assessing the risk of health impacts from radiation exposure. From a practical standpoint, the use of risk assessment for carcinogens has received broad and general endorsement. The early use of risk assessment of carcinogens relied heavily on replacing the uncertainties in the risk assessment process with very conservative assumptions to ensure that the risk would in no case be underestimated. As the practice of risk assessment has become widespread, considerable attention has been focused on improving the scientific basis for evaluating each step of the risk assessment process: the weight-of-evidence indicating likely carcinogenicity, the dose–response relationships, and the environmental exposures. Chemicals that are thought to cause health or ecological effects through threshold mech-

16 INSURANCE AND RISK MANAGEMENT FOR HAZARDOUS WASTE

anisms are also being evaluated by risk assessment approaches. More attention has been focused on the scientific relationships that underlie the characterization of suspect carcinogens and their dose–response relationships, including extrapolation from animals to humans and from high dose to low dose; far less attention has been focused on the exposure assessment, which can impact the outcome of the quantitative risk assessment by certainly as much as the assumptions in the dose–response extrapolation part of risk assessment. Recent advances in both areas, when applied to hazardous-waste-site risk assessment, can substantially alter the outcome of the site-specific risk assessments. This paper will provide an overview of scientific developments in risk assessment and describe how the use of these improved scientific data may alter the outcome of the standard Superfund risk assessment approaches.

1.2 Risk Assessment

1.2.1 An Overview of the Process

In 1976, EPA adopted the first policy for the use of risk assessment of toxic chemicals that were suspected to be human carcinogens and accompanied this policy statement with guidelines for the scientific risk assessment process (EPA, 1976; Albert et al., 1977). These guidelines were adopted in response to the need of a major regulatory agency to develop a means for regulating the presence of hundreds of suspected carcinogens in the environment under numerous environmental legislative statutes that had been adopted by Congress. In short, it was obvious that EPA could not regulate all the suspected carcinogens that were being identified in rapid succession as environmental contaminants to a zero-risk level, as had been the risk goal of the Food, Drug and Cosmetic Act's Delaney Clause. Such a goal had clearly been the objective of the strong environmental movement that characterized the first half of the decade of the 1970s. The adoption of guidelines for risk assessment, with the implication that EPA planned to accept residual risk as a regulatory policy, was initiated under the watchful eyes of the scientific community, the regulated community, and the environmental communities. In short, there was considerable skepticism about the approach as a basis for public policy because of the substantial scientific uncertainties, particularly in quantitative risk assessment. After the EPA action, several other endorsements followed. The Interagency Regulatory Liaison Group (IRLG) adopted similar scientific principles in 1979 (IRLG, 1979). These guidelines were followed by a report from the

National Academy of Sciences (NAS) that endorsed the use of risk assessment and provided descriptive terms for each step of the risk assessment process, which have now been adopted as a common vocabulary (NAS, 1983). In addition, the Office of Science and Technology Policy (OSTP) published similar scientific principles in 1985 (OSTP, 1985), and EPA updated its earlier guidelines (EPA, 1986a). In short, the application of risk assessment to toxic chemicals, both to evaluate scientific evidence that indicates that a chemical might be a human carcinogen and also to provide information as to the magnitude of current and anticipated public health impacts, has been endorsed and described in many forums and has also been the subject of discussion in many scientific conferences.

The process is generally described in four steps: hazard identification, dose–response modeling, exposure assessment, and overall risk characterization (NAS, 1983). In practice, over the last 12 years, the hazard identification step in risk assessment has relied on all the available human, animal, and/or in vivo or in vitro data to describe the weight-of-evidence that indicates that a chemical might be a human carcinogen. At various times, the weight-of-evidence has been stratified according to either the International Agency for Research on Cancer (IARC) criteria (IARC, 1982) or the more recent EPA stratification scheme (EPA, 1986a) for assigning a category to the weight-of-evidence. While these two categorical schemes are very closely related, the EPA scheme expands on the inadequate evidence labeled Category 3 in the IARC criteria to include three additional categories: C to indicate evidence that constitutes the category of *probable* carcinogen for humans; D to indicate inadequate testing; and E to indicate negative evidence.

Dose–response modeling has largely followed a linear nonthreshold hypothesis for low-dose extrapolation as a basis for defining a plausible upper limit on the risk, meaning that the risks are unlikely to be higher but could be considerably lower (Crump et al., 1977; Crump and Watson, 1979; Crump, 1981; OSTP, 1985; EPA, 1986a). This model relies on the possibility that any suspected carcinogen can induce cancer by a single-hit phenomenon and makes no distinction for different biologically based mechanisms of cancer induction. To date, other models that have variously been suggested for low-dose extrapolation from high-dose data have been empirically based models that seek statistically to define the best shape to the dose–response curve; they have not been based on data that seek to describe the biological events that lead to cancer.

Other assumptions, such as those used to extrapolate animal responses to humans, have been adopted in order to be protective where scientific information was lacking (e.g., surface area is often chosen as the conversion

18 INSURANCE AND RISK MANAGEMENT FOR HAZARDOUS WASTE

factor rather than body weight). Dose is assumed to be synonymous with exposure, unless there are data to the contrary. Other conservative assumptions have also been chosen, including, for example, the interpretation of the significance of benign tumors, which can lead to malignancy.

Exposure assessment likewise has followed a conservative trend. Generally, *maximum plausible levels* of chemical exposure have been used in risk assessments, sometimes in conjunction with *average exposure* estimates. An example of a frequently used conservative assumption is that an individual is exposed for a lifetime of 70 years unless there is evidence to the contrary. In practice, the overall risk characterization has relied on a ranking of the weight-of-evidence that placed considerable weight on any tumor response in animals and sought to describe quantitatively the risk to current or anticipated exposed populations as an *upper-bound risk* based on *maximum plausible exposure* estimates. The first decade of experience with carcinogen risk assessment has been studied both from the scientific standpoint and from the standpoint of the use of risk outcomes in public policy decisions (Anderson and CAG, 1983). In short, if scientists have been successful in the past describing risk assessment as upper-bound estimates reflecting maximum plausible exposures, then as better science is developed to fill the gaps of uncertainty, risk assessments should be expected to become less conservative.

1.2.2 Current Trends

Historically, protective assumptions replaced uncertainties; in some cases, uncertainty is now being replaced by improved scientific information. In the area of weight-of-evidence, fresh consideration is being given to the weighting of evidence at high dose and its appropriateness for low-dose weighting. For example, in the Carcinogen Assessment Group's risk assessment of ethylenethiourea (ETU) (EPA/CAG, 1977), the uniqueness of the observation of rat thyroid tumors was discussed in the context of a threshold, namely that these tumors resulted from suppression of thyroid activity only after the administration of a sufficiently high dose of ETU. Currently, the rat response is being examined to determine whether environmental exposure levels are likely to approach those that could be expected to elicit the rat thyroid tumor response; if not, it may be appropriate only to factor the mouse liver tumor response results into the weight-of-evidence determination for environmental exposure levels. Other chemicals are similarly being reviewed for their relevance to human exposure because of mechanism of action, tumor type observed, dosing levels used,

RISK ASSESSMENT ISSUES: INACTIVE WASTE SITES

or metabolic and pharmacokinetic differences between humans and laboratory test animals.

The improvements in dose–response modeling probably represent the most dramatic departure from practices of the last 12 years. There is a clear effort by regulatory agencies to seek a biological basis for the development of more accurate estimates of risks expected to occur at environmental exposure levels. This effort represents a substantially different approach from that of applying empirical formulas to estimate low-dose responses from high-dose data; rather the attention is focused on the importance of research data that may guide low-dose modeling efforts. Such an approach provides, at a minimum, an indication of the extent to which the plausible upper bounds may be overestimating risk for particular chemicals.

Early efforts to define more accurate estimates of risk began at EPA in early 1985 and have culminated in the development of a generic approach using a two-stage model. This model adapts the clinical observations of Moolgavkar and Knudson (1981) to parameters involving exposure to toxic chemicals. The effort was first undertaken by EPA's Risk Assessment Forum and was ultimately published in the *Journal of Risk Analysis* in early 1987 (Thorslund et al., 1987). Thus far, EPA has proposed two important decisions in line with the trend toward less conservatism in dose–response modeling. Both of these decisions were discussed in a recent *New York Times* article (Shabecoff, 1988). For example, the EPA's Risk Assessment Forum has recommended lowering the arsenic ingestion potency by approximately an order of magnitude (Levine et al., 1987; Moore, 1987) based on modifications in dose–response calculation methodology and better estimates of the exposure involved in the epidemiology studies that were the basis for the evaluation. There is a further consideration of reducing the arsenic ingestion potency by still another order of magnitude to reflect the fact that skin cancer caused by arsenic ingestion is less likely to lead to death than is lung cancer induced by inhalation. Considerations of the latter raise the issue as to whether or not treatability, survival, and severity should be routinely considered as a part of the risk assessment process, and in particular as part of the potency evaluation. In addition, EPA has proposed to downgrade the potency of dioxin based on several factors, but most importantly, on the use of the two-stage model of carcinogenesis for modeling the promoting activity of dioxin. This model indicates that the potency of dioxin may be less by two orders of magnitude or more, than the potency defined by the linear nonthreshold model for low doses (T.W. Thorslund and G. Charnley, in preparation). This work was prompted by recommendations of the EPA Science Advisory Board and is still under consideration (EPA/OPTS, 1986b).

20   INSURANCE AND RISK MANAGEMENT FOR HAZARDOUS WASTE

The two-stage model of carcinogenesis has also been applied to several other chemicals with similar outcomes. For example, the model has also been applied to chlordane and heptachlor, as well as methylene chloride (T.W. Thorslund et al., 1988, private communication). While the mechanisms in each case differ, the outcomes of the model most often indicate several-orders-of-magnitude lower potency at low dose than that predicted by the linear nonthreshold model at the plausible upper bounds.

Additional applications of the biological model have involved the polycyclic organic compounds. Past practices have used the potency of benzo(a)pyrene as a unit equivalency to all other potentially carcinogenic polycyclic organic compounds, greatly overestimating risk. This practice has continued in spite of the fact that comparative potency methods have been developed for other chemical classes, such as the dioxins. When assembled in the aggregate, several laboratory studies provide a more substantial basis for developing a comparative potency approach for PAHs (M.M.L. Chu and C.W. Chen 1984, unpublished; Thorslund et al., 1986). In addition, the shape of the dose–response curve for benzo(a)pyrene itself has been reevaluated. Benzo(a)pyrene is a genotoxic agent, as indicated by a linear rate of DNA adduct formation that parallels exposure. The tumor dose–response data do not parallel DNA adduct formation, however, but appear to fit a quadratic equation, indicating that two events are probably necessary to induce the response. EPA's initial cancer potency estimate for benzo(a)pyrene does not reflect this relationship. The comparative potency approach for other polycyclic compounds, together with a revised dose–response curve for benzo(a)pyrene, has been used to accurately predict tumor outcomes in bioassays of chemical mixtures, which is not possible using upper-bound estimates (Thorslund et al., 1986). Another example of a chemical that may require two events to produce a cancer outcome is benzene. Current investigations are examining the mechanistic data, which indicate that benzene causes chromosome damage that is thought to be responsible for the chromosomal deletions and rearrangements observed in leukemia patients. This relationship implies that, although linearity may establish a plausible upper bound on human leukemia risk from benzene exposure, a quadratic relationship may be more appropriate to estimate the actual risk. Should this turn out to be the case, the risk from low-dose exposure to benzene would be considerably lower than previously estimated (T.W. Thorslund and G. Charnley, 1988, private communication).

A great deal of attention is also being focused on the metabolic and pharmacokinetic data to estimate actual levels of chemical exposure to the target tissue. In extrapolating animal data to humans, the effective dose in the animal studies has always been assumed to be the dose that the

RISK ASSESSMENT ISSUES: INACTIVE WASTE SITES 21

animal was exposed to by route-administered-dose. As our ability to de-scribe the actual dose to the target tissue in the animal improves, so will our ability to extrapolate animal responses to humans. In addition, the importance of pharmacokinetic data to define the significance of human exposure in the environment is exceedingly important.

Less progress has been made in the case of threshold pollutants. While attention is currently focused on developing biologically based dose–response curves to better describe the threshold dose for disease causation, the majority of these chemically induced effects are still described by ap-plying safety factors to no-observed-effect levels (NOELs) from animals studies, or for some few chemicals, describing the effective dose for ob-servations in humans, e.g., lead. In either case, the results are uncertain and the outcomes subject to scientific debate.

Of equal importance, trends in exposure assessment research are also leading to improved estimates of population exposures, which provide a better foundation for current and projected exposures. Traditional prac-tices have relied heavily on generic models to describe exposure to human populations. EPA has developed generalized dispersion models for de-scribing air transport and similar generalized dispersion models for surface and groundwater. The overall impact of these dispersion models has been to provide conservative estimates of exposure.

The use of generalized models provides a practical approach for wide-spread exposure estimation by regulatory agencies because it would be highly impractical for a national agency to evaluate site-specific parameters for every source. For important cases, however, it is possible to estimate actual parameters that may refine the estimates obtained by generic mod-eling. An example is the risk assessment of the ASARCO smelter in Ta-coma, Washington, which was conducted by EPA (Patrick and Peters, 1985). The use of generalized dispersion modeling using the human ex-posure model (HEM) (which assumes a flat terrain and an immobile pop-ulation, and uses meteorological data from the closest weather station), when coupled with the dose–response curve, estimated a maximum indi-vidual risk of about 1×10^{-1} for populations living near the smelter. Sub-sequently, a local study was conducted that permitted the use of several site-specific assumptions including a more accurate description of the actual terrain, local meteorological data, and better emissions information. The result was the lowering of the exposure assessment and the overall risk by about an order of magnitude. This brought the risk into closer alignment with the limited monitoring data that was available for the ambient air.

The same phenomenon has been observed when comparing estimates that use generalized dispersion models for groundwater with estimates that rely on site-specific parameters. For example, in figure 1-1, the generalized

22   INSURANCE AND RISK MANAGEMENT FOR HAZARDOUS WASTE

dispersion model, the vertical horizontal spread (VHS) model using EPA default values overestimates the risk by a factor of 5.7 when compared to the results from the more complex equation, which incorporates measured site values (Domenico and Palciauskas, 1982; EPA, 1985a). Another important area that has sharpened exposure estimates and, practically, has lowered the outcome from exposure assessment by several orders of magnitude (and thus the quantitative risk assessment) has been considerations of bioavailability. For example, dioxin was originally assumed to be 100% biologically available in soil. Recent studies, however, have demonstrated that dioxin is only partially available, between 0.5% and .85% depending on soil type (Umbreit et al., 1986). In practice, it has been our experience that dioxin is mostly available in the range of 15-50% (P. Chrostowski, 1988, private communication). Dioxin in fly ash also was originally assumed to be up to 100% available. Recent studies have found that this is not correct; rather, dioxin in fly ash is biologically available between 0.1% and 0.001% (van den Berg et al., 1986). The bioavailability issue is now being commonly investigated in many different situations where the availability in soil and fly ash is important to the outcome of the risk assessment.

Although improving the scientific information available for site-specific exposure assessment tends to lower the overall outcome of the exposure

1.   VHS model using simple equation, EPA fixed default values

$$\frac{C}{C_0} = \text{erf} \left[\frac{Z}{2(DY)^{0.5}} \right] \text{erf} \left[\frac{X}{4(DY)^{0.5}} \right]$$

$$= 0.34$$

2.   VHS model using complex equation, measured site values

$$\frac{C}{C_0} = \frac{1}{4} \left[\text{erf} \left(\frac{Z + Z}{2(D_ZY)^{0.5}} \right) - \text{erf} \left(\frac{Z - Z}{2(D_ZY)^{0.5}} \right) \right]$$

$$\left[\text{erf} \left(\frac{X + X/2}{2(D_XY)^{0.5}} \right) - \text{erf} \left(\frac{X - X/2}{2(D_XY)^{0.5}} \right) \right]$$

$$= 0.06$$

Conclusion: EPA method overestimates concentration at exposure point by factor of 5.7.

Figure 1-1. Ground water modeling.

RISK ASSESSMENT ISSUES: INACTIVE WASTE SITES

assessment and thus the risk assessment, there are important exceptions. For example, a recent paper that addressed the issue of risk associated with inhaling volatile organic chemicals from contaminated drinking water during shower activity (Foster and Chrostowski, 1987) indicated that as much as half or more of the total body risk could be associated with the shower exposure rather than with the drinking water exposure. In addition, recent improved methods for modeling the actual deposition of particulate matter from stationary sources tend to raise the risk compared to the earlier EPA air-transport models, which assumed that both large and small particles bounced from the surface of the earth in very similar ways and were carried from the site by air transport. The more recent models take into account the fact that small particles deposit on the surface and are not so readily transported (Sehmel and Hodgson, 1979). Also, closer attention to chemical conversions may tend to raise or lower the risk; for example, trichloroethylene is converted under anaerobic conditions to vinyl chloride, which has a higher potency value by ingestion than does trichloroethylene (Parsons et al., 1984; Cline and Viste, 1984). Recognition of this conversion raises the overall risk assessment for circumstances that are appropriately evaluated by these methods.

Numerous other refinements in exposure assessment are also being incorporated in the risk assessment process, for example, use of human biological data to assist in exposure estimation, better descriptions of lifestyle for subpopulation groups, the use of statistical methods to describe likely exposure below detectable limits, and the use of pharmacokinetic data to describe the actual dose to target tissue. These developments rely on advancing research in multiple disciplines for use in the practical consideration of human exposure.

1.3 Applications to Hazardous Waste Site Risk Assessment

Waste site risk assessment practices have roughly paralleled the conservative (public health protective) approaches of risk assessment over the last dozen years. The majority of this experience has been gained from the investigation of Superfund sites according to the standard EPA Superfund Manual and related guidelines (EPA, 1985b, 1986c,d). In these investigations, a risk assessment is a formalized methodology applied to determine the potential for human health and environmental impacts associated with a site under the no action alternative or to evaluate the potential benefits from remedial alternatives.

24   INSURANCE AND RISK MANAGEMENT FOR HAZARDOUS WASTE

Generally, the initial step in conducting a risk assessment involves a review of all available environmental monitoring data at the site in order to select potential chemicals of concern on which the assessment will focus. At this step, chemical measurements with inadequate quality assurance/ quality control or chemicals that are present as part of the natural background may be rejected for inclusion in the assessment. The next step, hazard identification, involves identifying chemical-specific human health and ecological effects criteria. This may involve an evaluation of available data, including epidemiology, animal bioassay studies, and in vivo and in vitro studies. In the absence of human data to describe low-dose effects, the frequently used approaches for dose–response characterization are for *threshold* (noncarcinogenic) and *nonthreshold* (carcinogenic) effects. These approaches generate numerical health effects criteria to be used in the calculation of risk. While some guidance levels generally exist for most toxic chemicals, further scientific work may be warranted. Recent reconsideration by the EPA of potency factors for arsenic, dioxin, and polycyclic aromatic hydrocarbons are good examples. Following hazard identification, potential pathways by which human populations may be exposed under current or potential future land-use conditions are identified. An exposure pathway is composed of the following four elements: (1) a source and mechanism of chemical release to the environment; (2) an environmental transport medium (e.g., groundwater) for the released chemical, and/or a mechanism of transfer of the chemical from one medium to another; (3) a point of potential contact of humans or biota with the contaminated medium (the exposure point); and (4) an exposure route (e.g., ingestion) at the exposure point. All four of these elements must be present for a pathway to be considered complete. To evaluate exposure at an exposure point, the concentration of chemicals of concern must be evaluated. Many times these are actual measured concentrations; however, when they have not been measured, or in order to estimate future concentrations expected to occur either over a longer time (i.e., a 70-year lifetime) or at exposure points not previously investigated, environmental fate and transport modeling may be necessary. For Superfund sites, once concentrations of chemicals of concern at the exposure points have been determined, they are compared with *applicable or relevant and appropriate requirements* (ARARs). When ARARs are not available for all chemicals in all media, quantitative risk estimates are developed by combining the estimated intakes of potentially exposed populations (often derived using conservative assumptions regarding chemical concentrations, exposure duration, exposure frequency, and the efficiency of absorption in biological media of chemicals) with either existing health effects criteria or improved evalua-

RISK ASSESSMENT ISSUES: INACTIVE WASTE SITES

tions based on more recently available data and methods. Conservative assumptions are generally made in risk assessments to compensate for uncertainty and to explore the potential for adverse health effects using conditions that tend to overestimate risk, so that the final estimates will usually be near or higher than the upper end of the range of actual exposures and risks. Greater uncertainty in the site-specific data base generally leads to more extensive reliance on conservative assumptions; conservative (i.e., protective) assumptions are chosen to make certain that risks will not be underestimated. Because there is uncertainty, risk assessments generally do not present an absolute estimate of risk; rather, most risk assessments establish plausible upper bounds on risk to indicate the potential for adverse impacts. Thus, risk assessments are more useful where data are available to narrow uncertainties and to permit the most accurate descriptions of risk possible. In the absence of such data, conservative approaches that provide upper-bound risk estimates present clear guidance for the evaluation of low risk (i.e., even at the upper bounds the risks are low and therefore most often do not warrant regulatory attention), but are less instructive for remedial prescription where the social and economic costs are high.

1.3.1 Potential Pathways of Exposure to Contaminants

All pathways of exposure are considered: groundwater, surface water, soil, and air. The pathway that is most often of greatest concern is groundwater. For purposes of discussion, we will focus on this route as an example of an exposure-route evaluation.

To evaluate exposure to groundwater, standard intake assumptions are generally employed. These are that an average adult ingests two liters of water a day over a 70-year lifetime and that the average body weight over the exposure period is 70 kilograms, unless there are clear data to define alternative choices. For example, these assumptions can be arguably too stringent, or in some cases, such as outdoor workers in an arid climate, they may not be stringent enough. If the demographics and activity patterns of the population are known, more accurate intake assumptions may be used that will often diminish risks.

Additionally, inhalation exposures to contaminants in groundwater may occur through use of water in day-to-day activities such as cooking, bathing, washing of dishes and clothes, or showering. Dermal exposures are also possible. Although many of these exposures may be dependent upon individual water-use patterns, exposure through showering may be quantified

26 INSURANCE AND RISK MANAGEMENT FOR HAZARDOUS WASTE

using the model of Foster and Chrostowski (1987). For many volatile organics, quantification of the additional risks through inhalation of contaminants while showering may be similar to the risks associated with ingestion. In some instances, risks from all inhalation activities combined may be greater than those associated with ingestion, especially if the chemical involved is more toxic by the inhalation route (e.g., 1,1-dichloroethylene). Dermal exposures are generally small compared to ingestion or inhalation, although they may be substantial when chemicals that are absorbed with a high efficiency (e.g., dimethyl sulfoxide) are involved. Failure to assess these pathways could lead to groundwater risk-management decisions not protective of public health or associated with an inaccurate representation of liability.

In many locations discharges of groundwater into surface water bodies create additional potential pathways of exposure. This is a particular concern for water bodies that are of moderate size (i.e., that have sufficient flow to support aquatic life and are not so large in volume as to dilute concentrations of contaminants discharging in groundwater to insignificant concentrations). Contaminants that have high octanol-water partition coefficients (K_{ow}s) have a potential to bioaccumulate and generally are of particular concern. These chemicals may not only be toxic to aquatic life but may potentially cause risks to other organisms higher in the food chain or to humans that ingest fish from these surface water bodies on a regular basis. Additionally, surface water bodies may be used for recreational activities such as swimming, or for drinking water supplies; these may create additional exposure pathways to contaminants in groundwater.

An added consideration is needed to provide an assessment of anticipated exposures. For example, a change in local pumping conditions due to the installation of a new well (particularly one that has a high yield, such as an industrial or municipal well) may have an influence on contaminant migration; or groundwater that is not currently a drinking-water source may become a source in the future.

In evaluating potential exposure to contaminants present in groundwater, it is important to define the aquifer(s) to be evaluated. This may be a particular concern when a source area is underlain by more than one aquifer that may potentially be used for a water supply. The potential exposure associated with the use of each one of the aquifers may be evaluated individually. However, in many instances, particularly in areas with fractured bedrock, domestic wells are open to more than one water-bearing zone, and estimation of the potential future concentrations in these wells would be dependent upon the well construction. This type of information in some instances is not readily available, and is difficult to estimate for hypothetical future wells.

RISK ASSESSMENT ISSUES: INACTIVE WASTE SITES 27

EPA's groundwater strategy (EPA, 1986e) suggests that risks associated with potential water supplies should be evaluated as if the water supplies were actually in use. Thus, aquifers with natural mineral contents low enough for potability and with high enough potential yields are all considered potable whether in current use or not.

1.3.2 Approaches to Quantifying Exposure and Risks Associated with Contaminants in Groundwater

Site-specific information is essential in evaluating the potential risks associated with exposure to contaminants present in groundwater. For a screening approach, general hydrogeological parameters can often be obtained from available publications (e.g., Freeze and Cherry, 1979; Walton, 1985). Use of these general default parameters, however, could lead to an inaccurate portrayal of risk. However, to refine groundwater models with concomitant refinement of risk estimates, site-specific information with regard to the hydrogeological characteristics and potential source areas must be well defined. With regard to hazardous waste sites, there is frequently little or no information available regarding the history of chemical disposal in potential source areas; thus, conservative assumptions are generally made with regard to the nature, extent, frequency, and duration of chemical release to the subsurface.

Oftentimes estimation of concentrations of chemicals of concern in the groundwater begins with evaluating the transport of contaminants from the source, which is often soil, through the unsaturated zone. This can be done through a variety of approaches, ranging from a simplified steady-state soil-water partitioning model to time-dependent models that consider linear adsorption/desorption without accounting for dispersion (Enfield et al., 1982) or more complex compartmental numerical models that incorporate time-varying transport, advection, and dispersion, such as the Pesticide Root Zone Model (EPA, 1984). Applying these models to the same site may result in soil-pore water concentrations that may vary by as much as several orders of magnitude.

The output from the unsaturated zone models may then be coupled with groundwater models to predict concentrations of chemicals of potential concern at potential exposure points. The groundwater models may range from simplified mass-balance mixing models through analytical solutions of transport equations to complex three-dimensional numerical models. In some instances, simplified models used for screening purposes show that, even using conservative assumptions, the estimated concentration at a potential exposure point may not be associated with a risk and the conserv-

28   INSURANCE AND RISK MANAGEMENT FOR HAZARDOUS WASTE

ative assessment may be sufficient. However, refinement of conservative assumptions is often necessary to ensure that the evaluation is realistic and that remedial actions will not be undertaken needlessly.

An example of the importance of applying site-specific parameters to a hazardous waste site in the context of groundwater solute transport modeling can be instructive. For example, the VHS model (EPA, 1985b,c) is a steady-state groundwater model in which the only attenuation mechanism is vertical and horizontal spreading; the model neglects longitudinal dispersion and chemical degradation kinetics. Application of the VHS model to a particular site using EPA fixed default values resulted in a ratio of concentration at the exposure point to concentration at the source (C/C_o) of 0.34. However, a refinement of the model, incorporating vertical and horizontal dispersion coefficients and site-specific parameters, resulted in a C/C_o ratio of 0.06. Thus the generic EPA model overestimated the concentrations and risks at the exposure point by a factor of 5.7 (figure 1-1).

In many instances, the steady-state assumption is not applicable. At another site, site-specific groundwater modeling incorporating chemical decay and source decay illustrated that assumption that the observed concentration persists over 70 years could result in a substantial overestimation of the risk. For example, work completed by our scientists at a site in California indicated an upper-bound lifetime risk as high as 10^{-3} associated with ingestion of water containing trichloroethylene (TCE) in an aquifer (figure 1-2). This level is associated with a 70-year lifetime exposure via exposure to drinking water from the contaminated aquifer. Scenario 2 in figure 1-2 describes the decline in associated risk when hydrogeology models are applied to the site; the model assumes that the source of contamination has been removed. In table 1-1, the monitoring-well data are given, and table 1-2 displays the risk comparison over time given the ability to model the area. In this particular circumstance, remedial action was being considered that would cost in the million-dollar range and would require a number of months to install. If the hydrogeology models are correct, the theoretical risk could be lowered considerably over the first 18-month period, given the natural ability of the hydrogeology of the area to remove the contamination. Caution, however, should be exercised in assuming that the source has been removed, because recent publications indicate that in some circumstances some chemicals may remain entrapped in soil micropores and thereby provide a slow, diffuse release (Sawhney et al., 1988).

As the concentration of a particular compound present in groundwater may decrease through natural processes such as biodegradation, the total risk resulting from exposure to the groundwater may not necessarily de-

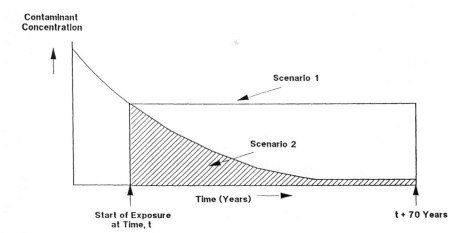

Figure 1-2. Scenario 1 assumes decay in contaminant levels until start of consumption and then a lifetime exposure to a constant concentration of the contaminant. Scenario 2 assumes decay in contaminant levels both before and during the period of exposure.

crease. Some compounds are transformed into more toxic compounds through biotransformation. For example, under anaerobic conditions, some halogenated aliphatics have been found to undergo reductive dechlorination (Bouwer et al., 1981; Kobayashi and Rittmann, 1982; Vogel and McCarty, 1985). The transformation is sequential, with, for example, tetrachloroethylene yielding first trichloroethylene (TCE) and ultimately vinyl chloride (Parsons et al., 1984; Cline and Viste, 1984). Trichloroethylene is categorized as a probable human carcinogen with a cancer potency factor of 0.011 mg/kg/day^{-1}, whereas vinyl chloride is a known human carcinogen with a cancer potency factor that is approximately two orders of magnitude greater than TCE by the ingestion route. Biotransformation is dependent upon a variety of factors such as availability of organic chemicals, oxidation/reduction conditions, availability of microorganisms, growth factors, toxicity, and inhibition. Consequently, the extent to which biotransformation may occur in groundwater at a particular site will vary, and the degree to which it occurs is difficult to quantify. Use of a risk assessment for TCE without considering the potential impact of vinyl chloride could result in an underestimation of risks at the site. This

30 INSURANCE AND RISK MANAGEMENT FOR HAZARDOUS WASTE

Table 1-1. Groundwater Monitoring Data

Well #	Sampling Date	[TCE] mg/l	Well Average [TCE] mg/l
1	1/1/85 1/20/85 2/2/85	.100 .144 } .124 .127	
2	1/1/85 1/20/85 2/2/85 2/8/85	<.005 1.10 1.98 } 1.27 2.00	
3	1/1/85 2/2/85	<.005 <.001 } .003	
4	1/1/85 2/2/85	.055 .080 } .068	0.36
5	1/20/85 2/2/85	.210 .177 } .194	
6	1/1/85 2/8/85	.510 1.40 } .96	
7	1/20/85 2/2/85	.160 .305 } .23	
8	1/1/85 2/8/85	.070 <.005 } .04	

could lead to risk management decisions from an inaccurate data base, which could be ultimately expressed as lingering liability even after cleanup had been accomplished.

1.4 Summary

In summary, it has been shown that risk assessments are generally conservative evaluations, primarily due to selection of assumptions to compensate for data limitations and uncertainties. The methods selected to estimate exposure and to quantify exposure-point concentrations may have a substantial effect upon the estimation of risk associated with exposure to contaminants in groundwater. Most often, more accurate data will provide risk assessment outcomes that are less conservative, although use of the shower model and chemical transformation to more potent chemicals can have the opposite impact. There is no question that the best hazardous waste site cleanup decisions must rely on the most accurate risk assessments

RISK ASSESSMENT ISSUES: INACTIVE WASTE SITES

Table 1-2. Risk Comparison Over Time.

Time of Exposure Initiation (Years)	[TCE] at Start of Exposure (mg/l)	Lifetime Upper Bound Cancer Risk	
		Scenario 1	Scenario 2
t_0	.36	1.1×10^{-4}	3.5×10^{-6}
1	.23	7.1×10^{-5}	2.2×10^{-6}
2	.14	4.5×10^{-5}	1.4×10^{-6}
3	.09	2.8×10^{-5}	8.8×10^{-7}
4	.06	1.8×10^{-5}	
5	.04	1.1×10^{-5}	
6	.02	7.2×10^{-6}	
7	.01	4.6×10^{-6}	
8	.009	2.9×10^{-6}	
9	.006	1.8×10^{-6}	
10	.004	1.1×10^{-6}	

possible; this fact stresses the importance of accurate initial site characterization. The implications of overestimating potential risks associated with contaminants present in groundwater or other media may result in implementation of expensive remediation that could have otherwise been less restrictive. Conversely, inaccuracies in predicting risk may also result in underestimations of exposure that may have far-reaching ramifications in the areas of public health protection and liability evaluation. Currently, far more attention is focused on costly remediation than on reducing the theoretical risk through better risk assessments. Substantial experience demonstrates that improved risk characterization is possible for most sites. This additional scientific effort is important to the process of distinguishing which sites require the greatest attention for remediation.

Acknowledgements

We want to thank Luci Henry, Dr. Anderson's Administrative Assistant, for assistance in preparing the manuscript, and Judy L. Fauls for typing the manuscript. In addition, we would like to thank Gail Charnley, Sarah Foster, and Todd W. Thorslund for their constructive comments and assistance.

32   INSURANCE AND RISK MANAGEMENT FOR HAZARDOUS WASTE

References

Albert, R.E., R.E. Train, and E.L. Anderson (1977). Rationale developed by the Environmental Protection Agency for the assessment of carcinogenic risk. *J. Natl. Cancer Inst.* 58:1537.

Anderson, E.L., and The Carcinogen Assessment Group (CAG) of the U.S. Environmental Protection Agency (1983). Quantitative approaches in use to assess cancer risk. *J. Risk Anal.* 3(4):277.

Bouwer, E.J., B.E. Rittmann, and P.L. McCarty (1981). Anaerobic degradation of halogenated 1- and 2-carbon organic compounds. *Environ. Sci. Technol.* 15:596-599.

Cline, P.V., and D.R. Viste (1984). Migration and degradation patterns of volatile organic compounds. Proceedings of the National Conference of Uncontrolled Hazardous Waste Sites, p. 217.

Crump, K.S., H.A. Guess, and L.L. Deal (1977). Confidence intervals and test of hypotheses concerning dose–response relations inferred from animal carcinogenicity data. *Biometrics* 33:437.

Crump, K.S., and W.W. Watson (1979). *A Fortran Program to Extrapolate Dichotomous Animal Carcinogenicity Data to Low Doses*. National Institute of Environmental Health Sciences, Contract No. 1-ES-2123.

Crump, K.S. (1981). An improved procedure for low-dose carcinogenic risk assessment from animal data. *J. Environ. Pathol. Toxicol.* 52:675.

Domenico, P.A., and V.V. Palciauskas (1982). Alternative boundaries in solid waste management. *Groundwater* 20:303.

Enfield, C.J., R.F. Carsel, S.Z. Cohen, T. Phan, and D.M. Walters (1982). Approximating pollutant transport to groundwater. *Groundwater* 20(6):711-722.

Environmental Protection Agency (EPA) (1976). Interim procedures and guidelines for health risks and economic impact assessments of suspected carcinogens. *Fed. Reg.* 41:21402.

Environmental Protection Agency/Carcinogen Assessment Group (EPA/CAG) (1977). *Preliminary Report on Ethylene Bisdithiocarbamate (EBDC)*.

Environmental Protection Agency (EPA) (1984). *Users Manual for the Pesticide Root Zone Model (PRZM), Release 1*. Prepared by Carsel, R.F., et al. for Technology Development and Application Branch, Environmental Research Laboratory, Athens, Georgia. EPA-600/3-84-109.

Environmental Protection Agency (EPA) (1985a). Final exclusions and final vertical and horizontal spread model (VHS). *Fed. Reg.* 50:229.

Environmental Protection Agency (EPA) (1985b). *Endangerment Assessment Handbook*. Prepared by PRC, Environmental Management, Inc., for Office of Waste Programs Enforcement, Washington, DC.

Environmental Protection Agency (EPA) (1985c). Hazardous waste management system; Identification and listing of hazardous waste; proposed rule and request for comment. *Fed. Reg.* 50(38): 7896-7900.

RISK ASSESSMENT ISSUES: INACTIVE WASTE SITES 33

Environmental Protection Agency (EPA) (1986a). Guidelines for carcinogen risk assessment. *Fed. Reg.* 51(185):33991.

Environmental Protection Agency/Office of Pesticides and Toxic Substances (EPA/OPTS) (1986b). *Report of the Dioxin Update Committee.*

Environmental Protection Agency (EPA) (1986c). Guidelines for the health risk assessment of chemical mixtures. *Fed. Reg.* 51:34014.

Environmental Protection Agency (EPA) (1986d). *Superfund Public Health Evaluation Manual.* Prepared by ICF, Inc., for Office of Emergency and Remedial Response, Washington, DC. EPA 540/186/060.

Environmental Protection Agency (EPA) (1986e). *Guidelines for Groundwater Classification under the EPA Groundwater Protection Strategy.* Office of Groundwater Protection, Office of Water, Washington, DC.

Environmental Protection Agency (EPA) (1987f). *Fed. Reg.* 50(38):7896-7900.

Foster, S., and P. Chrostowski (1987). Inhalation exposures to volatile organic contaminants in the shower. Presented at 80th annual meeting of the Air Pollution Control Association, New York, June 21.

Freeze, R.A., and J.A. Cherry (1979). *Groundwater.* Prentice-Hall, Inc., Englewood Cliffs, NJ.

Interagency Regulatory Liaison Group (IRLG) (1979). Scientific basis for the identification of potential carcinogens and estimation of risks. *J. Natl. Cancer Inst.* 63:243.

International Agency for Research on Cancer (IARC) (1982). Monographs on the evaluation of the carcinogenic risk of chemicals to humans. Supplement 4, *Chemicals and Industrial Processes Associated with Cancer in Humans.* Lyon, France, p. 7.

Kobayashi, H., and B.E. Rittmann (1982). Microbial remediation of hazardous organic compounds. *Environ. Sci. Technol.* 16:170A-183A.

Levine, T., A. Rispin, C.S. Scott, W. Marcus, C. Chen, and H. Libb (1987). *Special Report on Ingested Inorganic Arsenic: Skin Cancer; Nutritional Essentiality.* Draft for U.S. Environmental Protection Agency Science Advisory Board Review.

Moolgavkar, S.H., and A.G. Knudson (1981). Mutation and cancer: A model for human carcinogenesis. *J. Natl. Cancer Inst.* 66:1037.

Moore, J.A. (1987). *Recommended Agency Policy on the Carcinogenicity Risk Associated with the Ingestion of Inorganic Arsenic–Action Memorandum.* U.S. Environmental Protection Agency, Office of Pesticides and Toxic Substances, Washington, DC.

National Academy of Sciences/National Research Council (NAS/NRC) (1983). *Risk Assessment in the Federal Government: Managing the Process.* Prepared by the Committee on the Institutional Means for Assessment of Risk to Public Health, Commission on Life Sciences. National Academy Press, Washington, DC.

Office of Science and Technology Policy (OSTP) (1985). Chemical carcinogens: Review of the science and its associated principles. *Fed. Reg.* 50:10372.

34 INSURANCE AND RISK MANAGEMENT FOR HAZARDOUS WASTE

Parsons, F., P.R. Wood, and J. DeMarco (1984). Transformation of tetrachloroethene and trichloroethane in microcosms and groundwater. *Res. Technol.* p. 56.

Patrick, D., and W.D. Peters (1985). Exposure assessment in setting air pollution regulations: ASARCO, Tacoma, a case study. Presented at the Society for Risk Analysis annual meeting, Washington, DC.

Sawhney, B.L., J.J. Pignatello, and S.M. Steinberg (1988). Determination of 1,2-Dibromoethane (EDB) in field soils: Implications for volatile organic compounds. *J. Environ. Qual.* 17(1):149.

Sehmel, G.A., and W.H. Hodgson (1979). *A Model for Predicting Dry Deposition of Particles and Gases to Environmental Surfaces.* Prepared for the U.S. Department of Energy by Pacific Northwest Laboratory. PNL-SA-6271-REV 1.

Shabecoff, P. (1988). EPA reassesses the cancer risk of many chemicals. *New York Times.* January 4, p. A1.

Thorslund, T.W., G. Charnley, and E.L. Anderson (1986). *Innovative Use of Toxicological Data to Improve Cost-Effectiveness of Waste Cleanup.* Presented at Superfund '86: Management of Uncontrolled Hazardous Waste Sites, Washington, DC, December 1-3.

Thorslund, T.W., C.C. Brown, and G. Charnley (1987). Biologically motivated cancer risk models. *Risk Anal.* 7:109.

Umbreit, T.H., E.G. Hesse, and M.A. Gallo (1986). Bioavailability of dioxin in soil from a 2,4,5-T manufacturing site. *Science* 232:497.

van den Berg, M., M. van Greevenbroek, K. Olie, and O. Hutzinger (1986). Bioavailability of PCDDs and PCDFs on fly ash after semi-chronic ingestion by rat. *Chemosphere* 15:509.

Vogel, T.M., and P.L. McCarty (1985). Biotransformation of tetrachloroethylene to trichloroethylene, dichloroethylene, vinyl chloride, and carbon dioxide under methanogenic conditions. *Appl. Environ. Microbiol.* 49:1080-1083.

Walton, W.C. (1985). *Practical Aspects of Groundwater Modeling.* Second edition. National Water Well Association, Worthington, Ohio.

Discussant: Ruth Patrick

I'm going to comment on Betty Anderson's paper as it was written, which was pretty well summarized by what she said.

This is an important paper with a comprehensive approach for risk assessment issues associated with cleaning an active hazardous waste site. It points out the probability of underestimation or overestimation of effects from using the wrong model, the lack of site-specific data, and the lack of knowledge of the behavior of chemicals under various surface and subsurface environments.

For example, incorrect evaluation of the importance of various types of exposure may lead to a misunderstanding of the source of the risk. For example, inhaling volatile organic chemicals during a shower may form more than half the total risk. This may be more important than drinking the water, and it is certainly more important than dermal exposure.

Another example is the problem of toxicity of dioxin to humans to which Betty referred. Here, the use of the wrong model has overestimated the risk factor 5.7 times when compared with the results from more complex equations that incorporate measured site values. Lack of technical knowledge also has affected the perception of dose to dioxin. As she points out, we used to think it was 100% biologically available; we now know it is .1% to .001%. I would like to add that we have made a recent study of dioxin, and there is no definite proved case of death of a human from dioxin. There are cases of, for example, nausea, rashes, and things of this sort, and death in animals, yes, but not as yet death in humans — a fact that says there may be something wrong in the way we diagnose human effects.

Since site-specific characteristics are very important in cleaning up sites, one must understand the geohydrology of the site, the length of time the hazardous materials have been in the site, the characteristics of the soil and the geological materials in the site, and other chemicals that may be present. All of these will affect the toxicity that may be produced.

Another important field of investigation is the correct estimate of actual levels of chemical exposure to target tissues, which Betty referred to in her talk. As our ability to describe the actual dose-to-response for tissues in the animal improves, so will our ability to extrapolate animal response to humans — an ability which really isn't too good now in many cases.

When one considers exposure to hazardous materials from the use of groundwater, we find that our technical data is relatively sparse. The behavior of various hazardous materials in the ground is highly variable according to the environmental conditions that exist. This means that we must have data concerning soil and geological materials through which the chemicals pass in the unsaturated as well as the saturated zones. It is well known that the physical structure of the soil may affect the movement of contaminants and whether they are concentrated on soil particles and later released or whether they remain absorbed on soil particles.

The chemical characteristics of the soil are also important in determining the rate of movement and degradation of hazardous material. Biological organisms in the soil may have a great effect upon the toxicity of chemicals. They may as in many petroleum products greatly reduce any toxic effect by their metabolic activity. For some chemicals, depending upon whether this activity occurs in the aerobic or anaerobic conditions, the end product

36 INSURANCE AND RISK MANAGEMENT FOR HAZARDOUS WASTE

may be very different. As Betty said regarding trichloroethylene, under aerobic conditions these organisms render it rather nontoxic; under anaerobic conditions, vinyl chloride may be produced.

She discussed to some extent the effects of groundwater contamination of lakes. It is true that a large portion of the water of lakes is groundwater. And if there are fish in these lakes and people eat the fish, there may be a direct dose to man. However, there are many other effects that are very important about which we have little data. For example, it is well known that our forests and many of our commercial crops are very dependent upon the microflora for their productivity. And their productivity is greatly affected if these organisms are not functioning properly.

We know little about the effects of the leachate from waste sites upon these organisms. Furthermore, it has been estimated by Leopold and others that about one third of the flow of all of our streams under natural conditions is groundwater. If the groundwater entering a stream contains toxic concentrations of certain hazardous materials, the surface waters may become unfit for drinking or irrigation, particularly if these waters are used over a long period of time. In the past, we have tended to think of air, water, and land as separate components of the environment. We are now beginning to realize the seriousness of our mistake and the fact that they all interact.

It is most important that we do research and develop the technology and technical know-how to deal with these hazardous materials correctly. Otherwise, the importance of their toxicity may be greatly overestimated and in some cases underestimated. These conclusions affect not only the health of man and his environment, but also man's economic and social well-being.

Discussant: James Wilson

Dr. Anderson has, as we have come to expect, described in exemplary fashion the current process of health risk assessment and some of the changes that are taking place in that process. She has also pointed out some of the problems in application. I would like to expand on a couple of the things she said and add two minor points.

First: implicit in her remarks was how she, and I, and the others of us in the Society of Risk Analysis define health risk assessment. As we understand the process, the output of a risk assessment is part of a decision. Its purpose is to inform the decision making — for instance, what is to be done at a waste site. It helps to decide if any remedial action is needed

RISK ASSESSMENT ISSUES: INACTIVE WASTE SITES

beyond cleaning up the trash, and if so, how much and of what kind. So think about what we've said in terms of how the product of a health risk assessment is going to be used. It's part of decision making, but it's not the only input to decisions on what's to be done with hazardous waste.

One other minor point; Dr Anderson has described a number of interesting studies that her associates have been carrying out in the last few years. I'm sure she shares my regret that these have not yet found their way into the open scientific literature. When they do so, and if they receive the acceptance by the scientific community that I'm sure they will receive, they are going to make a significant contribution to progress in this field.

The two subjects that I'd like to amplify are our exposure assessment, and the changes in how hazards of low exposures to chemical carcinogens are carried out. I'm going to look at exposure assessment from the viewpoint of the hazardous-waste-generator camp, but I am going to comment on hazard assessment from the viewpoint of a scientist active in the field.

I've heard frequent complaints from people involved in certain cleanups that the EPA field staff makes very unrealistic exposure assessments. Typically, these people combine all the worst-case default assumptions, ending up with unrealistic scenarios like this one: the typical person exposed is a 145-pound newborn with the appetite of a teenage boy, who works hard in the sun in front of his house 24 hours a day for 70 years, while an optimally bad breeze wafts towards him the maximum concentration of pollutant from the site. That reminds me of the advice to keep your feet on the ground, your shoulder to the wheel, your head in the clouds, your eyes on the stars — and then try to work in that position.

Clearly, no one is described by that kind of exposure assessment. Nobody has all the characteristics that are used as default assumptions. Common sense says that that's wrong; mathematically it's wrong, too. Stringing together those assumptions in that way is approximately the equivalent of multiplying together 90th- or 95th-percentile values from a series of distributions, believing that doing so will give you approximately the 90th- or 95th-percentile of the final distribution. It doesn't work that way. Instead what you get is something far out into the tail of the final distribution, beyond the 99th percentile. Mathematically, the proper procedure is to characterize the distributions, multiply them together to obtain a final distribution, and then identify the properly conservative value. Otherwise, only the means (or other measures of central tendency) can be multiplied together in this way.

Now, it's pretty clear why this occurs. It's part of the phenomenon of "CYA." No bureaucrat has ever been fired for going with the most healthy conservative estimate of risk. It's easy to see why that should be so. Each

38 INSURANCE AND RISK MANAGEMENT FOR HAZARDOUS WASTE

of these parameters is sanctioned as a default value by the agencies, not only by EPA but by FDA and everybody else in the business; these parameters are to be used when there is no other information available. When there are no other data on which to base an assessment, there's no penalty assigned for using these default values. So naturally people do. It's only human nature.

There's really no easy solution to this problem. It is something we can't legislate away because it is a widespread and ingrained practice. It will help to make one small change in the EPA Exposure Assessment Guidelines: to make explicit something that is now implicit. (Several of us will be working on that over the next year or so.) It is implicit in those guidelines that data are to be preferred to default values; this point needs to be made explicit. In the meantime, any of you who come up against an exposure estimate that relies on this kind of bad mathematics and bad science should raise a ruckus. Make it painful somehow for the people who have done the assessment not to change it, and do whatever seems appropriate to make them realize that there are better methods available.

Now on cancer hazardous assessments: Dr. Anderson described innovative approaches. What most of them have in common is that they take advantage of something that's now being recognized in the scientific community. When I say now being recognized, we need to say that there is not yet agreement on how best to describe the phenomenon. However, a consensus is forming around the idea that cancer in animals can be caused by two fundamentally different kinds of processes.

Present methodology for estimating low exposure hazard assumes that there is only one kind of process that can operate, namely the direct reaction of a chemical with DNA. This kind of reaction requires that the hazard be proportional to cumulative exposure; we call this *low-dose linearity* or a *nonthreshold process*. It has been the standard for more than a decade.

The second kind of process, just becoming widely recognized, is indirect. It involves a chemical speeding up the normal process of cell division. Several mechanisms are now known by which this can occur. Speeding up cell division increases the chance that normal or background mutagens (e.g, radiation, cosmic rays) will cause DNA damage that is trapped, as it were before repair, becomes a mutation, and leads to cancer. There are at least some cases — called by toxicologists *hormonally mediated* — where that process is now almost universally agreed to have a true threshold of biological response. What this means is that exposures that do not increase risk can occur. These cases occur because of the body's ability to accomodate to small changes in physiology without altering the status of important processes like cell turnover. There is a normal variation in these processes, and strong forces exist to keep them within usual limits.

EPA has just proposed to regulate a certain class of compounds called *goitrogens*, which cause enlargement of the thyroid and thus thyroid tumors, by a threshold process. In effect, the agency proposes to treat these tumors as a consequence of chronic toxicity, and to regulate protecting against such toxicity. It looks to me as though EPA will receive the assent of its Science Advisory Board on this interpretation, and that EPA will proceed to act on that basis.

Induction of goiter in rats and mice apparently leads to tumors because this condition is accompanied by a continuing elevation in the rate at which the thyroid cells divide. It is becoming apparent that other kinds of indirect or *nongenotoxic* animal carcinogens also create conditions of elevated cell division rates in the target organs. Theory predicts just such phenomena and rationalizes what is observed.

I suggest that all of these nongenotoxic carcinogens, which act only by elevating the rate of cell division, will turn out not to be effective below some level of exposure. It will be a couple of years before we know, but that's the way it looks to me right now. This will mean that we can treat them like chronic toxins, as EPA proposes now to do with goitrogens.

However, there is another implication of this that is probably more important for the immediate assessment of hazards — namely the recognition that the traditional kind of carcinogen, which we call *genotoxic*, can also act by this second process. The second process only operates at high doses. Yet compounds are only tested at high doses. Because the effects of the two processes reinforce each other synergistically, we are likely to overestimate the hazard.

We estimate cancer hazard by applying some mathematical formula, some algorithm, to the tumor incidence data obtained from animal bioassays. EPA typically estimates something called an *upper bound*, derived from the slope of the curve describing incidence as a function of exposure. The standard bioassay does not provide enough information to allow one to separate the two effects; it only describes the result of the two acting together. Because of synergy, the result is much greater than the sum of the independent parts.

Thus, for example, the well-known carcinogen benzo(a)pyrene appears to be 10 times as potent when the algorithm is applied to high-dose data as it is when evaluated using only low-dose data. The same thing is true of another famous experimental carcinogen called 2-AAF, which induces liver tumors in mice.

We don't know yet how many traditional carcinogens operate by both processes and are thus less dangerous at low exposure than we thought. However, we suspect that it might be the majority.

Five years ago, I was at the University of Pittsburgh for a conference on the epidemiology of hazardous waste sites. The consensus of the epidemiologists there at that time was that a one-armed man would be able to count on his remaining fingers all of the sites at which human health injury from a hazardous waste facility would be demonstrated. I think we now know of two examples. Nevertheless, those epidemiologists said that despite the fact that they expected to see few injuries, the sites ought to be cleaned up.

I think that's the message that the American public has been sending to all of us involved: whether or not there is any health risk, the sites ought to be cleaned up. I suppose the challenge to those of us in the health risk assessment business is to make sure that the assessments used for cleanup decisions give the most reliable picture possible of the danger posed. The risk managers, the people who are going to be making the decisions on what to do at these sites, can then take all the other factors into account and come to reasonable conclusions about what is to be done.

2 INSURABILITY ISSUES ASSOCIATED WITH CLEANING UP INACTIVE HAZARDOUS WASTE SITES

Leslie Cheek, III

2.1 Introduction

In December, 1987, the General Accounting Office (GAO), Congress' investigative arm, released a report[1] whose Executive Summary states matter-of-factly:

> While still not fully understood, the extent of the nation's potential hazardous waste problem appears to be much larger than is indicated by EPA's [Environmental Protection Agency] inventory of sites. GAO now estimates, largely on the basis of EPA data, that as many as 425,000 sites may need to be evaluated, compared with about 27,000 in CERCLIS [Comprehensive Environmental Response, Compensation, and Liability Information System], of which a small portion is expected to become NPL [National Priorities List] sites.[2]

Although the report notes EPA's belief that "only a small portion of the estimated number of sites will actually be found to require cleanup,"[3] the nearly 16-fold disparity between the EPA's CERCLIS list and GAO's inventory is a shocking reminder of how enormous the waste cleanup task may be.

The Congressional Office of Technology Assessment (OTA) estimated

41

42 INSURANCE AND RISK MANAGEMENT FOR HAZARDOUS WASTE

in 1985 that cleaning up the 22,000 *inactive* sites then on the CERCLIS list would cost as much as $100 billion, and that the task would take 15 years to complete.[4] Another GAO study, also released in December, 1987[5] concluded that the process of cleaning up *active* sites permitted under the Resource Conservation and Recovery Act (RCRA) would not be complete until fiscal year 2025,[6] and would cost up to $22.7 billion.[7]

While it would be inappropriate to extrapolate to the GAO's estimate of the number of sites that may require cleanup (425,000) the OTA estimate of cleanup costs for 22,000 CERCLIS-listed sites, it seems clear that the cleanup process will last well into the next century and will entail several *hundreds* of billions of dollars.

Indeed, factors other than the sheer numbers of sites suggest that even estimates of this magnitude may be too low.

First, the average cost of cleaning up an inactive site will increase by between *three* and *five* times as a result of the cleanup standards[8] mandated under the Superfund Amendments and Reauthorization Act of 1986 (SARA). A prominent consulting firm estimated in March, 1986, that under the standards that ultimately became law, "(s)ite cleanup costs could be expected to rise by at least a factor of 2.6, and possibly a factor of five or higher."[9] As a result of these more stringent standards, the study concluded:

> Total Superfund program costs could jump from $16 billion, as estimated in the CERCLA 301 (a) study, to $39–81 billion. The estimated number of sites capable of being addressed by a $16 billion program would fall from 1,800 assumed by EPA to 300–700 sites. . .[10]

Congress chose to ignore the reduced number of sites capable of being cleaned up under its new standards; indeed, it established a series of timetables that is forcing EPA to vastly accelerate the multistep cleanup process. Section 116 of SARA ordered EPA to complete preliminary assessments for all sites on the CERCLIS list as of October, 1986, by year-end 1987; to complete all necessary inspections a year later, and all evaluations two years later; and to complete evaluations of all sites listed in CERCLIS after October, 1986, within four years of their listing.

In addition, Congress directed EPA to complete remedial investigations/feasibility studies (RI/FS) for NPL sites according to the following schedule:

- 275 sites by October 17, 1989;
- a total of 450 sites by October 17, 1990; and
- a total of 650 sites by October 17, 1991.

INSURABILITY ISSUES: INACTIVE WASTE SITES

Finally, Congress directed EPA to begin remedial action at a minimum of 175 sites by October 17, 1989, and at another 200 sites by October 17, 1991.[11]

Second, and more important, the types of facilities currently finding their way onto the NPL pose progressively more complex issues of liability, remedy, and transaction costs than do earlier additions to the list, and therefore are likely to require progressively greater expenditures of both public and private resources. Most numerous among these types of facilities are the so-called *nonhazardous* waste facilities regulated under subtitle D of RCRA. Here is what the GAO said about these facilities in late 1987:

> The estimates of nonhazardous waste, or subtitle D, facilities that may require cleanup are far less precise, although there appear to be more than were reported in 1985. Altogether, there are reported to be 261,930 nonhazardous waste facilities in the United States, both active and closed. These do not have to have EPA permits to operate, however, and only half of the 227,127 operating facilities are subject to any state permitting requirements. Although 58 percent of subtitle D operating facilities are reported to have inspections at least once a year, only about one-third were actually inspected in 1984, and only 5 percent had groundwater monitoring systems. Many of these facilities existed before hazardous waste disposal was regulated, and any of them could be receiving hazardous wastes from companies or households that generate unregulated small quantities. EPA and the states have already found serious contamination problems at some of these types of facilities, including 184 subtitle D landfills on the NPL.
>
> For these reasons, EPA and state officials suspect that hazardous waste may be present, in some amount, at virtually all of the estimated 261,930 subtitle D facilities. Of these, 70,419 facilities, by their nature, have a high likelihood of being hazardous waste sites. As of 1984, 35,622 facilities received hazardous wastes from small quantity generators, i.e., those facilities that generated 1,000 kilograms of hazardous waste or less a month. Another 32,941 were establishments (locations that include one or more facilities) that were reported closed as of 1984, and therefore, because of their age, were most likely accepting hazardous wastes before the disposal of hazardous waste was regulated. In addition, 1,856 are facilities that EPA classifies as open dumps because they pose a reasonable probability of adverse effects on health or the environment.[12]

These "nonhazardous" facilities include most of the nation's municipal landfills, some of which contain enormous volumes of waste. Cleanup actions at these sites will involve hundreds, if not thousands, of PRPs, will entail remedies of staggering cost; and will require litigation of stupefying complexity.

While the GAO study makes it clear that there is no way of predicting how many "nonhazardous" facilities will ultimately end up on the NPL,

44   INSURANCE AND RISK MANAGEMENT FOR HAZARDOUS WASTE

the fact that nearly one out of every five sites *currently* on the NPL is a nonregulated landfill suggests that the number will be huge.[13]

The current average cleanup cost per CERCLA site is already more than $10 million, when EPA and PRP legal expenses and other transactions costs are added to the $9.2 million average EPA remedy cost.[14] If, as has been responsibly estimated, the SARA cleanup standards boost this average by anywhere from three to five times the current per-site cost, each NPL site cleanup in years to come will cost between $30 million and $50 million.

If, out of a universe of 425,000 potential NPL sites, a mere 15,000 are ultimately listed for cleanup, the resulting cost will range from $675 *billion* to more than $1 *trillion*.

It strains credulity to believe that Congress will force American business to incur liabilities on this scale to deal with an environmental problem that ranks nowhere near the top of any rational set of public health priorities. However, until public policymakers recognize the need to allocate resources in a cost-beneficial manner, business will be saddled with a growing bill for cleaning up inactive hazardous waste disposal sites.

The immediate challenge to business in these circumstances is to make the best of a bad situation. In enacting SARA, Congress emphatically rejected the business community's argument that it makes little sense to finance the biggest public works project in the nation's history on a case-by-case basis, utilizing what has been termed in other contexts as the most complicated, expensive, arbitrary, and unpredictable legal system ever devised. At least until the next reauthorization of CERCLA/SARA, and perhaps beyond then, business must live with the fact that every nontax dollar that ultimately find its way into the cleanup process must be pressed through a legal and bureaucratic sieve that pits government against its business taxpayers, business against business, business against its insurers, and insurers against their reinsurers, leaving behind a sorry residue of huge legal fees, vast wastage of judicial time and resources, interminable delays, and venomous cynicism.

While conflict is inherent in the adversarial nature of CERCLA'S liability scheme, the intensity of that conflict has steadily escalated as both PRPs and their insurers have recognized the ruinous costs that CERCLA, as amended by SARA, could impose on them. For many corporations and their insurers, the fight to avoid or shift CERCLA liability is, quite literally, a fight for corporate life. As will be seen below, the stakes in CERCLA are so huge that both PRPs and their insurers have taken to launching preemptive legal strikes against each other, in attempts to learn, in advance of any legal determination of CERCLA liability, who will be required to pick up the tab.

INSURABILITY ISSUES: INACTIVE WASTE SITES 45

The megatrial recently concluded in the *Shell* waste cleanup coverage case is but a faint harbinger of what is to come. In their complexity, their expense, and their waste of time, the coverage suits are, by themselves, a compelling argument for change. Regardless of their outcomes, they demonstrate beyond cavil that there has to be a better way to fund the costs of cleaning up America's hazardous wastes.

2.2 Armageddon Now: Waste Cleanup Coverage Litigation

When the Federal 8th Circuit Court of Appeals ruled on February 26, 1988, that standard-form Comprehensive General Liability (CGL) insurance policies do not provide coverage for waste generators who must reimburse EPA for CERCLA cleanup costs or are ordered to do the job themselves,[15] the American Insurance Association (AIA) characterized the decision as "a victory of major importance."[16] Said AIA President Robert E. Vagley:

> The federal courts have now made clear that industrial polluters are not going to be allowed to shift the cost of cleaning up their hazardous waste under the Superfund programs to their insurers.
>
> Insurance companies did not agree to accept hazardous waste generators' burdens of complying with environmental requirements, did not charge premiums for that risk, and entered into contracts that clearly did not cover this kind of expense. Two federal appeals courts have confirmed those conclusions. We believe the issue should now properly be regarded as settled.[17]

While it is true that the *NEPACCO* decision adopted the analysis of the only other Federal Circuit Court decision on the issue, *Maryland Casualty Co. v. Armco, Inc.*[18] and that the Supreme Court refused to review that decision, the issue is far from settled. A review of *state* court decisions on the identical issue reveals only that insurers and PRPs have achieved the judicial equivalent of a Mexican standoff.

Moreover, the Supreme Court's refusal to grant *certiorari* in *Armco* does not necessarily mean that the high court approves of the result in that case. The best proof of this caveat lies in the court's serial refusals to review the widely divergent Circuit Court results in three major asbestos disease coverage cases in the early 1980s.[19]

Indeed, it appears from the escalating proliferation of huge, preemptive declaratory judgment actions relating to insurance coverage for waste cleanup obligations that the issue is not only unsettled, but is fast becoming the PRP-insurer answer to the regional lottery.[20]

The opening paragraph of a February 1, 1988, *Business Insurance* magazine article headlined *Superfund Unleashes Flurry of Coverage Suits* announced that attorneys for both policyholders and insurers agree that "litigation over insurance coverage for government-ordered hazardous waste site cleanups may rival asbestos-related coverage lawsuits in terms of time, expense and complexity."[21]

At this writing, the most recent of the megadeclaratory judgment actions was that filed in December of 1987 in Massachusetts Superior Court by United Technologies Corp. (UTC) and six of its subsidiaries against 240 of their first-party property and third-party liability insurers from the past 37 years. The UTC suit seeks defense and indemnification coverage for 102 off-site pollution cleanup claims and 36 on-site claims in 26 states.

The other major actions include Westinghouse Electric Corp.'s May, 1987, New Jersey claim against more than 140 of its insurers for coverage of cleanup costs at 74 hazardous waste sites and the costs of defending hundreds of bodily injury claims by customers alleging exposure to toxic substances in the company's products; and Shell Oil Co.'s 1983 suit against 270 of its liability insurers for coverage costs at waste sites in Colorado and California.

Of the megasuits, only the Shell case has been concluded. But enough other miniactions have been decided to make it anyone's guess as to how the larger actions will turn out.

About all that can be said for current state case law on the cleanup coverage issue is that jurisdictions noted for the generally pro-plaintiff bias of their judiciaries have tended to side with policyholders[22] while states with more conservative legal traditions have tended to support insurers' views.[23] A recent review of developments in this burgeoning field accurately noted that "liability for past handling, transportation, storage, treatment and disposal of hazardous wastes at currently inactive facilities (is) the most controversial environmental law issue of the decade."[24]

Indeed, the controversy is already so well developed that the major players have established institutions to enhance their positions on coverage issues: EPA's PRP demand letters now contain boiler-plate requests for information on the recipients' insurance coverage for the preceding 50 years; site-specific PRP steering committees are as happy to oversee coverage warfare with their insurers as to do battle with EPA over the apportionment of cleanup liability; the carriers have organized the Insurance Environmental Litigation Association (IELA) to assure nationwide coherence of position among insurers and provide *amicus curiae* support in key coverage cases; and the Chemical Manufacturers Association (CMA) recently established its own coverage litigation unit.

The environmental-coverage litigation community is now large enough

INSURABILITY ISSUES: INACTIVE WASTE SITES 47

to support a thriving trade press and a booming conference business. And law firms with both PRP and insurance defense clients are bidding up the prices of experienced coverage litigators.

In other words, a new industry has been born, ready to serve a widening circle of customers as the CERCLA net is cast ever more broadly and the stakes at issue grow daily larger. A future paved with never-ending legal fees stretches toward a horizon that grows more distant with each addition to the NPL.

Given the theoretical coverage litigation possibilities inherent in EPA actions at multigenerator NPL sites, it is small wonder that the legal fraternity is bullish on this corner of environmental practice. It is not at all outlandish to picture the following scenario developing out of the cleanup of an NPL-listed urban landfill used by 1000 PRPs for 10 years:

If each of the PRP's had only three primary liability insurers during the landfill's life, as many as 3000 insurers could become defendants (or plaintiffs!) in coverage actions; they would face litigation involving as many as 10,000 individual insurance contracts.

If each of the primary carriers had but five excess limits, umbrella, or reinsurance carriers on these 10,000 risks, as many as 50,000 additional suits might result from disputes among these entities, which in turn might face litigation involving *their* retrocessionaires!

The 10-year period could span the years during which the insurance industry modified its CGL policies from an accident to an occurrence basis, and added the pollution exclusion to the latter form, thereby injecting the mind-numbing possibility of as many as 100 *million* separate cross-claims among the primary carriers alone, as the insurers of all 10,000 involved contracts each seek to pin liability on the underwriters of policies other than their own.

If the excess limits or reinsurance carriers refuse to follow form as a result of decisions in these cases, the number of potential lawsuits climbs quickly into the *billions*, all as a result of the cleanup of one of hundreds or perhaps thousands of similar landfills.

What is to prevent the different (or perhaps many of the same) parties to cleanup at other landfills from repeating the same pattern of litigation, in the hope of different outcomes? The recent history of both tort and insurance contract law suggests strongly that no American common-law precedent, no matter how hoary with age or tradition, has a half-life longer than it takes a trial court judge to write an opinion overruling it.

From the standpoint of insurers, the volatility of the common law was most startlingly demonstrated in the trio of cases arising out of the cleanup of a Jackson Township, New Jersey, landfill[25] and in the February, 1987,

48 INSURANCE AND RISK MANAGEMENT FOR HAZARDOUS WASTE

decision of a New Jersey Superior Court in *Summit Associates, Inc. v. Liberty Mutual Fire Insurance Co.*[26]

Until the *Summit Associates* case, the most alarming precedents for insurers were those growing out of the contamination of wells serving 97 families by seepage of hazardous wastes from a landfill owned and operated by Jackson Township, New Jersey. In combination, these decisions destroyed the efficacy of the CGL's pollution exclusion, converted the CGL's per *occurrence* policy limit into a per *claim* policy limit, and overturned centuries of common (and insurance contract) law by awarding damages for the mere *possibility* of future harm. Only the last of these decisions has been modified on appeal.

New Jersey's courts have simply written the word *sudden* out of the phrase *sudden and accidental* in the CGL pollution exclusion, thereby eliminating the *temporal* distinctions that underlie the exclusion and converting it into a matter of the insured's *intent* in performing the act that gave rise to the pollution incident. The *Jackson Township* court dismissed the key phrase in the exclusion as follows:

> When viewed in the light of the case law cited, the clause can be interpreted as simply a restatement of the definition of 'occurrence.' The injury was 'neither expected nor intended.' It is a reaffirmation of the principle that coverage will not be provided for intended results of intentional acts but will be provided for the unintended results of an intentional act.

Having found insurance coverage for a transparently non-sudden occurrence under contracts patently intended to deal with sudden *and* accidental events, New Jersey jurists next went to work on the application of the term *occurrence* to the facts in the Jackson township case. The *American Home* court found that the occurrence was not the seepage of wastes from the landfill that contaminated the wells, but rather the separate contamination of *each* of the wells. Looking at the language in the occurrence definition stating that "continuous or repeated exposure to substantially the same general conditions shall be considered as arising out of one occurrence," the court found that each of the 97 plaintiff families had been exposed to different conditions, in that "they ingested different quantities of contaminated water; different V.O.C's, at different times, and each family's duration of exposure varied. . . ." Thus, with a stroke of the judicial pen, the insurer's liability went from X to 97X.

The *Ayers* court did to the CGL's coverage of bodily injury what the *American Home* court did to the definition of occurrence: it ignored it, and awarded $13.4 million in damages to the 360 individual plaintiffs, not

INSURABILITY ISSUES: INACTIVE WASTE SITES

one of whom even alleged bodily injury, as follows: $8.2 million for the creation of a medical surveillance trust fund; $5 million for "loss of quality of life"; and $200,000 for "emotional distress." (The court also awarded more than $2 million to cover the cost of hooking the plaintiffs' houses up to the municipal water supply system.) On appeal, only the $8.2 million in medical surveillance funds was deleted from the award.

Happily, not all courts have dealt so cavalierly with insurers' efforts to enforce the temporal aspects of the pollution exclusion. In *Great Lakes Container Corp. v. National Union Fire Insurance Co.*, 727 F.2d 30 (1st Cir. 1984), for example, the court found no ambiguity in the exclusion:

> We agree with the District Court that when the policy is read against the complaint, there is no ambiguity and exclusion (f) applies. The government has alleged that Great Lakes is liable because pollution and contamination of the soil, surface and subsurface waters has taken place as a concomitant of its regular business activity. Property damage resulting from such activity falls squarely within the language of exclusion (f). There is no "occurrence" within the meaning of the policy alleged, nor any allegation of a sudden and accidental discharge.

Similar results were reached in *American States Insurance Co. v. Maryland Casualty Co.*, 587 F. Supp. 1549 (E. E. Mich., 1984) and *Barmet of Indiana, Inc. v. Security Insurance Group*, 425 N.E. 2d 201 (Ind. App., 1981).

The ability of courts to paralyze the will of the insurance industry to provide environmental liability coverage in the future by reinterpreting or, worse, totally ignoring the plain language of past contracts between insurers and insureds was best illustrated in the *Summit Associates* case.

Summit Associates, Inc., purchased a piece of property in Edison Township, New Jersey, unaware that it had previously been used by the township as a sewage treatment facility. When workmen discovered toxic wastes on the property, Summit was ordered to remove 150 tons of sludge and 50,000 gallons of liquid waste from the ground, at a cost of some $438,600. Summit filed a claim with its commercial multiperil insurer, Liberty Mutual, for its cleanup and removal costs.

Liberty denied the claim on the basis of two explicit exclusions in its contract — one precluding coverage of any pollution damages unless the occurrence giving rise to those damages is both "sudden and accidental," and the other precluding claims for damage to property owned by the insured.

The New Jersey court declared both exclusions "ambiguous," and thereby found a way to construe them "liberally in favor of the insured." "Sudden and accidental," said the court, really means "neither expected

50 INSURANCE AND RISK MANAGEMENT FOR HAZARDOUS WASTE

nor intended from the standpoint of the insured," and thus the pollution exclusion is inapplicable. As for the owned property exclusion, the court reasoned as follows:

> ... the underlying Public Policy in this area is quite clear when the potential for damage to the health, safety and welfare of the people of this State must outweigh the express provisions of the insurance policy in issue. As a result, the exclusion clause in the policy which pertains to excluding coverage where the damage is to the policy holders [sic] land, must be held inapplicable where the danger to the environment is extreme.

Not only did the court cite no precedent whatsoever in support of this proposition, it also went on to explain why it had chosen to vitiate a contract heretofore found perfectly consonant with public policy:

> ... The question that arises is what party will bear the burden of the cost of the clean-up in a situation where the landowner does not have the resources to pay for the cost of the clean-up? Certainly, to impose such clean-up costs on government agencies would certainly create an undue burden on taxpayers, who should not be forced to assume such a burden in cases involving private landowners. . . .
> ... A precedent must be set to provide coverage for the case where the private landowner is ordered to undertake the necessary clean-up. Thus, exclusions denying coverage for damage to property owned by the insured should not be applied under these circumstances. . . .
> This policy must control over the plain meaning doctrine in situations such as that presented by this case, because of the nature of the case, the potential damage which may result, and the cost which may be imposed upon a landowner.

Consistent with this whole-cloth formulation, the court ordered Liberty to reimburse Summit for all its cleanup costs, prejudgment interest, *and* $37,000 in attorneys' fees.

What the New Jersey court really said was this: "We don't give a damn what an insurance policy that was negotiated in good faith between two contracting parties says when somebody other than the insurer might have to pay a loss plainly excluded from the policy's coverage. We're going to make the insurer pay whenever we find a 'public policy' rationale for doing so."

No insurer's lawyer reading this decision is ever going to recommend that his client write commercial multiperil policies for New Jersey risks with even the remotest potential for pollution loss, because there would be no assurance whatsoever that the terms and conditions on which the decisions to insure and the prices of the coverage were based would be respected by the state's courts.

INSURABILITY ISSUES: INACTIVE WASTE SITES 51

From the standpoint of PRPs, however, the *Jackson Township* and *Summit Associates* cases must appear as beacons in a landscape benighted by decisions like those in the *Nepacco*, *Armco* and *Great Lakes Container Corp.* cases.

The fact that there are strongly worded precedents for both camps' views, and the fact that many of these precedents reversed earlier or lower-court holdings, have simultaneously heightened both camps' anxiety and provided new incentives to litigation. The absence of a clear trend in these cases has given both sides no choice but to continue, like Iran and Iraq, a war whose original cause became pointless in the escalating bloodshed.

2.3 Apocalypse Tomorrow: The Socioeconomic Fallout From Unchecked Waste Cleanup Coverage Litigation

Just as neither Iran nor Iraq won their conflict, neither PRPs nor insurers will win the war over the insurance coverage of waste cleanup obligations. Indeed, the consequences of either side winning are so frightening as to make the current stalemate, frustrating though it may be, preferable to a decisive outcome.

If the majority of courts were to decide that waste cleanup costs are covered by CGL policies (with or without pollution exclusions), the resulting exposure would almost certainly bankrupt every major liability insurance carrier in this country, and many of their foreign reinsurers as well.

By the same token, if that majority were to rule that waste disposal is not the sort of fortuitous event insurance was intended to handle, then thousands of businesses would face huge and wholly unanticipated retroactive liability for decades of routine and entirely lawful commercial practice. Many of these businesses would be forced to close, others would have to assume crippling debt, and all would be hamstrung in their ability to compete.

The American property–casualty insurance industry is so huge (with annual premiums in excess of $160 billion, assets of more than $374 billion, and some 500,000 employees working for more than 3500 companies)[27] and so universally unpopular (insurance comes into play only when things go wrong) that it is not hard to understand why some courts have assumed, niceties of contract language to the contrary notwithstanding, that insurers can more easily absorb the costs of Congress' *post hoc* remedy for 200

52   INSURANCE AND RISK MANAGEMENT FOR HAZARDOUS WASTE

years of unsafe hazardous waste disposal than can the business and governmental entities that make up the community of PRPs.

Nor is it particularly surprising that some courts seem to assume that insurers have virtually limitless reserves that can be tapped for unforeseen liabilities, even those as large as cleanup costs under CERCLA as modified by SARA.

The irony of these assumptions is that the insurance business is no more clairvoyant than any other business, and, like all other enterprises, prices its products and projects its reserve needs on the basis of past experience extrapolated into the future, using reasonable loss development and inflationary trend factors.

Upon close analysis, it is illogical to assume that insurers of general liability contracts written in the decades prior to CERCLA's 1980 enactment could have foreseen that enactment and its attendant economic consequences and built these costs into their prices and reserve calculations. It is beyond credulity to believe that liability insurers were any better prepared than the rest of American business for Congress' decision to cram retroactive, strict, joint and several liability for waste cleanup down their collective throat.

The simple fact is that prior to the most recent policy years, insurers had no experience whatsoever with waste cleanup costs, and thus had *no* reserves for CERCLA liabilities established under any policy written prior to its enactment.[28]

Under normal circumstances, anticipated reserve requirements are built into the prices that insurers charge in given lines of insurance, and actual reserves are established on a case-by-case basis, with the amounts involved taxed against the premiums collected during the year in which the accident giving rise to the case occurred. In this manner, the industry is usually able to determine, reasonably soon after the fact (three years in such short-tail lines as automobile and homeowners', five or more in the long-tail general liability line and its medical malpractice and product liability sublines), whether its original anticipated reserve requirements were accurately calculated.

If these requirements were not correctly estimated, neither past nor future policyholders can be taxed for the deficiencies involved.

Insurers cannot go back to holders of occurrence-based policies long ago presumed closed and say, "Because recent changes in the law made your premiums 600% inadequate, we are now billing you for the balance." These policyholders would rightly say (and, indeed, repeatedly have argued in their coverage briefs) that they paid their money precisely to relieve themselves of the burden of such unanticipated liabilities. Some have gone

INSURABILITY ISSUES: INACTIVE WASTE SITES

even further, arguing that even though insurers could not have foreseen the enactment of CERCLA, they nevertheless intended to cover the liabilities that statute created.

Nor would insurance regulators or the imperatives of competition permit insurers to say to future claims-made policyholders, "Because retroactively imposed liabilities made the prices we charged for occurrence-based policies in the 1960s and 1970s too low, you will have to make up the shortfall." As noted above, state-approved rating and statistical plans permit only limited reflection of past rate inadequacy in future insurance prices. And even if full recoupment were permitted, insurers seeking it would quickly lose market share to carriers without such burdens from the past. Finally, the entire point of claims-made policies is to bring their pricing closer in time to the loss-and-expense experience of the insured class by eliminating the endless retroactive effect of the occurrence-based policy form. Claims-made insureds would rightly reject pricing practices that incorporated experience from an occurrence-based past.

Similarly, rate regulatory statutes require that future insurance prices be based primarily on actual loss and expense experience of current and recent insureds, not on speculation about losses and expenses that may arise under policies from the distant past. Here again, the exigencies of competition for ongoing business would supplement regulatory strictures against speculation in insurance pricing practices.

Given the regulatory and practical restraints on insurers' ability to reach either backward or forward to accumulate reserves for huge unanticipated exposures, the industry would have but two means of attempting to fund these costs — profits from its ongoing business and the surplus built up during its 200-year history. Neither alternative, nor the two in combination, would come close to meeting the potential exposure; worse, either would doom many of the industry's major competitors.

In a growing economy, insurance capacity must increase in order to keep pace with the demand for coverage of new lives, enterprises, homes, and possessions, and with the steady increase in the cost of goods and services for which insurance pays. Increased capacity is a function of the capital a free enterprise market is willing to commit to the insurance function. The insurance business can attract and retain capital only if it can provide a competitive return on invested funds. If its profits are eliminated or radically curtailed to create reserves for unmeasurable CERCLA exposure, capital will be withdrawn from the business.

The resulting shrinkage of insurance surplus would first create escalating shortage of needed coverage, and price increases as demand exceeded supply and insurers sought to rebuild surplus through improved profita-

54   INSURANCE AND RISK MANAGEMENT FOR HAZARDOUS WASTE

bility. Higher prices would accelerate the flight of superior risks to such alternatives to commercial insurance as captive insurers and risk retention groups. This adverse selection process, in turn, would exacerbate insurers' loss and expense experience, thereby precipitating still more availability problems and still higher prices.

As was the case during the aftermath of the 1980–1984 commercial-insurance price war, insurers would tend to reduce their writings in their least profitable lines first. Recently, the least profitable lines have been those covering the commercial, professional, and governmental entities most profoundly affected by continuing rapid changes in common-law tort liability. The capacity recovery after the price war has restored market stability in most of these lines. An ongoing shortage of capacity in these lines would create severe dislocation in the provision of vital goods and services, since many providers could not risk operating without insurance protection.

As profits disappeared, the stock prices of investor-owned insurers would decline, and these insurers would find it both difficult and expensive to raise additional capital. Mutual insurers would have to finance any additions to surplus entirely out of policyholder premiums and dividends.

Should CERCLA-related losses and expenses exceed any funds diverted from profits and additions to surplus, insurers would be required to dip into surplus, a step with profound implications for insurance availability and cost. For every dollar withdrawn from surplus, an insurer must curtail its premium writings by at least two dollars, in order to maintain an appropriate ratio between premiums and surplus.[29] If invasion of surplus were to become widespread within the industry, severe shortages of capacity would soon become evident even in relatively profitable lines.

Invasion of surplus is an insurer's last and most dangerous defense against insolvency. If only 10% of an insurer's surplus is invaded in a given year, that insurer must reduce its premium writings by 20%, 30%, or more, thereby sharply curtailing its cash flow and further limiting its profit-making potential. A series of years of similar surplus impairments can bankrupt an insurer in less than five years.

The surplus of the entire property-casualty insurance industry is currently about $110 billion. As large as this figure seems in the abstract, it is the foundation upon which the industry's $200 billion in premiums is written, not just for contracts that may be called upon to respond to CERCLA-related claims, but for every other policy the industry writes as well. Expropriation or erosion of this surplus would threaten the entire industry, and deprive millions of Americans of the protection they need in their personal and professional activities.

INSURABILITY ISSUES: INACTIVE WASTE SITES 55

In an individual case involving perhaps a few hundreds of thousands or even several millions of dollars, a judge who is determined to ignore insurance contract language denying a policyholder coverage for a CERCLA-related claim might find it easy to rationalize the small chip his decision to find coverage will take out of the edifice of the American insurance business. But because, in virtually all states, insurance contract law is judge-made (as opposed to statutory) law, that single decision may be dispositive of hundreds of other cases and thousands of pending claims. Furthermore, the decision may be adopted as precedent by other jurisdictions, compounding its effect on insurers' exposure. Thus, a single decision can easily create literally billions of dollars in unanticipated and unreserved exposure for hundreds of liability insurers.[30]

Of course, as noted earlier, insurers appear to be winning as many of these cases as they lose (perhaps more), meaning that the unanticipated liabilities at issue in those cases may have fallen on enterprises or governmental entities as ill-prepared for them on an individual basis as the insurance industry is in the aggregate. Indeed, for some individual policyholders, these liabilities may have been catastrophic.

Over time, however, the ostensible disparity in economic power between insurers and policyholders may lead a majority of courts to shift cleanup costs to insurers, particularly as the profile of the typical PRP changes.

The very early CERCLA cost-recovery cases (e.g., Conservation Chemical Co.) were brought for sites at which the PRPs were well-heeled members of the Fortune 500. Only now are these cases reaching the second tier of CERCLA-spawned litigation: EPA and the PRPs have settled the differences between and among them, and the PRPs are in the process of trying to hand the legal and cleanup bills to their insurers.

In these cases, with giant industrial corporations squaring off against huge insurers, there is less temptation for the courts to take the David vs. Goliath position that the insurance policies at issue were contracts of adhesion forced on helpless policyholders by omnipotent insurers, or that sophisticated corporate risk managers had a "reasonable expectation" that their policies would cover waste cleanup costs.

But the current generation of CERCLA cases is sweeping progressively smaller entities into the PRP net, and future generations of cases, involving vast urban landfills, will bring the wonders of CERCLA liability to every mom-and-pop entity that ever disposed of hazardous wastes. What impact will the changing demography of the PRP community have on judicial trends in waste-related coverage litigation?

In a colloquy published recently in the *Docket*, the journal of the Amer-

56 INSURANCE AND RISK MANAGEMENT FOR HAZARDOUS WASTE

ican Corporate Counsel Association, the author of this paper and Washington environmental attorney David B. Graham debated this question as follows:

> GRAHAM: . . . (T)hink about the future: Congress has given the cleanup of hazardous wastes its highest environmental priority, with what poll after poll shows to be overwhelming public support. You and I both know how expensive the cleanup job will be, and we both know that only a fraction of the necessary money will come out of the Federal treasury and from the Superfund taxes imposed on industry. Where is the balance to come from?
>
> Do you believe that the majority of judges, who read the same polls and newspapers that we do, are going to let words in insurance forms that numerous courts have found to be ambiguous stand in the way of mobilizing funds for a popular cause? Do you think that these judges will allow PRPs to file for bankruptcy when there is any plausible basis for finding insurance coverage for the PRPs' Superfund obligations?
>
> CHEEK: These judges are sworn to uphold the law, not to ponder notions of what is or is not a worthwhile reason to find insurance coverage where none exists.
>
> Moreover, assuming for the sake of argument that judges make decisions on the basis of socio-economic, rather than legal considerations, the state law equivalent of bankruptcy is also available to insurers, who not only provide jobs but also play a vital role in protecting the economic security of millions of households and businesses. I have to believe that judges understand that they cannot indefinitely pluck feathers from the insurance goose without threatening its survival, regardless of the popularity of the cause at issue.[31]

In many jurisdictions, trial and even appellate judges are elected public officials, and since, in such common-law fields as insurance contracts, the law is what the latest decision says it is, it is at least a strong possibility that locally elected trial judges (and locally selected juries) will tend to side with local employers who provide jobs to local voters in contract disputes with large insurers domiciled elsewhere.

But while elected local, and perhaps appellate, judges may have no political choice but to side with local employers, appointed State Supreme Court justices and Federal District and Circuit Court judges are in positions in which they can consider both unpalatable prongs of the Hobson's choice that CERCLA has spawned — namely, which is it to be: the serial destruction of every local economy in America or the piecemeal dismemberment of the nation's property-casualty insurance industry?

It is this writer's view that either extreme, and any point in between these extremes, is unacceptable. The cleanup of hazardous waste is a worthwhile objective, but as this analysis makes clear, continlued reliance on the judicial branch of government to finance history's costliest public works

INSURABILITY ISSUES: INACTIVE WASTE SITES

project on a case-by-case basis, using the most transaction-cost-intensive legal system in the world to extract billions of dollars in retroactive taxes from entities who could not have foreseen them and thus will be bankrupted by them, will become increasingly dangerous to this nation's economy.

2.4 False Hope: Judicial or Legislative Resolution of Insurance Coverage Issues

Rationality suggests that neither the courts nor Congress will permit the Apocalypse described above to occur. Politics suggests otherwise. It is this writer's considered opinion that the Supreme Court will not inject itself into a judicial fray revolving around entirely state-law issues, and that for Congress, environmental politics, like love, means never having to say you're sorry.

Most businessmen are conservative, and have instinctively welcomed the current administration's attacks on judicial activism and its careful appointments of strict constructionists to the federal bench. Under normal circumstances, PRPs and insurers alike would likely cheer for less federal judicial intervention in state law matters, and would wholeheartedly embrace the notion that there is no Federal common law.[32]

The impending apocalypse in hazardous waste-related coverage litigation, however, may remind both insurers and policyholders of Tom Wolfe's epigram: "A liberal is a conservative who's been arrested."[33] Both may find themselves longing for a single, definitive United States Supreme Court ruling on what is and is not covered under insurance contracts in waste cleanup actions, issues historically governed entirely by state, not federal, law.

Such judicial relief is unlikely, for a number of reasons. First, the Supreme Court has already had, and has wordlessly declined, several opportunities to rule on quite similar issues in the key asbestos disease coverage cases of the early 1980s.[34] These cases had enough procedural hooks for a grant of *certiorari* — a split among the Circuits being chief among them — but were missing the substantive hook — a question of federal law. All of the cases had arisen under state law, and the federal courts had taken the cases under their *diversity of citizenship*[35] rather than *federal question*[36] jurisdiction. Although the Supreme Court did not articulate its reasons for denying *certiorari* in the asbestos disease coverage cases, the absence of a federal question was undoubtedly a crucial factor.

Second, as a general matter, the Supreme Court has been reluctant to override state law in the absence of federal legislation on the same issues.[37] Thus, it is unlikely, in the absence of any federal statute dealing directly with insurance contract interpretation, that the Supreme Court would be willing to craft, out of whole cloth, a federal common law of insurance contract interpretation.[39]

Fourth, were Congress to attempt to resolve the problem by statute, its rules would not be given preemptive effect by the Supreme Court unless Congress were specifically to override all state insurance contract law. The Court has held that preemption of state law cannot be implied from a federal statute dealing with the same matter.[40]

Finally, the Court is as politically sensitive as the other branches of the federal government, and would undoubtedly welcome any excuse to avoid having to choose between the two horns of the Hobson's dilemma that the waste cleanup coverage issue represents. A definitive rule on the issue would mean economic ruin on a spectacular scale, regardless of which choice the Court made.

The same political considerations, and others, militate against a legislative resolution of the coverage issue, short of a redrafting of CERCLA that would relieve individual PRPs of the retroactive liability imposed under the original 1980 statute.[41]

It is a truism that Congress, as an essentially reactive institution, will not act on a particular problem until it is forced by circumstance to do so. And when Congress does act, it does so by doing the bare minimum necessary to placate the interests involved. With the apocalypse of waste cleanup coverage litigation some years down the road, the crisis that typically moves Congress to action has yet to occur.

Moreover, Congress tends to act only when a political consensus has been reached on both the need for action and the nature of the action needed. With the outcomes of the current waste cleanup coverage cases looking more or less like the judicial equivalent of a Mexican standoff, a consensus on the need for legislative intervention, much less a consensus on what form that intervention should take, is obviously remote.

Consensus itself implies a middle ground, with all affected interests surrendering some of their objectives in order to achieve the most important of their goals. The current scorched-earth, all-or-nothing approach to the cleanup coverage issues by both PRPs and insurers, illustrated most vividly by the dozens of preemptive megasuits now taxing judicial resources across the land, suggests that it will be a long time before the affected interests will be able or willing to reach a compromise on their differences that they could present to the Congress for legislative endorsement.

And if consensus among the affected interests is indeed an essential prerequisite for Congressional intervention, it would seem unlikely, under current circumstances, that any legislative solution would provide either

PRPs or insurers with the relief that both ultimately desire — predictable, finite obligations that can be factored into their business planning without undue hardship. This is so because PRPs and insurers are not the only parties necessary to a viable consensus.

Meeting PRPs and insurers' ultimate objectives would entail throttling back on the engine that is driving the cleanup coverage litigation, namely CERCLA's tort law-based liability scheme. The business community's lack of success in tempering either the original 1980 enactment or the 1986 reauthorization of CERCLA suggests, if anything, that the political momentum behind CERCLA may be impossible to halt, much less reverse.

First, although what many PRPs did in decades past could hardly have been described as pollution under then-prevailing definitions, and although the CERCLA tax system is inconsistent with it, there is an undeniable rough justice in the principle ostensibly underlying CERCLA: the polluter pays. Any proposed legislative resolution of the cleanup coverage disputes that even appears to be at odds with this principle will earn the immediate opposition of the professional environmental organizations and their legions of highly motivated supporters.

Second, for reasons of institutional pride alone, it would be extremely difficult for Congress effectively to admit that it erred in both 1980 and 1986, as it would be forced to if any relief were to be granted to PRPs and their insurers. Having twice embraced a fault-based approach to the funding of a societal need, Congress would find it hard to concede that it made the same mistake twice, even if the evidence to this effect were overwhelming.

Third, as a purely fiscal matter, Congress would be hard-pressed to justify giving the cleanup of hazardous wastes a higher priority for general revenue funding. The continuing federal budget deficit has already forced Congress to reduce its commitment to national objectives of far greater political importance than waste cleanup; Congress would be unlikely to voluntarily exacerbate its serious resource allocation problems solely to provide relief to politically unpopular segments of society.

Fourth, but for their cataclysmic socioeconomic implications, it would be easy for the opponents of change in CERCLA to argue that the PRP-insurer disputes are nothing more than private quarrels whose resolution is not necessary to the public interest. Just as many state legislators look upon the debate over the future shape of American tort law as but another round in the fight between plaintiffs' lawyers and defendants' insurers, so might Congress be convinced that waste cleanup coverage litigation is nothing more than an internecine squabble among members of the business community the outcome of which is irrelevant to the goals of CERCLA.

60   INSURANCE AND RISK MANAGEMENT FOR HAZARDOUS WASTE

Fifth, until the ranks of the PRP community are swollen with the small businesses, school boards, sanitary districts, counties and municipalities who will increasingly feel the sting of CERCLA liability, the interests advancing reform — mostly large corporations and major insurers — would be unable to overcome the combination of inertia and environmental political power behind the status quo.

Finally, Congress has already rejected one major effort to overhaul CERCLA (AIA's 1985 proposal)[42] and an attempt by a small group of insurers to resolve the coverage issues through an amendment to SARA.[43]

AIA's proposal was condemned by the business community for its substitution of higher taxes for CERCLA liability at NPL sites, excoriated by the Administration for its insistence on fair apportionment of post-1980 CERCLA obligations, and attacked by the environmental community for its fault-blind approach to dealing with the consequences of pre-1980 waste disposal. Indeed, the storm of criticism was so loud that the AIA was unable to find a Congressional sponsor for its draft legislation.

The insurers' proposed amendment to SARA, developed after the broader AIA proposal died a premature death, was never formally introduced. Then-chairman Robert T. Stafford (R–VT) of the Senate Committee on Environment and Public Works, who was sympathetic to the amendment, was unable to persuade any other member of the Committee to cosponsor it and decided that it would be wiser, under those circumstances, not to offer the amendment than to propose it and risk having its rejection used by its opponents to further their contrary interests.

Assuming, *arguendo*, that the foregoing analysis is correct, and that neither judicial nor legislative resolution of the waste cleanup coverage disputes is either feasible or imminent, there remains the third branch of the federal government, the Executive, whose power in this context resides with EPA. What prospect is there for relief through the administration and bureaucratic interpretation of CERCLA and SARA?

At this writing, EPA is showing encouraging signs that it actually wants to make CERCLA and SARA work. The agency has aggressively utilized its authority to enter mixed funding (fund and PRP resources) agreements to expedite cleanup[44] and has instructed its regions not to make the perfect the enemy of the good in fashioning cleanup remedies.[45]

But EPA is subject to Congressional oversight, and the Senate Environment and Public Works Subcommittee on Superfund and Environmental Oversight, chaired by Senator Frank R. Lautenberg (D–NJ), last year held a hearing to attack EPA for failing to require attainment of Maximum Contaminant Level Goals (MCLGs), rather than the less stringent (and, of course, vastly less expensive to achieve) Maximum Contaminant Levels (MCLs) under The Safe Drinking Water Act in selecting remedial actions.

INSURABILITY ISSUES: INACTIVE WASTE SITES

In a blatant attempt to establish a *post hoc* legislative history for SARA's cleanup standards, the Subcommittee Chairman attacked EPA's statutory defense of its choices (adequate protection of human health and the environment; cost-effectiveness) as inconsistent with Congressional intent.

The same Subcommittee recently held another hearing during which it was suggested that EPA was spending too much of its CERCLA tax revenue on remedial actions, and not recovering enough from PRPs.

The tenor of both hearings suggests that some members of Congress do want CERCLA/SARA to work, and that they see their political interests being better served by hamstringing EPA's efforts to make the system work than by encouraging them. Obviously, EPA's insistence on gold-plated cleanups to MCLG standards would increase PRPs' willingness to litigate the choice of remedy, and thereby slow down the cleanup process. Similarly, EPA's refusal to commit its own resources to fund orphan shares at NPL sites would increase the reluctance of PRPs to settle with the Agency.

Should hearings of this nature dissuade EPA from exercising what little discretion Congress has left it,[46] as little relief can be expected from the Executive Branch as from Congress and the Supreme Court. And should Congressional bullying prevent EPA from making the best of a flawed system, the ensuing bureaucratic failures will almost certainly be used to justify still more Draconian legislation when CERCLA is reauthorized again in 1990–1991.

And even if EPA succeeds in administering the system flawlessly, there are limits to the role it can play in resolving PRP–insurer conflicts. Certainly, expedited and rational cleanups would reduce initial transaction costs for both PRPs and their insurers on matters of mutual concern (e.g., apportionment of liability and selection of the remedy), but they would also accelerate the pace of secondary litigation while doing nothing to resolve PRP–insurer differences, which arise under an entirely different body of law over which EPA has no jurisdiction.[47]

Finally, from insurers' perspective, EPA has effectively exacerbated the PRP-insurer conflict by including in its PRP demand letters requests for information on the past 50 years of the recipients' insurance coverage. The effect of these requests has been to encourage PRPs (who hadn't already decided to do so) to seek insurance coverage of their CERCLA obligations.

2.5 Peace in Our Time: Toward a Voluntary Solution

Both PRPs and their insurers have a vital economic stake in the prompt cleanup of hazardous waste disposal sites: the faster these threats to public

62   INSURANCE AND RISK MANAGEMENT FOR HAZARDOUS WASTE

health and the environment are removed or neutralized, the fewer will be the claims that members of the public have been or will be harmed by exposure to them.

As expensive as cleaning up these sites may prove to be, such potential costs pale in comparison to the possible economic consequences of the revolution now under way in the nation's courts in toxic tort cases. Consider these recent examples from both state and federal courts:

- An Illinois jury awarded only $1 in compensatory damages each to plaintiffs alleging a variety of injuries from the spilling of a spoonful of dioxin in a tank-car derailment, but nevertheless slapped the defendant with $16.2 million in punitive damages.[48]
- A Federal Circuit Court, allegedly interpreting Mississippi law, held that plaintiffs who can demonstrate a greater than 50% medical probability that they will contract cancer from exposure to asbestos may recover damages for their risk of future disease.[49]
- The same court upheld an award for a plaintiff's fear of getting cancer in the future, even though there was no proof of a medical probability that he actually would develop the disease.[50]

As extraordinary as the outcomes in these cases are, they are but the forerunners of a vast new toxic tort jurisprudence founded on the "expert testimony from a small group of professional witnesses who call themselves 'clinical ecologists,' despite the fact that their views have been repudiated by the medical establishment."[51]

Yale Law School Professor E. Donald Elliott recently described the impact of clinical ecology on toxic tort cases as follows:

> Only several years ago, most knowledgeable lawyers thought that it would be very difficult to win chemical exposure cases under traditional principles of tort law; except where exposure to a toxic substance causes a rare disease with virtually no other known causes, conventional science generally cannot make the showing traditionally required by tort law: namely, that it is more likely than not that a particular plaintiff [']s illness was caused by exposure to a particular substance.
>
> Testimony from the clinical ecologists has effectively overruled this rule of law, dramatically changing the balance of advantage between plaintiffs and defendants in toxic tort cases. For a price, certain clinical ecologists will testify that exposure to even very small amounts of a wide range of chemicals suppresses the immune system, thereby weakening the body's ability to ward off disease and making the plaintiff vulnerable to virtually all diseases known to humankind, including many such as "nervousness" and "malaise" that present only subjective symptoms.

INSURABILITY ISSUES: INACTIVE WASTE SITES

The opinions of the clinical ecologists on these matters are generally rejected by conventional scientists, who question their methods and also emphasize the natural variability and reserve capacity of the immune system. Both the American Academy of Allergy and Immunology and the California Medical Association have issued official statements repudiating clinical ecology as unscientific.

Despite its marginal status as science, clinical ecology is increasingly important in toxic tort litigation because it gives plaintiffs' lawyers important strategic and economic advantages. The economic value of a toxic tort case to a plaintiffs' lawyer is heavily influenced by the number of claimants in the case, since the "going rate" for settlements is from $10,000 to $100,000 per plaintiff. If a plaintiffs' lawyer bases her case on conventional science, the number of claimants who can be joined in the suit is limited to the small subset of exposed persons who actually suffer from the particular diseases that the chemical in question has been shown to be capable of causing in animal tests or epidemiological studies. With a clinical ecologist on board as an expert, however, the plaintiffs' lawyer can sue, and probably get to the jury, on behalf of everyone who was (or conceivably might have been) exposed to the substance, on the theory that whatever happens to ail them was probably caused by the suppression of their immune systems by chemicals.[52]

As the number and pace of CERCLA cleanups accelerate,[53] so will the number of private suits alleging harm from exposure to the chemicals at the sites involved. And to the extent that larger numbers of courts are willing to admit into evidence what Professor Elliott tactfully calls "marginal science,"[54] the number of successful plaintiffs is certain to increase.

While the economic impact of waste-site-related toxic tort suits is obviously impossible to forecast, the trend in the common law appears to be toward the elements of a federal cause of action proposed during the 1984 Congressional consideration of CERCLA reauthorization[55] whose *annual* costs were estimated by AIA consultants at between $300 million and $56 billion.[56]

The proposed federal cause of action for injuries resulting from the disposal of hazardous substances was defeated in the U.S. House of Representatives on August 9, 1984, by a margin of only eight votes out of the 408 cast.[57] Similar provisions in what ultimately became SARA were avoided at the cost of preempting state personal injury statutes of limitation based on exposure (as opposed to discovery of the resulting harm).[58]

It is widely anticipated that proposals for a federal cause of action will surface again when Congress turns its attention to reauthorizing CERCLA/SARA in 1990 or 1991. While the common-law developments described above would suggest that codification would be unnecessary to the evolution of remedies for injuries alleged to have resulted from disposal of hazardous substances, it will undoubtedly be argued that the benefits cur-

64 INSURANCE AND RISK MANAGEMENT FOR HAZARDOUS WASTE

rently available to plaintiffs in only a minority of states ought to be universally applicable in toxic tort cases.

Also on the horizon for both PRPs and insurers are the mammoth costs certain to be involved in state and federal *parens patriae* claims for damage to natural resources under section 107(f) of CERCLA as amended by section 107(d) of SARA.[59] Designated federal and state officials, acting "on behalf of the public as trustees for natural resources under this Act and section 311 of the Federal Water Pollution Control Act,"[60] are required to "assess damages for injury to, destruction of, or loss of natural resources,"[61] and their assessments are given "the force and effect of a rebuttable presumption . . . in any administrative or judicial proceeding under this act or section 311 of the Federal Water Pollution Control Act."[62]

Here again, there is no realistic basis upon which to estimate the potential economic implications for both PRPs and their insurers of CERCLA/SARA liability for natural resource damages. But to the extent that the process of cleaning up dangerous waste sites is delayed, groundwater contamination can only grow worse, and removing contaminants from groundwater can be both time-consuming and extremely expensive.

Moreover, a number of court decisions suggest that, in addition to *parens patriae* actions under CERCLA, state officials may also assert both equity and state tort claims for groundwater contamination.

For example, courts in both New Jersey[64] and Michigan[65] have held that the state's interest in its natural resources allows it to maintain actions to prevent injuries to the environment, and also to sue for compensatory damages if the environment is in fact harmed. Of concern to insurers in both cases was their additional finding that these damages constituted "property damage" within the definition thereof in the relevant liability insurance contracts.[66]

In sum, the potential liabilities for both site-related toxic tort and natural resource damage claims are certainly large enough to constitute a major incentive to both PRPs and their insurers to expedite the cleanup of dangerous waste disposal sites. The scale of these potential liabilities also suggests the wisdom of reducing the transaction costs in the cleanup process in order to conserve resources needed for defense against related tort and other claims.

Thus far, every proposal for reducing the transaction costs and unpredictability of CERCLA/SARA liability has contemplated favorable Congressional action, which, as the analysis above suggests, is unlikely in the foreseeable future.[67] Moreover, the deep divisions within the business community over the CERCLA tax structure, and the schism between PRPs and their insurers, would doom any reform effort, even if the political conditions for change were otherwise present.

INSURABILITY ISSUES: INACTIVE WASTE SITES

65

It seems reasonable to conclude, therefore, that those upon whom CERCLA/SARA liabilities now fall must first resolve their own differences if they are ever to persuade Congress that a better system is needed. It also seems reasonable to conclude that the process of rapprochement should begin immediately, both because of its necessary complexity and because of the likely consequences of further delay.

Although the machinery developed by asbestos producers and their insurers to resolve their differences — the Asbestos Claims Facility — has fallen apart,[68] the mere creation of that facility demonstrates that it is possible for businesses and their insurers to develop procedures for resolving extremely complicated coverage issues. In addition, even if what is left of the facility ultimately collapses, the lessons learned from its dissolution can be applied to the development of a means of resolving the even more complicated issues over coverage of waste cleanup liability.

The fatal miscalculations in the initial dissolution of the Asbestos Claims Facility seem to boil down to two assumptions — first, that the character of asbestos disease claims would remain the same and would gradually decline in number; and, second, that the number of insurers, excess limits and umbrella carriers, and reinsurers initially subscribing to the facility would constitute a sufficient critical mass to bring carriers of the missing coverage layers aboard.

Unfortunately, both assumptions proved unfounded. The profile of claimants against the facility changed dramatically (from heavily exposed shipyard and insulation workers to lightly exposed workers from asbestos-using industries such as tire manufacture) and sharply increased, rather than gradually decreased, in number. The missing carriers not only stayed out of the facility; at least one reinsurer has alleged that its ceding carrier's participation in the facility "materially changed the risks covered" in the reinsuring agreement and has sued to void its coverage.[69]

As complicated as the issues were in the creation of the Asbestos Claims Facility, their complexity is dwarfed by that of the issues to be resolved in the waste cleanup coverage disputes.

In the asbestos disease coverage disputes, there was no question that bodily injury had occurred in workers exposed to asbestos, and that the economic consequences of that injury were compensable under the producers' liability insurance policies. The key issue was which policies were to respond in particular cases: which insurer in the continuum from first exposure to manifestation of disease should respond to an individual worker's claim?

As costly as the decision of the Federal Circuit Court of Appeals for the District of Columbia in the *Keene Corp. v. INA*[70] case has proven to

66 INSURANCE AND RISK MANAGEMENT FOR HAZARDOUS WASTE

be to insurers, the triple trigger theory developed by Judge Bazelon vastly simplified the process of apportioning liability among each producer's insurers: since every insurer in the continuum was theoretically fully liable for each individual worker's injury, dividing the responsibility according to the proportion that each insurer's aggregate cumulative limits bore to the particular producer's total limits of coverage was not only logical, but also served to spread each insurer's exposure over a longer period of time, thereby easing potential cash-flow problems and preserving assets for the payment of future claims.

Apportioning the producer's liability proved more complex, since the frequency with which they had been sued and the duration of their presence in particular markets and regions all had to be factored into the Facility's cost-sharing formula.

As noted above, the subscribers' decision to open the Facility was predicated on two critical assumptions. While it is easy on hindsight to criticize these assumptions, the Facility's organizers had so many short-term problems to solve that it is hard to fault them for failing to anticipate all of their long-term difficulties. Moreover, their failures contain valuable lessons for any effort to resolve coverage disputes in the waste cleanup context.

Indeed, given the extraordinary complexity of the issues in the waste cleanup coverage imbroglio, it is fortunate that the effort to organize the Asbestos Claims Facility was undertaken. For in addition to all the temporal issues present in the asbestos disease coverage cases (e.g., trigger of coverage), the waste cleanup coverage disputes involve questions not present in the asbestos disease context, such as whether there was an occurrence giving rise to property damage within the scope of the coverage. This complexity is compounded by the presence of a pollution exclusion in many of the policies at issue.

However, the fact that many disputes between PRPs and their insurers are settled short of litigation or short of actual judgment in litigation suggests that, complex as they are, the issues in these disputes are not intractable. Indeed, the current prevalence of cleanup coverage litigation may in part reflect the relative inexperience of the insurance community in CERCLA litigation generally, and insurers' concomitant lack of familiarity with the cooperative efforts common in the PRP community.

At least to this writer's knowledge, insurers have yet to organize, for themselves, any counterpart to the site-specific steering committees that PRPs routinely utilize not only to apportion liability among themselves, but also to negotiate, with EPA or other enforcement agencies, the nature of the cleanup remedy.

INSURABILITY ISSUES: INACTIVE WASTE SITES

It may be necessary for insurers to develop a sufficient level of comfort in resolving the issues that divide the insurance community before they will be willing to entertain the development of similar forms of cooperation between themselves and PRPs.

Fortunately, insurers have a tool — reservation of their rights under their policies — that would enable them to experiment both with different approaches to inter-insurer cooperation at individual sites and, ultimately, with similar approaches to insurer–PRP cooperation. As cooperative experience on a site-by-site basis is accumulated, insurers and PRPs alike might then want to establish procedures or an institution for a more generalized resolution of their disputes.

The reservation-of-rights device effectively allows insurers to work with their policyholders in circumstances in which the insurers believe that their obligations to defend and/or indemnify their policyholders are unclear, but that cooperation with these policyholders may be in the insurers' best interests regardless of their ultimate obligations. CERCLA cleanup claims are obvious examples of the kind of situation in which the reservation-of-rights device would be useful, if not crucial.

Its use would enable insurers of PRPs at a given site to develop, for example, an apportionment of their potential liability, both as to defense costs among themselves, and as to indemnity between themselves and their PRP policyholders, all without conceding that they owe those policyholders defense or indemnity duties.

Behind their reservation-of-rights shield, the insurers could begin to realize the efficiencies and transaction cost savings that their policyholders (insured or self-insured) routinely realize through their participation in site steering committees. For example, if the site is one used by 50 PRPs for 10 years, and each PRP had three different insurers during that period, a possible 150 insurers might be found to owe the PRPs a duty to defend and/or some reimbursement of their cleanup obligation.[71]

Without any kind of joint effort, the primary insurers would each have to hire counsel to deal with each other; perhaps other counsel to deal with their policyholders, excess limits and umbrella carriers, and reinsurers; and experts to advise them on the work being done by EPA or other environmental authorities and the PRPs on the cleanup remedy.

If, on the other hand, the insurers were to agree to a formula by which their potential defense and indemnity costs would be divided should they be found liable for them, they could eliminate the need of each individual company to retain counsel to deal with the PRPs,[72] and they could hire a single set of experts to monitor the technical work on the cleanup remedy. Instead of each of 150 insurers having to mount a multifront defense at

68 INSURANCE AND RISK MANAGEMENT FOR HAZARDOUS WASTE

each company's cost, the insurers, assuming equal potential exposure, would be liable for only 1/150 of a joint defense.

Consider the cost savings possible in even this limited form of cooperative effort. Assume that a competent team of legal and technical consultants would cost each of the 150 insurers $100,000; that's $15 million to fight off a liability that might be imposed anyway under policies whose prices in no way contemplated that policyholder dollars would be used to fight those who paid them, rather than claimants against those policyholders!

Against this, consider each insurer's 1/150 share of the same team cost of $100,000 — less than $700 per carrier to achieve the same result, less than 0.7% of what each carrier would otherwise have to pay.[73] The same cost-sharing formula, when applied to defense costs, and perhaps to indemnity, might ultimately cost the individual insurers less that they collectively might pay to resist their policyholders' claims *in toto*.

The economies of scale that insurers would experience in dealing with their common problems would, in this writer's judgment, lead naturally to a desire to further these economies through cooperation with their policyholders on matters in which both insurers and PRPs have a mutual interest, such as the apportionment of PRP liability and the nature of the cleanup remedy.

This cooperation might initially take the form of expanded steering committees, with representatives of both PRPs and their insurers, since both would have a potentially similar economic stake in low transaction costs, rational apportionment of cleanup liability, and cost-effective cleanup remedies. Moreover, those insurers who were ultimately found legally responsible (or willing to assume some liability without litigation) would be able to avoid the additional cost that would otherwise be necessary to apportion and measure their liability, and would incur lower losses as a result of shared expenses and apportioned responsibility.

Over time, the mechanics of PRP-insurer cooperation might be reduced to formulae applied by each steering committee to the circumstances at each site; or codified in procedures administered on an ongoing basis by a neutral party trusted in both communities; or institutionalized (like Clean Sites, Inc. or the ill-fated Asbestos Claims Facility).

Of course, the mechanics of PRP–insurer cooperation are much easier to picture than the legal understanding upon which joint activity would rest. And no legal understanding will be either possible or permanent unless both sides perceive it as being fair and evenhanded under all circumstances. Finally, and most important, both sides will have to believe that the economic consequences of compromise will be more favorable than those of intransigence.

While the judicial picture on cleanup coverage liability issues is murky, and is likely to be for some time, the decisions to date suggest that there may be grounds for reconciliation on a number of fronts.

First, on the matter of insurers' obligations to defend their policyholders against the demands of environmental authorities, the courts appear to be saying that this aspect of insurers' duties is broader than the duty to indemnify. The majority rule seems to be that even if only one of a series of allegations, prayers, demands, etc. in a complaint is arguably within the scope of a policy's coverage, the insurer is obligated to defend its policyholder against all of them.[74]

Second, on indemnification issues, the judicial pattern is less distinct, but some clear threads seem present.

If a PRP generator's wastes damage third-party property, the courts usually find coverage, assuming they can get past the pollution exclusion's "sudden and accidental" limitation and that they can find an occurrence during the policy period that gave rise to the damage.[75]

On-site cleanup demands on the PRPs property have given the courts a much harder time, not only because of the pollution and owned-property exclusions and the possible absence of an occurrence, but also because of the character of the typical EPA demand, which sounds more in equity than in law and does not seek damages so much as injunctive relief. While some courts have recharacterized the cleanup demands as suits for damages,[76] others have denied coverage for injunctive relief.[77] With respect to the owned-property exclusion, the most persuasive line of reasoning is that unless property other than the PRPs' has been contaminated, there is no coverage.[78]

These threads suggest the possibility that the insurance community might be willing, eventually, to provide PRPs with defense in every CERCLA cleanup claim, and to provide indemnity where PRPs' waste has actually caused third parties' bodily injury or property damage, if PRPs in return agree to be responsible for the removal or treatment of wastes on their own property or on the property of others if these wastes have not actually damaged that property.

The term "actually damaged" is meant to imply a distinction between traditional insurance concepts on the one hand and the mere presence of wastes or the contamination of resources in which private parties have no interest on the other hand.

Liability insurance contracts are written to respond to some event that results in property damage or bodily injury to third parties for which common-law money damages may be claimed. Yet many CERCLA actions are brought primarily to prevent an event from occurring, by requiring PRPs to treat or remove wastes that may not have caused harm to anything

70 INSURANCE AND RISK MANAGEMENT FOR HAZARDOUS WASTE

or anyone but for which damages might be claimed. The courts have tied themselves in intellectual and semantic knots trying to find in cleanup demand or cost-recovery cases either a covered event, someone to compensate, or damages: Pleas for injunctive relief have been recharacterized as suits for damages, the state's inchoate interest in trees and often unusable groundwater has been reclothed as real property ownership, and the mere presence of the wastes has been recast as an occurrence.[79] The results of these cases are predictably absurd, and insurers believe that PRPs must acknowledge this if there is to be any lasting accommodation between the two groups.

Insurers believe that it is unjust to apply, to *before-the-fact* administrative orders aimed at preventing harm, insurance contracts designed to respond to *after-the-fact* demands for money damages resulting from discrete events. For example, if the brakes on an auto insurance policyholder's car need repair, or if a tree in a homeowners policyholder's yard looks as though it is about to fall on his neighbor's house, neither policyholder would dream of demanding that his auto insurer fix his car's brakes to prevent him from having an accident, or that his homeowners insurer cut down the tree before it falls.

If insurance policies were to provide such maintenance-type coverage, insurers would be creating what they call a "moral hazard" for themselves — the possibility that policyholders would deliberately neglect routine maintenance in order to take advantage of their insurance coverage.

While insurers can readily understand the public benefits of cleaning up on-site waste in order to prevent harm, as insurers they cannot, by accepting in one context liability for the cost of such prophylactic non-events, create a claims environment that would, in all other contexts, lead policyholders to claim that liability insurance intended to protect them from the unpredictable consequences of their negligence is also intended to be a sort of all-purpose maintenance contract against the predictable ravages of use and the passage of time.

PRPs, on the other hand, believe that judicial observance of the distinction insurers draw between liability and maintenance would leave the bulk of the indemnity burden in typical cleanup actions on the PRPs. For site owners and operators, most, if not all, CERCLA costs are associated with cleanup on their own property. They argue that, to the extent the cleanup serves to prevent damage to off-site property of third persons, the insurer is merely being asked to pay now what it would have paid, but for the cleanup.

When insurers protest that this reading of their contracts forces them to argue that their policyholders were deliberately irresponsible in the handling of wastes, in that the policyholder's argument seems to assume that environmental damage will be an inevitable, and therefore expected, consequence of their waste disposal activities, the PRPs respond by pointing out that, under CERCLA, the courts are saying that it doesn't matter either that a PRP handled wastes responsibly or that his conduct did or did not cause harm. They argue that CERCLA liability attaches to a PRP's *status*, whether as generator, transporter, or facility owner, rather than to the consequences of his *conduct*. And if liability insurance is intended to answer for any legal liability imposed on the policyholder, they ask, why should the policyholder's conduct, other than willful or deliberate malfeasance, have anything to do with whether he is covered or not?

Insurers respond to this question by pointing out that they have been forced to focus on the consequences of their insureds' conduct because most of the linchpins of coverage under general liability policies — an occurrence (in policies without the "sudden and accidental" qualification of that term) that gives rise to third-party bodily injury or harm to property for which money damages are sought — are quite often entirely absent in Superfund actions.

PRPs also argue that the kinds of distinctions insurers might prefer to draw are unrealistic in the cleanup context. They ask, for example, what happens when there is both on-site and off-site property damage. Would insurance cover both the cleanup of the off-site damage and the on-site costs of preventing further off-site migration of wastes? They go on to assert that, even if these distinctions could be adequately articulated, it would be hard to forecast with the needed predictability how their use would play out on the bottom line at each particular site.

It is not the purpose of this paper to propose a perfect model for resolving cleanup coverage issues that divide the insurance community and the PRPs, but rather to suggest avenues that might be explored toward that end. Only experience can dictate the shape of any ultimate agreement. And even if the ideal solution were set to paper, these crucial questions would remain: Where would the insurers and the PRPs get the money to carry out their ends of the bargain? And how would the agreement be enforced against any and all parties to a particular site cleanup?

Insurers would likely have to fund their share mostly out of surplus, although some of the defense costs could be expensed as unallocated loss adjustment expenses. But the savings inherent in any cost-sharing agreement would so soften and stretch out the ultimate hit on insurers that the industry would probably be able to add enough to surplus from profits on its ongoing business to be able to fund, on a hand-to-mouth basis, its part of the bargain.

72   INSURANCE AND RISK MANAGEMENT FOR HAZARDOUS WASTE

PRPs, for their part, could build their expected costs into the prices of their products and services as a matter of prudent business planning. Here, too, cost-sharing would both soften and stretch out the hit on the business community.

Assuring universal participation in any global agreement on waste cleanup insurance coverage would depend on the extent to which its original signatories constituted a critical mass of those implicated at each site. Achieving such a mass may prove easier in the finite universe of liability insurers than in the almost infinite universe of PRPs.

Critical mass, as the experience of the Asbestos Claims Facility suggests, is that number of insurers and PRPs needed to convince nonsignatories at a particular site that their interests will be better served by participating in the agreement than by going it alone. Here, too, only experience will determine what constitutes a critical mass.

In conclusion, the consequences of failure to achieve an agreement are so dire as to make it imperative that, no matter how difficult the legal and technical issues dividing PRPs and insurers might be, men and women of good will begin now to develop a lasting, voluntary, marketplace solution to an important environmental problem — financing the cleanup of this nation's hazardous wastes in an efficient, effective, and equitable manner.

Notes

1. *Superfund: Extent of Nation's Potential Hazardous Waste Problem Still Unknown.* United States General Accounting Office, Rept. No. GAO/RCED-88-44, December, 1987.

2. *Op. cit. supra* n. 1 at 2-3.

3. *Id.* at 13.

4. See *Superfund Strategy*, Office of Technology Assessment, 1985, at .

5. *Hazardous Waste: Corrective Action Cleanups Will Take Years To Complete.* United States General Accounting Office, Rept. No. GAO/RCED-88-48, December 1987.

6. *Id.* at 2.

7. *Ibid.*

8. See §121, Pub. L. No. 99-499, October 17, 1986.

9. *Cost Implication of Changes in Superfund Cleanup Standards*, a study prepared by Putnam, Hayes & Bartlett, Inc., for the American Insurance Association, March 20, 1986, at 1.

10. *Ibid.*

11. §116, Pub. L. 99-499, October 17, 1986.

12. *Op. Cit. supra* n. 1 at 15-16.

13. *Id.* at 13. Of the nearly 1000 sites on the NPL, 184 are RCRA SUBTITLE D landfills.

14. *Op. cit. supra* n. 9 at 5.

15. See *Continental Insurance Companies v. Northeastern Pharmaceutical & Chemical Co.* (NEPACCO), _____F.2d.

INSURABILITY ISSUES: INACTIVE WASTE SITES 73

16. *Insurers Win Major Pollution Coverage Dispute.* American Insurance Association Press Release No. DC 13, February 26, 1988, at 1.

17. *Id.* at 1-2.

18. _____F.2d _____(4th Cir. 1987), *cert. den.,* _____U.S. _____(1988)

19. Insurance Co. of North America v. Forty-Eight Insulations, Inc, 451 F. Supp. 1230 (E.D. Mich. 1978), aff'd 633 F.2d 1212 (6th cir. 1980), aff'd on reh'g., 657 F.2d 814 (6th Cir.), cert. denied, 454 U.S. 1109 (1981); Porter v. American Optical Corp., 641 F.2d 1128 (5th Cir.), cert. denied sub nom. American Casualty & Surety Co. v. Porter, 454 U.S. 1109 (1981); Keene Corp. v. Ins. Co. of North America, 667 F.2d 1034 (D.C. Cir. 1981) cert. denied, 455 U.S. 1007, reh'g denied 456 U.S. 951(1982).

20. See, e.g., UTC sues 240 insurers for pollution cover. *Business Insurance*, February 1, 1988, at 28.

21. Superfund unleashes flurry of coverage suits, Business Insurance, February 1, 1988, at 1.

22. See, e.g., Summit Associates, Inc. v. Liberty Mutual Fire Insurance Co., N.J. Super. (Law Div., Middlesex County)Docket No. L-47287-84 (Feb. 25, 1987); Jackson Township Municipal Utilities Authority v. Hartford Accident & Indemnity Company, 186 N.J. Super 156, 451 A.2d 990(1982)

23. See, e.g., Waste Management of Carolina v. Peerless Insurance Co., 315 N.C. 688, 340 S.E.2d 374, reh. denied, 316 N.C. 386, 346 S.E. 2d 134 (1986); Transamerica Insurance Co. v. Sunnes, 77 Ore. App. 136, 711 P.2d 212(Or. Ct. App. 1985), review denied, 301 Or. 76 711 P.2d 631 (1986).

24. R.D. Chesler, M.L. Rodburg, C.C. Smith, Jr., Patterns of Judicial Interpretation of Insurance Coverage for Hazardous Waste Site Liability, 18 Rutgers L.J. 9 (1986). see also S.L. Birnbaum, T.R. Newman, I.A. Sullivan, W.J. Wright Jr., "Hazardous Waste Litigation: CGL Insurance Coverage Issues" Second Annual Insurance Litigation Institute at 3 (Prentice Hall Law & Business, 1988)

25. Jackson Township Municipal Utilities Authority v. Hartford Accident & Indemnity Co., 186 N.J. Super 156, 451 A.2d 990(1982); Township of Jackson v. American Home et al., Docket no. L-29236-8 (Law Div., Aug. 31 1984); Ayers et al. v. Township of Jackson, 189 N.J. Super. 561, 461 A.2d 184 (1983); Rev'd in part and aff'd, in part, 106 N.J. 557, 525A.2d 287 (1987)

26. N.J. Super. (Law Div., Middlesex county), Docket No. L-47287-84 (Feb. 25, 87).

27. 1987-88 Property/Casualty Fact Book, Insurance Information Institute, at 5-6 (1987).

28. The same reserve deficiency existed when asbestos disease claims accelerated dramatically in the late 1970s. With between 40,000 and 50,000 cases already pending, and new claims coming in at the rate of 1,500 per month, these liabilities threaten the financial viability of the Asbestos Claims Facility, and perhaps that of some of its members as well. Here, the results in insurance coverage cases have typically maximized the relief sought by policyholders.A similar phenomenon looms with respect to the industry's potential exposure for the costs of asbestos removal from public buildings. The number of buildings involved suggests that in the asbestos disease cases: an EPA survey found potentially dangerous asbestos in 511,000 office and other commercial buildings, 208,000 apartment houses, and 14,000, federal buildings. These are in addition to 33,000 school buildings found to contain friable asbestos.

29. The premium-to-surplus ratio of an insurer is a key indicator of its solvency. Most insurers are comfortable with a ratio of between 2 to 1 and 3 to 1, although higher ratios are found among companies with preponderantly,"long-tail" business.

30. This was the case in Keene Corp. v. INA, Supra n. 19, which held that any insurer of a business which exposed workers to asbestos is liable for the consequences of that exposure

74 INSURANCE AND RISK MANAGEMENT FOR HAZARDOUS WASTE

if it furnished coverage to the business at any point in the continuum from the worker's first exposure to the manifestation of his or her disease. See also In re Asbestos Insurance Coverage Cases, Judicial Council Coordination Proceedings No. 1072 (S.F. Sup. Ct., Dept. 9, 5/29/87).

31. Superfund defense and cleanup costs: An insurer-policyholder counsel colloquy. Docket, vol. 6, no. 1, American Corporate Counsel Association (Winter, 1988), at 9/

32. See Erie Railroad Co. v. Tompkins, 304 U.S. 64 (1938).

33. See Wolfe, Tom (1987). The Bonfire of the Vanities. Farrar, Straus, Giroux, New York.

34. See supra n. 19 and accompanying text.

35. 28 U.S.C. §1332(a).

36. 28 U.S.C. §1331 (a).

37. See Younger v. Harris, 401 U.S.37 (1971). See also Friendly, Federalism: A Foreword, 86 Yale L.J. 1019(1977)

38. The McCarran-Ferguson Act of 1945, 15 U.S.C.§ 1011-1015, provides that general Federal statutes are not applicable to the business of insurance unless they specifically so provide. Insurers historically have discouraged Federal involvement in insurance matters.

39. Upon occasion, the Court has adopted State Law as Federal law in particular cases, in effect replicating, at the Federal level, the varieties of State law approaches to given issues. See e.g., Mordan v. C.G.C. Music, Ltd., 600 F Supp. 1049 (1984), 804 F. 2d 1454 (9th Cir. 1986); U.S. v. Kimbell Foods, Inc. 440 U.S. 715 (1979).

40. See, e.g., Silkwood v. Kerr McGee Corp., 464 U.S. 238 (1984), in which the Court ruled that punitive damage claims under State Law were not impliedly preempted by pervasive Federal regulation of the nuclear industry.

41. See American Insurance Association, 1985. Proposal to Reform and Expedite Cleanup Under Superfund, which recommended that pre-1980 disposal problems be dealt with under a fault-blind, tax-funded, public works system.

42. See supra n. 41 and accompanying text.

43. Known among business lobbyists as the "silver bullet" the amendment, advanced by a coalition including Travelers Liberty Mutual, CIGNA, Crum & Forster and the Alliance of American Insurers, would have eliminated coverage of CERCLA cleanup liability under any pre-1980 insurance policy that did not specifically provide it.

44. See § 122(b)(1) of CERCLA, as amended by SARA; and "Superfund Program; Mixed Funding Settlements" 53 Fed. Reg. 8279-85 (March 14, 1988).

45. See, generally, §121, Pub. L. 99-499, which establishes CERCLA cleanup standards. The section vests EPA with discretion to select remedial actions which meet its general standards, but fail to meet standards under specific environmental statutes, under certain circumstances. See, also "Policy Shift Is Urged for Toxic Cleanup Fund" New York Times, March 22, 1988, at A23.

46. See supra, n. 11 and accompanying text.

47. Indeed, EPA's sole involvement in insurance issues is concentrated in section 108 of CERCLA which directs the Agency to establish financial responsibility requirements for those subject to the statute.

48. Kenner v. Monsanto, 112 III.2d 223, 492 N.E. 2d 1327 (1987).

49. Jackson v. Johns-Manville Sales Corp., 781 F. 2d 394, 413-15 (5th Cir.), cert. denied, 106 S. Ct. 3339 (1986).

50. Dartez v. Fibreboard Corp., 765 F.2d 456 (5th Cir. 1985).

51. E. Donald Elliot, "Toward Incentive-based Procedure: Three Approaches for Reg-

INSURABILITY ISSUES: INACTIVE WASTE SITES 75

ulating Scientific Evidence" Working Paper #76 (unpublished), Civil Liability Program, Center for Studies in Law School, at 5 (footnotes omitted) (March 1988) (quoted with permission of the author and hereinafter cited as Elliott).

52. Elliott, supra, at 6-8 (footnotes omitted). Professor Elliott reports that the "going rate" for one of the leading clinical ecologists' testimony is $20,000 per plaintiff, and that 83 percent of the claims currently pending against Exxon seek damages for harms other than clinically diagnosable physical injury.

53. See supra, n. 11 and accompanying text.

54. Elliott, supra, n. 51 at 5.

55. See, generally, Title II of H.R. 5640, 98th Cong. 2d sess. LL(1984). Section 203 of this bill did not rule out claims for pain and suffering from and individual's fear of experiencing injury, illness or death; it might become ill. See also aa 1114(b) (1) (C) of H.R. 4813, 98th Cong. Ist Sess. (1983), which created a presumption that a plaintiff's harm was caused by exposure to waste site chemicals if a jury found that "exposure to such hazardous substances has a reasonable likelihood of causing or significantly contributing to death or to a personal injury or illness of the type suffered by the applicant." The subparagraph specifically permitted the introduction of "immunological studies" into the evidence in support of the presumption. Ronald E. Gots, M.D. Ph.D., President of the National Medical Advisory Service, at page 5 of his March, 1984, Response to H.R. 4813, prepared for the Crum & Forster Insurance Companies, said that the bill "would formulate a presumption of causality upon a framework of unresolved scientific disputes built upon a foundation of disparate, controversial, irrelevant or scientifically meaningless data."

56. See Jim J. Tozzi and Charles W. Chesler, "The Federal Cause of Action: An Estimate of Its Costs" (Sept. 20, 1984).

57. 130 Cong. REC. H 8854-55 (daily ed. Aug. 9, 1984).

58. See Pub. L. 99-499, § 112 (d) (2) (1986).

59. See Pub. L. 99-499, § 107 (d) (1986).

60. Pub. L. 99-499, § 107 (d) (2) (A) and (B).

61. Ibid.

62. Pub. L. 99-499, § 107 (d) (2) (C).

63. 43 C.F.R. Part II, 51 Fed. Reg. 27674-27753 (1986).

64. Lansco, Inc. v. Environmental Protection Dept., 138 N.J. Super. 275 (Ch. Div. 1975), aff'd 145 N.J. Super. 433 (App. Div. 1976), certif. den. 73 N.J. 57 (1977)

65. United States Aviex Co. v. Travelers Insurance Co., 125 Mich. App. 579 (Ct. App. 1983).

66. Insurers should also ponder the implications of the logic in these cases, which involved off-site damages to property other than that of the policyholders, for possible liability for on-site damages under first party property insurance contracts. Cf. Riehl v. Travelers Insurance Co., Civil Action No. 83-0085 (W.D. Pa., August 13, 1984), rev'd 772 F. 2d 19 (3d Cir. 1985). The Court of Appeals reversed a District Court holding that the insurer was obligated to defend and indemnify its insured against CERCLA cleanup liability for contamination of the insured's own property, notwithstanding the fact that the contamination had caused no third-party bodily injury or property damage in the traditional sense.

67. Indeed, the SARA experience suggests that the next reauthorization of CERCLA could well make its liabilities more, not less, onerous.

68. See, Fifth producer leaves asbestos claims facility. Business Insurance, April 11, 1988, at 2.

69. Ibid.

76 INSURANCE AND RISK MANAGEMENT FOR HAZARDOUS WASTE

70. Supra n. 19.

71. This figure of course would not include the policyholders' excess limits carriers or umbrella carriers, or any of the insurers' reinsurers. These carriers may also want to participate in collective arrangements, if only to assure that primary dollars are wisely and efficiently spent.

72. PRPs at some sites have banded together to sue their insurers for defense and indemnity. A joint insurer response, whatever the outcome, would be cheaper for all the insurers and would enhance their leverage in either litigation or negotiation.

73. These economies of scale would be even more spectacular if insurers other than the primary carriers became part of the cost-sharing arrangement.

74. See, e.g., Independent Petrochemical Corp. v. Aetna Casualty & Surety Co., 654 F. Supp. 1334 (D.D.C. 1986); Shapiro v. Public Service Mutual Ins. Co., 19 Mass. App. 648, 477 N.E. 2d. 24 (1985); Shapiro v. American Home Assurance Co., 616 F. Supp. 960 (D. Mass. 1985).

75. Buckeye Union Ins. Co. v. Liberty Solvents, 17 Ohio app. 3d 127, 477 N.E.2d. 1227 (1984).

76. New Castle County v. Hartford Accident and Indemnity Co., No. 85-436JLL (D.C. Del., November 2, 1987); Solvents Recovery Service of New England v. Midland Ins. Co., No. L-025610-83 (N.J. Super. Ct., Law Div., Union County); Consolidated Rail Corp. v. Certain Underwriters at Lloyds, No. 84-2609 (E.D. Pa., 6/3/86); U.S. Avoex Co. v. Travelers Ins. Co., 336 N.W. 2d 838 (Mich. App. 1983).

77. Maryland Casualty Co. v. Armco, supra; Mraz v. Canadian Universal Ins. Co., 804 F.2d 1325 (4th Cir. 1986); CPS Chemical Co. v. Continental Ins. Co. (No. l-060537-84 N.J. Sup. Ct.), Mealy's Lit. Rpt., 11/10/87.

78. Atlantic City Municipal Utilities Corp. v. CIGNA companies, No. A-1320-84t7 (N.J. Super. Ct. App. Div., Dec.19, 1985); E.C. Electroplating Inc. v. Federal Ins. Co., No. L-06-2919-85 (N.J. Super. Ct., Feb. 18, 1986). But see Summit Associates v. Liberty Mutual Ins. Co., supra n. 26.

79. New Castle County v. Hartford Accident & Indemnity Co., supra; ("damages" given ordinary meaning and covers cleanup costs); Broadwell Realty Services v. Fidelity and Casualty Co., A-5301-85 (N.J. Super. Ct., App. Div., May 23, 1986, Supp. June 24, 1986) (demands to remedy damage to groundwater and surface water are covered by insurance policy); Solvents Recovery Service Of New England v. Midland Insurance Co., No. L-025610-83 (N.J. Super., Law Div. Union County)(in a request to appeal an interlocutory order finding coverage, Hartford Ins. Co. argues that no actual environmental harm has been demonstrated from hazardous substances spill. See Mealy's Litigation Reports, Vol. 52, p. 4053, March 24, 1987.).

Discussant: Kenneth Abraham

This is a marvelous paper, full of all the wit and insight that we've come to expect of Leslie Cheek. In Virginia where I live, Thomas Jefferson is a hero, among other things for his versatility. He could argue a case, play a violin, design a building, dance a minuet. Les Cheek is one of a handful of modern Jeffersons that I'm lucky enough to know. He can deliver a speech, lobby a senator, draft a statute, interpret an insurance policy, write

INSURABILITY ISSUES: INACTIVE WASTE SITES 77

an academic paper, and finally, it must be said, pull the wool over your eyes in such a charming and moderate way that you feel a little like the lady from Kent in the famous limerick:

There once was a lady from Kent,
Who said she knew what it meant,
When men took her to dine,
Gave her cocktails and wine,
She knew what it meant, but she went.

And in reading this paper I almost went too.

The argument of the paper is that the Superfund regime is a financial disaster — that neither the Congress, the courts, nor the EPA is going to be able to do anything about it in the near future, and that therefore the business and insurance communities had better get together and settle their differences or both may go down in flames.

I agree with much of this argument. The Superfund regime is a disaster, for both the business community and the insurance industry. Of course, so is the way in which hazardous waste was disposed of in this country for the 35 years between the end of World War II and 1980, and I include the way in which the insurance industry came to a late recognition of the problem. I share Les Cheek's doubt that the courts or the EPA can solve the problem in any fundamental way. But he argues that because Congress isn't about to solve the problem, the business and insurance communities should sit down, as the Bible suggests that the lion should lie down with the lamb, and work together to solve their problems. Now I suppose that if Congress won't act, it behooves business and insurers to cooperate rather than to fight tooth and nail, case by case, as they do now. Unfortunately, when the lion lies down with the lamb, the lamb usually doesn't get very much sleep.

I'd like to talk a bit about the political economy of this problem, because it seems to me that if there isn't sufficient political consensus for Congress to act to reform the Superfund regime, then agreement among PRPs and their insurers isn't likely either. Les says that the Congress won't admit it was wrong in 1980 and 1986, that it won't use general revenue for cleanup and that it views the issue as an internecine struggle between PRPs and insurers. But it seems to me that if the PRPs and insurers did in fact reach an agreement about how to allocate their responsibilities, Congress would be able to ignore these problems. It's the absence of an agreement by these two groups about how to allocate responsibility that precludes legislative reform, and that is likely to preclude PRPs and insurers from agreeing in any other forum as well.

78 INSURANCE AND RISK MANAGEMENT FOR HAZARDOUS WASTE

Let me take Les's two most prominent explanations for his prediction that Congress won't act. First, he says that Congress won't admit that it was wrong to adopt CERCLA and SARA. But I believe Congress will admit that it's wrong whenever it gets something out of it — the first rule of legislation, after all, is that legislatures are shameless. Second, general revenue need not be used for cleanup — much more extensive surcharges could finance a public-works cleanup program that paid little attention to cost recovery on a site-by-site basis — but the insurance industry surely would have to pay a sizable portion of these surcharges. Based on the politics of the situation, I'd say that the insurance industry would have to pay between 20% and 40% of cleanup costs, or about $50 billion between now and the turn of the century. The money would have to be generated from payouts under existing liability policies or from new taxes. I don't hear the insurance industry agreeing to that kind of legislative compromise, and I predict that during the CERCLA reauthorization, a few years from now, you won't hear that kind of proposal either.

Now, I'm not suggesting that as a matter of equity the insurance industry should contribute that much. I'm only suggesting that it will probably take that to get a deal struck, and that the industry won't offer that much. My evidence is drawn from the proposal for private cooperation by the most reasonable of all proponents of the insurance industry's interests, Leslie Cheek. What does he suggest PRPs and insurers agree to do?

He would have insurers agree to defend all cleanup suits and to indemnify the costs of actual off-site damage, in return for PRP agreement to bear both the cost of off-site cleanup where there is no property damage and the cost of all on-site cleanup. Well, first of all, once you have that kind of agreement, defense costs will fall dramatically; and since the vast majority of Superfund costs are going to be incurred for remedial action where what the industry defines as property damage has not occurred, the industry will bear a very small percentage of total cleanup costs as well. When PRPs cost out the Cheek proposal, they'll see that it leaves them holding the bag. Maybe they should be left that way, maybe not; but they aren't likely to agree to do so. In his characteristically elegant way, Les admits this. He says, "Of course the mechanics of PRP–insurer cooperation are much easier a picture than the legal understanding upon which joint activity would rest." Even Jefferson couldn't have matched Cheek's ability at understatement.

Let me conclude by stating what I think are the implications of what I've said. I think we have to distinguish three different issues:

(1) Is the Superfund cleanup program too ambitious? That is, should we be spending less on cleanup? Both PRPs and insurers think we should

INSURABILITY ISSUES: INACTIVE WASTE SITES

be spending less, but on that question they are likely to be long-term losers.

(2) Aside from that question, how should we allocate responsibility for the money that we will spend on cleanup? Here Les takes no explicit position. I have argued that his proposal implies that the insurance industry should bear a smaller proportion than I think is necessary for legislative or private consensus to be reached, that at least will reduce transactions costs.

(3) How can we reduce transaction costs? The answer to this is obvious — there has to be either legislative or private agreement on a formula that will allocate responsibility without incurring transactions costs on a case-by-case, site-by-site basis. Each side is going to have to feel some pain if an acceptable agreement is to be reached. If the transactions costs are as high as everyone says they are, then there ought to be enough savings available to make an agreement possible without either side feeling too much pain. Only the lawyers and consultants will hurt if such an agreement is reached because the vast majority of transactions costs are the hours on their bills. On the other hand, if the transactions costs aren't high enough to induce the two sides into an agreement that captures savings on transactions costs, then proposals for reform in order to reduce transactions costs ought to be viewed more as camouflage. In reality, such proposals should be understood less as efforts to reduce transactions costs and more as attempts to attain a better deal on substantive liability than the current regime seems to afford the party making the proposal. Thus, each side proposes an enlightened way for everyone to save transactions cost that just happens to save that side more money than it saves anybody else!

I hope that we do see a cooperative agreement reached, not only because I share the view that there are sizable transactions cost savings to be had, and not only because such an agreement will produce quicker cleanup of hazardous waste sites, but also because then Les Cheek will turn his attentions onward, and enrich us all with his views about the next problem on the horizon.

Discussant: Cornelius Smith

I consider Mr. Cheek's article to be exceptional. However, rather than discussing the insured versus insurer hazardous-waste-site liability arguments, I'd like to direct your attention to an article on the subject listed by Leslie Cheek in his bibliography. It appeared in the Fall 1986, Rutgers Law Journal, and was authored by me and Michael Rodburg and Bob Chesler of the New Jersey firm of Lowen, Stein, Sandler. It is a primer

80 INSURANCE AND RISK MANAGEMENT FOR HAZARDOUS WASTE

on the Comprehensive General Liability (CGL) policy waste site coverage issue. It attempts to be intellectually objective about the background decisions that were rendered before Superfund. This is very important because most articles written on this subject refer only to the latest cases. Our Rutgers article also looks at cases of the 1940s, 1950s, 1960s, and 1970s. How did the courts treat historic groundwater contamination cases under the original Comprehensive General Liability policies? Our article addresses such subjects and then attempts to analyze them in the context of the Superfund issues of today. Although the article is a couple of years out of date, it's easy enough to bring it up to date.

I endorse the underlying theme of Les's article that the insureds and their insurers should make every effort to resolve their differences promptly. And, in order to do so, I think they've got to use alternative dispute resolution techniques and possibly binding arbitration on the key generic issues. It is only then, really, that insureds and insurers representing the PRP community can try to minimize the transaction costs of dealing with the government and each other, including allocation of responsibilities.

I differ with Les on a great number of his characterizations of the judges' rulings in cases discussed in his article. I would like to point out that the draftsmanship of key CGL policy terms — those of the old *caused by accident* policies, the *occurrence* policies and then the later ones in the 1970s with the pollution exclusion clause — was not a prime example of clarity and precision. Underwriters, the drafters of those CGL policies, deserve to pay the price for their ambiguity. Key words like *sudden* and *accidental* were left totally undefined by these policies. The pollution exclusion clause, which is often relied upon by the underwriters, contains an exception-within-an-exception. And, if something is an exception-within-an-exception it becomes a double negative; a double negative becomes a positive. So why is the insurance industry surprised that judges have interpreted the pollution exclusion clause totally out of existence by finding it to be merely a restatement of the old occurrence policy language?

Additionally, the rules of judicial construction allow every judge to interpret anything that is ambiguous against the drafter. These are standard-form policies for the most part, so the insurance industry bears the burden of anything that is ambiguous. Judges also try to give effect to the reasonable expectations of the insured. In this regard, I was very happy to see that Les has said, "The policyholders rightly argue that they paid their CGL premiums" — I'm paraphrasing here: "during the 1950s and the 1970s" — "precisely to relieve themselves of the burden of such unanticipated liabilities."

INSURABILITY ISSUES: INACTIVE WASTE SITES

Well, that really is the point. Let us step back a moment here and examine the basic nature of these policies. *Comprehensive general liability* — what do these words mean to the average businessman buying a policy? They mean the equivalent of a major medical policy or a business umbrella policy to cover all the types of potential liabilities that can't readily be foreseen and that will protect his business assets. Particular types of liabilities, like employees' coverage, fire and theft, etc. are specifically excluded by CGL policies because you can go out and buy particular types of insurance for those exposures. So the bottom line is that a CGL policy is an *everything-else* policy. And Superfund liability, a retroactive statutory liability that was not foreseen by either the insureds or their insurers, is arguably an example of an everything-else liability that should come within CGL policy coverage.

These CGL policies were written in the 1950s, 1960s, and 1970s when Superfund liability was unforeseen. Who's going to bear this burden in a 1980 controversy between an insured, who paid the premium to rightly get out of unforeseen liability, and the insurance company, who took the risk of issuing a comprehensive general liability policy to cover unforeseen risks of his business? The insurance company may not have calculated its premiums correctly — a lot of businesses don't calculate the future correctly when they go into business. It is not surprising that judges came to the conclusion that the burden should be borne by the insurance company rather than the policyholder.

I suggest that a broad group be formed to address these important issues. It should be composed of representatives of major PRPs and of major CGL insurers, both of the primary and excess layers. Each of the interest groups should hire one law firm or one voice to speak for them to simplify negotiations.

First of all, an effort should be made to identify all the generic issues that arise in the insured/insurer Superfund cases. That's a critical first step — to identify them, not resolve them. Then an effort should be made to try to negotiate a compromise, whether it comes out percentagewise or however you do it, on each one of those generic issues.

I then propose that the issues that are left unresolved be sent for binding arbitration before a quality panel composed of persons drawn from the Center for Public Resources list or from similar sources. The ultimate resolutions of these generic issues, whether they be by compromise or by alternative dispute resolution, should be embodied in an agreement that both insurers and PRPs can sign. What they would basically agree to is to abide by this decision wherever their cases came up, on a case-by-case basis, no matter what state law governs.

That's a concrete proposal. I'm not saying how it should come out. I think that, to a certain extent, it would be a roll of the dice. But I think there is room to minimize the transaction costs between the insurance industry and insureds, which I feel is an important step to take. I don't believe that those transaction costs are as large as the transaction costs between PRPs and the government, or the costs resulting from hazardous waste site cleanup overkill. If you look at the total cost, over $100 billion, I really don't think insured/insurer transaction costs a major percentage. But I think they are an issue.

These transactions are, however, resource-intensive, and they divert PRPs from dealing with the critical issues of minimizing transaction costs with the government and each other and from getting on with these clean-ups. That's what we really want. We want to clean up these hazardous waste sites cost-effectively as soon as possible. If we end up fighting among ourselves rather than getting on with the job, we deserve the criticism of Congress and of other interested parties.

3 RISK MANAGEMENT ISSUES ASSOCIATED WITH CLEANING UP INACTIVE HAZARDOUS WASTE SITES

James Seif and Thomas Voltaggio

3.1 Defining Terms

A meaningful discussion of risk management at hazardous waste sites requires a clear understanding of the terms. Risk management is a process that is part of a chain of activities that is used in formulating government policy. This chain can generally be shown as follows:

Risk Assessment → Risk Management → Risk Communication

Each of these links must be employed to successfully manage governmental programs that deal with hazards. Risk is defined in Webster's as a possibility of a loss or injury. In a regulatory context, such as in EPA, risk is usually the danger of injury to human health or welfare, or to the environment. This risk is manifested in numerous ways. The mode to be considered in this paper is the risk due to exposure of humans or the environment in general to hazardous substances. Risk assessment is the method that we use to define the probability of harm coming to an individual or a population as a result of exposure to a substance or a situation. This assessment uses a base of scientific research and is usually quantitative.

Risk management is a public process that is used to decide what to do

84 INSURANCE AND RISK MANAGEMENT FOR HAZARDOUS WASTE

where risk has been determined to exist. The process must factor in benefits, cost of control, and any statutory framework for control.

Risk communication is the process by which we inform the population of the risk, the assessment of that risk, and how we intend to manage that risk. An excellent, scientifically valid assessment and a brilliantly derived management process can easily be construed as a failure unless risk information is communicated effectively to the public.

3.2 Societal Factors

Risk management is an extremely volatile issue in today's society. There are several factors that appear to contribute to this phenomenon. First, there is a new scientific awareness and public interest in health and fitness. One needs only to count the number of health spas and fitness equipment stores to get an indication of how much this society values its health. There is a decrease in the faith and confidence that the public puts in science and technology. Examples of catastrophes like the Challenger, Bhopal, Three Mile Island, and Chernobyl bring to light the uncertainties in the ability of science to control risk effectively.

Chemical hazards are frequently in the news. This constant repetition of problems or calamities tends to maintain the focus, and results in people believing that risks are becoming unmanageable.

There is considerable disagreement among knowledgeable scientists as to the risks of chemical exposure. The public cannot understand the differences among the positions of the experts and tends to assume the worst.

These factors tend to decrease the faith that the public puts into the risk management decisions made by government. Effective risk communications require the recognition of these issues.

3.3 Necessary Elements of Risk Assessment and Management

In order to provide meaningful information regarding risk, we must be sure to include the following necessary elements.

First, risk calculations must be expressed as distributions of estimates, not as fixed numbers that can be manipulated without regard to what they really mean. Distributions reflect the uncertainty that is inherent in risk assessment science. Additionally, better tools are needed to explain the meaning of probability distributions to the public.

RISK MANAGEMENT ISSUES: INACTIVE WASTE SITES 85

Second, the public must be informed of the assumptions that underlie the analysis and the management of the particular risk. If we use very conservative assumptions, then this must be communicated so that the public understands the differences in probability between the assessment conditions and reality.

Third, we must communicate clearly that risk reduction is our business, not cost–benefit analyses. Cost must be included in any risk management analysis; however, the goal of risk management is the balancing of risk against risk by using cost as one factor to decide which risks can be deferred and which must be addressed immediately.

3.4 Risk Management

There are two major elements in risk management: priority-setting and making choices. Priority-setting is extremely important for an agency like EPA. We have many statutes with differing requirements and philosophies. We have limited resources and many constituency groups to whom we must be responsive. These competing interests require that the agency set its priorities effectively. The use of risk management allows the priority-setting process to be based on the principal of providing for the greatest degree of risk reduction using the available resources.

Making choices is the Agency's mission — in the final analysis, this is what we are paid to do. In the hazardous substance cleanup field, it involves the selection of the appropriate remedy to render a site safe. In the hazardous-waste-permitting field, it is the conditions of operations that a facility must meet to allow it to operate safely. In making choices, the Agency normally balances the resources available (either to the Agency or the permittee in the above examples) with the risk reduction that will result from the action taken. The risk reduction is a function of the health and nonhealth benefits that will accrue and the confidence that we have in being assured that the choice made will bring about the results that we anticipate.

Factors that must be considered in all risk management processes are comparability and consistency. Risk management choices for similar problems should exhibit these qualities unless there is a good reason not to do so. One reason may be a difference in philosophy between problems with similar risks but regulated by different statutes. Some statutes are technology-based and others are risk-based. Some require both to be factored in. The use of technology or a requirement to be consistent with other relevant statutes will tend to skew a choice from comparability and

86 INSURANCE AND RISK MANAGEMENT FOR HAZARDOUS WASTE

consistency with similar choices. In these cases, risk communication is necessary to properly explain the seeming dichotomy.

There are several areas in which EPA is concentrating its efforts to improve risk management. These are obtaining better and more consistent information bases, using more varied forms of risk management tools, and strengthening the role of communications. In Region III, we are using a concept of MERITS. This is a program of managing for environmental results. Apart from the normal Agency program initiatives, projects are proposed and prioritized based upon an assessment of risk for the environment. Resources are identified and used to fund the project.

3.5 Cleaning Up Inactive Hazardous Waste Sites — Superfund

Cleaning up inactive hazardous waste sites is a portion of the Agency's program of dealing with the hazardous substance problems in the country. There are several major areas of the program. One area is the control of chemical production under the Toxic Substance Control Act. Another is the control of use of chemicals under the pesticides laws or the asbestos programs. Still another is the prevention of chemical exposure from accidents under Community Right to Know and Emergency Preparedness law. The two major laws that deal with hazardous chemicals, however, are the Resource Conservation and Recovery Act (RCRA) and the Comprehensive Environmental Response, Compensation and Liability Act (CERCLA or *Superfund*). The major distinction is that RCRA deals with the current handling of hazardous waste by operating facilities, and Superfund deals with the cleanup of past mismanagement of the handling of hazardous waste.

The major innovation in the Superfund law is the ability of the government to arrange directly for cleanup using money from a trust fund set up under the law. The current statute provides for $8.5 billion for implementation of the program. Enforcement provisions allow for authority to order responsible parties to clean up and also to recoup costs back into the trust fund for money that the EPA spends when it arranges directly for cleanup.

3.6 Removal Program

The Superfund statute provides for a mechanism to clean up acute, short-term, immediate risks to public health or the environment. The mechanism

RISK MANAGEMENT ISSUES: INACTIVE WASTE SITES 87

is called a removal response. It is used for transportation accidents, spills or air releases of hazardous substances, acute threats such as storage of deteriorating and incompatible hazardous substances, etc. A removal response usually involves immediate response to *stabilize and contain* the hazard; it is not normally a cleanup program. It is meant to reduce the threat sufficiently so that further study can be made prior to any large-scale cleanup without posing unacceptable short-term risks.

In keeping with the concept of short-term containment and stabilization, Superfund limits these responses to 12 months or $2 million, although exemptions are provided for special circumstances. To facilitate quick response, there are few administrative burdens to overcome prior to activation of trust-fund money. Cleanup is carried out by contractors under the supervision of EPA personnel. These contractors have been competitively selected and are available for cleanup for a period of several years on a standby basis. When called upon, they are available in a matter of hours. EPA has performed approximately 800 removal actions since the inception of Superfund. This has resulted in significant reduction of risk to human health and the environment.

3.7 Remedial Program

The remedial program provides the framework for an organized response to inactive hazardous waste sites. It consists of several steps. These are discovery and assessment, prioritization, investigation, and cleanup. These steps are the same if a responsible party performs the cleanup or if the Superfund trust fund is used.

Discovery and assessment start by organizing an inventory of sites that have the *potential* of environmental or public health risk. The Agency started with a list of 500 potential sites in 1980. This list has mushroomed to over 30,000 in 1988. Sources of these potential sites are citizen complaints, state records of hazardous waste activity, legal notification, etc. Once the inventory is established, it requires a program of assessment to determine the sites that require further review and the sites that can be deferred. This assessment program consists of a two-stage process of document review (called a Preliminary Assessment) and on-site investigation and sampling (called a Site Investigation). The sampling effort is limited and geared to finding if a release of hazardous substances to the environment has occurred or can occur. A further area of investigation is the determination of the means by which the hazardous substance can impact human health or the environment. The presence of a large amount of

88 INSURANCE AND RISK MANAGEMENT FOR HAZARDOUS WASTE

highly toxic chemicals in the middle of the Gobi Desert has a different risk than a smaller amount of less toxic chemicals in a recharge area for a sole-source aquifer.

Once the assessment is completed, a prioritization process begins. This process uses a mathematical quasi-risk model called the Hazard Ranking System, or HRS. The HRS builds on the assessment information and factors in such items as waste quantity, toxicity and persistence of the chemical, the route that the chemical takes to impact the environment, etc. The result is a numerical score based upon the actual or potential risk to human health and the environment. This will be discussed in detail later in the paper.

Once a site has been scored, it is compared against other sites. If it presents a certain level of risk (currently 28.5 using the HRS), then it is published in the Federal Register on the National Priority List, and work may proceed to fully characterize the site to determine the appropriate cleanup method. At this point, responsible parties are located and asked to perform the work. If they refuse, EPA will perform the work and seek recoupment of the funds expended when the cleanup is completed. In either case, the cleanup process is the same.

The next step is the determination of the scope and degree of contamination at the site. All the sampling up to that point has been to determine if there is a problem or not. The purpose at this point is to fully characterize the situation; the amount of hazardous substances, the extent to which they have migrated or can migrate, the hydrogeological regime, and many other factors that will allow the Agency to determine the appropriate cleanup remedy.

Once the site is characterized, a comparison of feasible remedies will be made and will be summarized for public review and comment. At that point, a decision will be made as to the appropriate remedy for the unique circumstances of the site. This remedy decision is a significant risk management decision and will be explained later in this paper.

When the remedy is selected, the responsible parties are again given an opportunity to perform the remedy. As before, if they decline, the Agency will perform the cleanup and seek recoupment of those funds. Plans and specifications are then prepared and bids are received for the cleanup. Unlike the removal program, contracts are awarded individually for remedial projects. Because of the greater cost of remedial projects and their lack of the same type of urgency as removal projects, it is believed that fixed-cost or unit-price contracts are more cost-effective.

The remedy is then implemented, followed by an extended operations and maintenance phase. In few situations do we find a remedy completed

RISK MANAGEMENT ISSUES: INACTIVE WASTE SITES 89

with no need for continuing activity. Usually, groundwater pumping and treating, extended monitoring, or similar activity will continue for years. This fact is very important to communicate to the public. While the risk is significantly reduced by the cleanup, continued work must continue to ensure the effectiveness of the remedy.

3.8 Risk Management Decisions in Superfund

As one follows the Superfund process outlined above, it becomes apparent that several key risk management decisions occur during the phases of Superfund response. We will outline the following in subsequent sections of the paper:

(1) Removal prioritization
   (A) On-Scene Coordinators
   (B) Agency for Toxic Substances and Disease Registry
(2) Removal cleanup levels
   (A) Action Memoranda
   (B) Exemption request criteria
(3) Remedial prioritization
   (A) Hazard Ranking System
   (B) National Priority List
(4) Remedial cleanup levels
   (A) Feasibility studies
   (B) Selection of remedy process

3.8.1 Removal Prioritization

Because of the need for quick response in a removal situation, the risk assessment and risk management processes are less formal than in remedial situations. An assessment will be made by special Agency employees called On-Scene Coordinators (OSCs). The OSC is responsible for assessing and directing cleanup work in acute emergency situations. The OSC will sample at the site to determine the existence of a threat and will call upon the health professionals at the Agency for Toxic Substances and Disease Registry (ATSDR) and EPA toxicologists to assist him in determining the sites that must be prioritized. It is usually the judgment of the OSC and his management that determines the prioritization of sites for removal work.

90 INSURANCE AND RISK MANAGEMENT FOR HAZARDOUS WASTE

The major factor considered is the immediacy of the threat to human health. Usually this threat is to an individual drinking water supply, or is a threat of fire and explosion or direct contact with toxic chemicals. It is usually ATSDR who provides an expert opinion on the seriousness of these threats.

Most of the sites handled by the removal program are obviously significant threats, and little controversy occurs unless resources are stretched too thin to handle all the sites that need immediate attention. Fortunately, this does not happen frequently.

3.8.2 Removal Cleanup Levels

The removal program documentation showing the reasons for the need for response and the explanation of the steps to be taken to clean up is called an Action Memo. This document gives a brief history of the site, the basis for determining that a threat exists, the method of stabilization or containment, the costs and time required to effectuate the action, and the levels of contaminants that the cleanup will reach. This last item is the risk management decision that is made for the action. As stressed before, the purpose of the removal program is not total site cleanup, but rather a containment or stabilization. Of course, if the action is small enough, it may be more cost-effective to complete the cleanup than to leave it to be assessed by the remedial program. This occurs frequently in projects such as transportation accidents or sites where a small number of drums are involved. For many removals, however, contaminants will remain on site to be assessed further.

Generally, the criterion for completing a removal action will be the determination of the level of contamination that will not pose an acute or short-term risk to human health or the environment; therefore, many times, contaminants remain on site that may present long-term or chronic health risks. The time frame usually used for short-term risks is three to five years, since that is a very conservative time frame for a site to be assessed in the remedial program. This issue is usually the most difficult one to communicate to the public at a removal action.

Outside forces play a large role in the risk management process for these actions. The community relations program of EPA requires extensive involvement of the public in all aspects of the removal action, but especially the extent of cleanup. Responsible parties also play a role in commenting on appropriate cleanup levels. Finally, elected officials at the local, state, and federal level play active roles in this process. It is usual for EPA to

RISK MANAGEMENT ISSUES: INACTIVE WASTE SITES 91

be in the middle with the responsible party on one side, the public on the other, and elected officials arrayed at many different points.

At times, removal actions need more than $2 million or 12 months to complete. There are exemption requests that can be approved by our national office. There are two criteria that may be used to seek an exemption. One criterion is that continued response is immediately required, there is still an immediate risk, and assistance would not otherwise be forthcoming. The other is that continued response actions are appropriate and consistent with future remedial actions. The latter criterion is new, and policy is still being formulated on its implementation. The risk management process for exemptions balances the need for additional cleanup against the availability of trust-fund money and the desire to see larger cleanups being funded under the remedial program with its more cost-effective contracting procedures. While multimillion dollar exemptions have been approved, they are rare. Most removal actions terminate below $1 million, and exemptions rarely go over $3–4 million.

3.8.3 Remedial Prioritization

As explained previously, the Hazard Ranking System and the National Priority List are the formal EPA processes for remedial prioritization. The original Superfund law in 1980 required a site to be listed on the National Priority List for remedial construction money to be allowed to be used. Because remedial cleanups can run into the tens of millions of dollars, this insured that the trust fund money would be utilized at the worst sites first. Conversely, if a site does not present sufficient risk to score high enough using the Hazard Ranking System, then the site cannot be listed on the National Priority List and remedial response cannot be conducted. Therefore, the risk management process for remedial prioritization is the Hazard Ranking System and National Priority List process.

As sites have been assessed and investigated, they are scored using the Hazard Ranking System. The analysis of the chemical constituents, and of hydrogeological and other pertinent data, is assembled and a worksheet is prepared. Various weights are given and a raw score is calculated. This is normalized to a 0–100 scale, and the result is used to determine if a site is eligible for listing on the National Priority List. Currently, a site is eligible if it scores more than 28.5 using the Hazard Ranking System.

A frequent question asked is: "Why 28.5 ?" The original reason was that in the first Superfund statute, EPA was required to generate a list of at least 400 sites for the initial National Priority List. EPA assembled all the HRS evaluations at the time and ranked them from highest to lowest. The 400th site scored around 28.5! Since that time, EPA has made an evaluation of whether or not using other values would result in better risk management. We have failed to find a better alternative.

The Hazard Ranking System considers two types of release mechanisms and three types of pathways to the environment. A release can be actual (or observed), or potential. An actual release is one that is measured by scientific sampling and is at least three times above background levels for that chemical in the environment. A potential release is one that could occur depending upon future events. For a potential release to be scored, there must be a pathway from the source to the target population, and the target population must be near enough to be impacted. A potential release scores less points than an actual release.

The three pathways for contamination to reach the target population are the groundwater, the surface water, and the air. Each of these pathways is evaluated and scores are generated. An additional factor is proximity to a sensitive environment. Rules of thumb for a high score with this model are as follows:

(1) There needs to be a drinking water intake or a number of wells within one to three miles of the site. Therefore if a site is in an urban environment with city water, then the groundwater pathway will probably not score highly.
(2) The same is true for surface water.
(3) There must be proximity of a target population for an air release to score. No potential air releases are allowed in the model due to the difficulty of assessing such a situation.
(4) The hydrogeology must be clear, especially when aquitards are suspected.

The Hazard Ranking System is undergoing a major revision as a result of a requirement in the latest amendments to the Superfund law. A proposal is expected this year. Changes projected are the addition of an air-potential route, more emphasis on environmental effects, and the addition of a direct-contact route. Outside interests also play a strong role in the risk management decisions of remedial prioritization. The public is involved in the National Priority List process in that the list is an Agency rulemaking action. This means that the list is proposed in the Federal Register. Comment is solicited and a docket is maintained for all to see the comments. A response is prepared and the list is then promulgated by appearing again in the Federal Register, along with a listing of all significant comments and

RISK MANAGEMENT ISSUES: INACTIVE WASTE SITES 93

the Agency's response. Considerable interest is also shown by responsible parties and elected officials. Since decisions whether to list or not have significant financial consequences, a great deal of time is expended and effort is made toward communicating the bases of our recommendations. Frequent meetings occur, and the community relations program is very active during this stage.

3.8.4 Remedial Cleanup Levels

The determination of remedial cleanup levels is probably the most time-consuming and difficult of the risk management decisions in the Superfund program. Because of the financial consequences, a very rigorous procedure is outlined in the program requirements. The general name for this procedure is the remedial selection of remedy process.

Before the remedy can be selected, a feasibility study is prepared. This study takes all of the data generated in the investigation of the scope and extent of contamination at the site and factors in a number of feasible alternatives that could result in satisfactory levels of cleanup. The cleanup alternatives vary from a no-action alternative, to contamination containment options (such as capping the site), to highly complex treatment and destruction options (such as incineration or chemical treatment). The study contains descriptions of technologies, costs, time lines, implementability analyses, legal constraints, technological constraints, and other pertinent factors. The purpose of the feasibility study is to provide the decision maker with all reasonable options from which to choose a remedy. The feasibility study should be neutral in order to allow the decision maker the widest flexibility.

When the feasibility study is completed, the entire project study is provided to the public, elected officials, and the responsible parties for their review and comment. Meetings are held and explanations are given. Comments from these reviews are assembled and collated and provided to the decision maker, who in most cases is the EPA Regional Administrator.

The documentation for the remedy selection is provided in a document called the Record of Decision. This is prepared by EPA and includes a summary of remedial studies, a description of the feasible alternatives, with their costs and time frames, a summary of the comments made by the public and others, and a description of the alternative that has been selected by the decision maker.

Current policy is to base the selection of remedy on the use of nine criteria for analysis. These criteria are:

94   INSURANCE AND RISK MANAGEMENT FOR HAZARDOUS WASTE

(1) Overall protection of human health and the environment
(2) Compliance with applicable or relevant and appropriate requirements of other environmental laws (ARARs)
(3) Long term effectiveness and permanence
(4) Reduction of toxicity, mobility, or volume
(5) Short-term effectiveness
(6) Implementability
(7) Cost
(8) State acceptance
(9) Community acceptance

Some of these criteria are explained further below.

Many of these criteria are required by the latest amendments to the Superfund law. The major changes from the original legislation are the requirement for ARARs, long-term permanence, and reduction of toxicity, mobility or volume.

In the previous legislation, the major criteria were overall protection, implementability, cost, and acceptance by the state and community. These criteria tended to make the remedy selection process highly risk-based, with the reality check provided by cost and acceptance. For the first several years, remedies tended to contain the waste in place via capping or similar means, or provide for excavation and off-site disposal. Remedies tended to cost an average of $5 million, and this resulted in a relatively high acceptance rate by states, communities, and responsible parties.

When the Superfund amendments arrived with a requirement for ARARs, long-term permanence, and reduction of toxicity, etc., remedy costs escalated significantly. While not enough time has passed to develop an average remedy cost under the new requirements, many of the remedies are in the $10–50 million range. Remedies now require treatment of waste in most cases. Incineration, biodegradation, and chemical fixation are the leading technologies selected thus far. In many cases, acceptable risk reduction would allow a containment remedy, yet the Congressional mandate for permanence and toxicity reduction seems to require more expensive treatment alternatives. Countering this is the need still to be cost-effective. One can see the dilemma facing the decision maker.

While the public, responsible parties, and elected officials were very active during the public comment period, the selection of remedy process is meant to occur with little outside involvement. The process assumes that the decision maker assembles all the technical information from the feasibility study, the comments from the public, responsible parties, and elected officials, the Agency response to the comments received, and the

RISK MANAGEMENT ISSUES: INACTIVE WASTE SITES 95

nine criteria. He then is expected to select the remedy that best balances all the criteria — surely a formidable task!

Once the remedy is selected, the Agency communicates this decision to the public, along with an explanation of the rationale for the choice. This communication phase is very important to the continued working relationships of the public and other interested groups.

It would be foolish to state that this process works as smoothly in practice as in theory. There is a great deal of involvement by responsible parties, the public, elected officials, national program offices, contractors, auditors, and any other group affected by the remedy selection. It is difficult to keep focused on the program goals while attempting to balance these competing forces. Nevertheless, the process seems to work, thanks to dedicated staff and decision makers. The risk management process for selection of remedy is probably the most excruciating that a regional EPA official makes, yet it seems to fulfill that Agency mission to keep risk assessment and risk management procedures the centerpiece of decision making.

3.8.5 Deletion

A final area of risk management is the decision whether to delete (i.e., remove) the site from the National Priority List after it has been cleaned up. This process requires that once a site has had the selected remedy implemented satisfactorily and a decision made, no further fund-financed response is necessary. The decision is submitted to the Federal Register as a proposed action and comment is solicited. When the comment period ends, a responsiveness summary is prepared and presented to the decision maker, who in this case is in our national office in Washington. The final decision is promulgated in the Federal Register as a final Agency action. The risk management process for deletion has not been a controversial process due to the similarities of approach with earlier processes in the remedial program.

3.9 Insurance and Responsibility

The responsibility for cleanup of inactive hazardous waste sites lies with the responsible parties. If they are unwilling or unable to perform the cleanup, then the Superfund law provides a mechanism for the government to arrange for cleanup itself, using trust fund money. There has been much discussion as to whether a responsible party should agree to clean up itself

96 INSURANCE AND RISK MANAGEMENT FOR HAZARDOUS WASTE

or wait until the Agency cleans up and presents its bill. In the early days of the program, few cleanups were performed by responsible parties. Over the past three to four years, more and more cleanups have been performed by such parties. This is a promising trend and is expected to continue. Obviously each site cleaned up by a responsible party frees up money to be used to clean up another site that does not have a willing or able responsible party.

There are a great many details that must be arranged in order for a responsible party to agree to clean up. The Superfund amendments devote much language to the requirements for such agreements. Any agreement must be enforceable, provide for the same general process of risk management as Agency cleanups, and provide for many other legal issues that are not the subject of this paper. A general rule for a responsible party is to take the risk assessment, risk management, and risk communication process that the Agency has and adhere as closely as possible to it.

Indemnification of cleanup contractors is a major problem faced in the cleanup of inactive hazardous waste sites. The problem stems from the inability of these firms to obtain insurance and hence the request to the government to indemnify cleanup contractors so work may proceed. The Superfund amendments addressed this issue partially with a provision to provide for indemnification under certain circumstances. These procedures are being developed and will be issued soon. The problem with the Superfund amendments is that it does not allow for indemnification of state liability in strict liability states. This problem is currently under review by the Agency. In the meantime, cleanups are being performed and very few, if any, cleanups have been halted due to this issue.

3.10 Conclusions

The cleanup of hazardous waste sites under Superfund involves each link of the risk chain: risk assessment, risk management and risk communication. The program was developed as a risk-based process, and there has not been a need to modify an existing process to incorporate the Agency initiatives in this area. Risk management decisions are incorporated throughout the process at major points, and they reflect the Agency guidance. The risk management process involves heavy reliance on the risk assessments provided in program operation, significant public input throughout the process, and feedback of the decisions to affected people. The balancing of the statutory bias for treatment and toxicity reduction against the pure risk factors in remedy selection provides the greatest challenge to the decision maker in the program.

References

Public Law 96-510, *Comprehensive Environmental Response, Compensation and Liability Act of 1980*, 94 STAT.2767-2811, December 11, 1980, and amendments.

Environmental Protection Agency (1982). National oil and hazardous substances contingency plan. *Fed Reg* 47(137): 31180-31243, and amendments.

Environmental Protection Agency (1984). Risk assessment and management: Framework for decision making. EPA/600/9-85/002.

Discussant: Jack Schramm

Jim Seif's paper, I think, is an excellent presentation of the Superfund program's details. There is a lot of stopping and evaluating and communicating and choosing and starting the cycle over again.

There is a lot of opportunity for the competing forces that he writes about to assert themselves, one against the other and against EPA, too. One would expect, under such circumstances, that theory and practice often pass one another in the night, as Jim Seif points out. Inexorably, however, the program does move forward. The toughest decision for the regulator, Jim notes, is the selection of a remedy. This happens after — and only after — a lot of weeding out has been done (risk assessment, if you will) through the Hazard Ranking System (HRS) and listing on the NPL. And may I say parenthetically that the HRS process is front-end-loaded with a lot of very conservative, subjective judgments about what constitutes particularly a *potential* hazard as distinguished from *actual* hazard. Of course, this makes the later selection of remedy decision even more difficult, not to say controversial. But be that as it may. . . . It is that remedy decision that I want to focus on principally here, because that's where the rubber hits the road. It's what corrective action is all about. Cleanup levels, costs, fund balancing, and the philosophy of Superfund are all bound up in that decision. It seems, indeed, a tough decision to make; but is it that tough in reality?

Let's step back a moment and look at the evolution of environmental law and policy over the last 20 years. The Clean Air Act and the Clean Water Act of the early 1970s, plus their later amendments, were based on establishing healthy and aesthetically pleasing ambient conditions for sur-

98 INSURANCE AND RISK MANAGEMENT FOR HAZARDOUS WASTE

face water and air. The standards were implemented through a system of plans and permits that, by and large, permitted "insignificant" degradation, because zero discharge was not technically practicable — although that goal was never abandoned and is still being paid lip service.

Waste management facilities, it is fair to say, are treated differently. Zero discharge as a concept is alive and well in the regulation of hazardous waste. I believe this is so — some pop psychology — because we often drink the water beneath our feet and we can't see what's going on down there. So, the groundwater becomes the resource to be protected.

Yet, there is no Clean Groundwater Act. Instead, what we got after the Air and Water Acts were the Safe Drinking Water Act, RCRA, and CERCLA. Under those Acts, in effect, waste management facilities are not permitted a degradation zone around areas of potential impact.

With redundant controls for landfills, for instance, zero discharge is an operative concept again. (The contrast is interesting because no one expects to be able to breathe the air at the stack or to drink the water from the outfall.) The fear of waste is pervasive. Zero discharge is not risk management; it is zero risk.

At this point all I am suggesting is that we have some perspective and understand the social and political context of waste management in this country.

What actually happens at a remediation site when the cleanup of a release is technically practical is that risk assessment and management are not truly practiced by the regulators. This is so because cleanup levels are preestablished; i.e., you clean up the groundwater to the MCLs even though no one may be drinking the water now or in the foreseeable future.

In other words, risk assessment is employed in setting the generic standard, but is not employed, nor is risk managed, at the site-remediation level.

Meeting the groundwater protection standards at the unit boundary involves seeing what's down there, looking at the MCL tables, and then designing a remedy to meet those numbers. The standard drives the technologies. You either pump and treat, do source controls or institutional controls, or some combination of these.

(Permit me this aside: During the MCL v. MCLG debate, someone commented that if the MCLG standard, which is the more stringent of the two, won out as the cleanup target, people would be moving to Superfund sites because that's where the cleanest water in America would be!)

So, the fact of the matter is that the cleanup level — the standard — is preestablished. That's not so bad so far as it goes because our industry, for one, is always looking for the certainty of a number, both for our site remediation work and for our own facilities as well.

RISK MANAGEMENT ISSUES: INACTIVE WASTE SITES 99

The problem is that that preestablished number would be operative in the Gobi Desert if it were located in the U.S.A. True, attenuation is an optional remedy, but it principally exists on paper; it really isn't practised in fact. Risk assessment and management should be made to focus on who in fact is at risk.

I'm not here arguing that the rural people should receive less protection than urban folks. What I am suggesting is that these judgments need to be more sophisticated, that groundwater mapping be done, and data accumulated, and that localized, risk-based ambient groundwater standards be established.

That is what's missing in our groundwater protection efforts today. We need to bring more rationality to the process. Absolute numbers in the hands of a young engineer lacking in experience and perspective can themselves be wasteful and counterproductive.

If we are more rational, the concepts of *cost* and *protection* might find some common ground. That becomes more and more significant as we consider the numbers in Leslie Cheek's paper.

The next point is one of departure, but it relates to groundwater and needs to be made because this debate may be played out in the Congress as it considers some of the comprehensive groundwater protection bills now pending in some of its committees.

What usually happens when people think there are too many chemicals floating around the environment is that the ratcheting down is usually applied against the commercial waste industry (and we deal with only about 4% of the hazardous waste generated in America). And believe it or not, we don't mind being held to a high level of accountability. We're regulated within an inch of our lives, and that's okay.

What we are bound to ask, not only in the name of equity, but also in the name of sound environmental policy, is this: Where is the same level of accountability for agricultural chemicals, for 17 million septic tanks, road de-icing salts, urban runoffs and other nonpoint sources? There are a lot of other sources that perhaps are much more pervasive and serious threats to the groundwater than subtitle C facilities, as attested by many neutral observers.

UST (underground storage tank) regulation is a step in the right direction. And subtitle D regulation will address a broad array of threats — if municipalities are held to the same standards as commercial facilities!

But action there doesn't excuse inaction as to those other sources.

In the final analysis, proactive legislation and regulation will be infinitely more effective in every way than a reactive Superfund program. So an effective groundwater strategy should be factored into an integrated hazardous waste strategy — a strategy that clearly addresses, assesses, and

100   INSURANCE AND RISK MANAGEMENT FOR HAZARDOUS WASTE

manages risk right where it really begins to affect public health and the environment.

Discussant: Paul Portney

Let me start by complimenting Jim Seif and Tom Voltaggio on their paper. For those of you who are not familiar with the remediation process under Superfund, the paper is an excellent description of the steps EPA must go through to initiate a cleanup at a site. Because their paper was largely descriptive, I would like to put in starker contrast a number of issues that they obliquely touched upon that have to do with risk management, and several issues they didn't touch upon. I want to emphasize that some of what I'll have to say here is not necessarily in reaction to material that was in the paper. The reason I want to do this is because there are certain important issues about Superfund that we have not touched on yet in the conference. It is in that spirit that I will try to raise these issues. I think I'll basically do so in the form of a number of questions that I hope we'll have a chance to address later.

The one overarching theme I have in my comments is that the discussion so far has focused largely on who will pay for cleanups. Whether or not the tab to be paid is a $100 billion or more or less, a question we should begin to think about is whether or not anybody should pay for cleanups at sites where the benefits don't seem to justify the cost.

I'd like to begin by identifying a number of issues, some of which ran through Jim and Tom's paper, that I think are the key to understanding the risk management options with respect to inactive hazardous waste disposal sites.

First of all, with respect to some of the comments that Betty Anderson made, I think it is important that we understand, and understand soon if possible, whether or not current risk assessment techniques are really adequate for the design and operation of a risk-based cleanup system. That is to say, are the techniques such that we really can derive careful rankings of which sites pose the greater risks to health and the environment, and then order cleanup priorities on that basis? If, as I hope and believe, the answer to that question is tentatively "Yes" how does the cost-effectiveness of Superfund cleanup actions compare to cost-effectiveness in terms of health and environmental protection associated with other environmental regulatory programs? How does it compare with other nonenvironmental regulatory programs? How does the cost- effectiveness of health protection from Superfund cleanups compare with other nonregulatory programs, such as national health insurance, women's, infants, and children's nutritional programs administered by the Department of Health and Human Services, etc.?

In other words, if we are going to spend a lot of money to clean up Superfund sites, we certainly ought to know whether or not that money can provide more protection for health, or more protection for the environment, through other environmental programs or through other government budgetary programs that possibly don't have anything to do with environmental protection.

Next, how can we stop action on an emergency removal, planned removal, or remedial action site, short of cleaning it up to an eat-the-dirt level, if we decide that further action isn't worth it? Is it politically acceptable or even possible to make a decision that we're going to stop at a certain point because the risk reductions associated with further cleanups at the site just don't merit spending an additional $1 million, $2 million, $5 million, or $10 million? An observation I'd like to make here pertains to Leslie Cheek's paper. He talked about the Superfund program becoming a public works program. I once heard it said that the easiest way to make a capitalist out of a socialist is to give him $20. I think it's also the case that the easiest way to make a proponent of the Superfund program out of somebody who heretofore had been an opponent is to point out that his/her district has four or five companies that will benefit from Superfund cleanup spending. I think this is tremendously important, not to mention the possible consequences for the economy of a $20 billion or greater expenditure of money for Superfund cleanups. We all want to have an economy in the United States that is capable of doing more than renting video tapes and delivering hot pizzas to each other.

The question is, if cleanups on the order of magnitude that we're discussing endanger the competitiveness of domestic industries and do not provide benefits commensurate with those costs, then I don't believe we should go ahead with this program. If these cleanups provide benefits that are equal to or greater than the cost associated with them, then we all ought to be willing to live with the economic consequences so long as we understand them clearly.

Finally, let me do a little multiplication to elaborate on the numbers that we heard thrown around here today. There are currently approximately 1000 sites on EPA's National Priorities List. The best estimate we have now is that under the 1986 SARA, an average cleanup will cost $20–30 million. Thus, it will take $20 billion to clean up those 1000 sites. But EPA itself has recognized that there are 30,000 sites on the so-called CERCLIS list. Those are sites that could eventually make their way onto the

National Priorities List. If, for some reason, we decided that we ought to go after all 30,000 of those sites, at $20 million per site, that's $600 billion. I'm not saying that it's likely that all of those 30,000 sites will become the targets of $20-million Superfund cleanups, but nevertheless one has to begin to do this kind of arithmetic in thinking about this program.

Finally, I mentioned 30,000 sites that EPA has identified. EPA is routinely castigated by the General Accounting Office, which says that according to its accounting, there may be as many as 400,000 sites that could eventually qualify for remedial action under Superfund. Four hundred thousand sites times $20 million per site is $8 trillion. By comparison, the U.S. gross national product is currently $4.5 trillion per year. Again, and I want to be very clear about this, I'm not suggesting that we will find 400,000 sites where we will decide that we ought to spend $20–30 million to clean them up. I will furthermore say that if we were really confident about our ability to do cost–benefit analysis, and if we had found that many sites and had found that the benefits of cleaning up 400,000 sites at $20 million per site were worth more than $8 trillion, then we ought to go ahead and do it. But I'm trying to give you some idea of the possible eventual magnitude of the Superfund program, given the way some people are beginning to count sites and given what appears to be the average cost of cleaning up sites under a 1986 SARA.

As I say, none of these issues I presented are ones that Jim Seif and Tom Voltaggio highlighted explicitly; but I think a number of these issues ran through their paper. At the very least, their paper prompted me to think about some of these other issues and present them today.

II MANAGING EXISTING WASTE SITES

Part II is devoted to the management of existing waste facilities both in the United States and abroad. Two of the papers are written by Europeans who have had extensive experience in developing risk assessment procedures and insurance programs for hazardous waste facilities. The third paper on risk management issues discusses the lessons learned from the implementation of the RCRA program since its enactment in 1976.

Risk Assessment Issues. Peter Schroeder describes a procedure that provides a risk profile of an existing waste facility by characterizing the probability and severity of hazards. Risk assessment techniques are utilized to determine where specific hazards lie on a two-dimensional probability–severity grid and to suggest risk reduction measures. This approach enables insurance premiums to reflect the potential risk associated with the waste facility.

Insurability Issues. Baruch Berliner and Juerg Spuehler discuss issues associated with insurability of existing hazardous waste facil-

103

ities. Berliner points out that hazardous waste risks pose serious challenges to the insurance industry due to great uncertainties about the probability and consequences of the risks and exogenous factors such as liability rules. In the second part of the paper, Spuehler shows how different types of pooling arrangements in Europe have provided insurance against pollution problems. He then proposes a liability life policy, an innovative insurance arrangement that provides coverage for the facility from its opening to the time it may close, including the post–closure period. Premiums will vary depending on the nature of the facility.

Risk Management Strategies. Richard Fortuna focuses on changes in RCRA over time to provide a perspective on risk management strategies. He feels that a strict liability standard, direct and indirect economic incentives, and a corporate conscience are necessary for industry to adopt preventive and protective waste management technologies. Fortuna feels that risk assessments should be used to establish national minimum performance standards for existing waste facilities, but argues against the use of site–specific risk assessments given the scientific uncertainties that currently exist in characterizing hazards.

4 MANAGING EXISTING HAZARDOUS WASTE FACILITIES: RISK ASSESSMENT ISSUES

Peter Schroeder

4.1 Introduction

Risk assessment does not make a product safe or a waste less hazardous. But it should identify the problem and lead to solutions.

In the last 10 years, considerable efforts have been made not only to develop more sophisticated and safer products but also to tackle the problem of wastes from the many different manufacturing processes by the application of sophisticated technologies. But we still cannot say that we use the same knowledge in this field as we use to develop new products.

Since the waste we suffer from today is the last step in the life cycle of a product or process, we apparently did not anticipate its associated problems. Therefore today's efforts to control the unpleasant situation can be qualified as an after-the-fact approach. The question has to be raised why we did not incorporate all the forward-looking techniques developed in such technologies as aircraft, space, nuclear, communication, quality, and safety to solve the problems in the design phase to prevent today's catastrophic waste problems.

In this context, the risk assessment process would have become a vital part or even a key to the solution. By definition, risks are assessed or

106　INSURANCE AND RISK MANAGEMENT FOR HAZARDOUS WASTE

quantified loss potentials. The losses have not occurred yet, so if we know what could happen, why do we not prevent it from happening? It is much cheaper to do a thing right the first time than to correct it later.

Whatever the causes for many large accidents were, many could have been prevented for a fraction of the costs for their compensation, as is indicated in table 4-1.

Seveso could have been prevented by a change of the chemical process to a low-temperature process or by changing the steam piping. The suspension of the Hyatt walkway was designed properly, but a change of the suspension rod arrangement during installation led to a significant reduction of the load-bearing capacity. In Bhopal, a number of safety features were out of service, and in Schweizerhalle a properly designed sprinkler system, fire walls, and stop valves in the sewer system would have prevented the disaster.

So the lack of foreseeability and life-cycle thinking have played a significant role in most large accidents. With reference to pollution, this is even more dramatic because of the long delay time between emissions and damages. So the traditional after-the-fact approach becomes more severe and is further enhanced by the complex reactions among the ecosystems. As a consequence, control measures have been initiated too late with a limited effect, because the control efforts were governed to limit the damages and not to prevent them.

4.2. Emissions versus Control Efforts over the Life Cycle

Since data is hardly available, a qualitative review of emission intensity and the associated safety efforts to control them are outlined in figures 4-1–4-4. According to OECD statistics,[1] 90% of the chlorofluorocarbon CFC-12 manufactured between 1960 and 1985 was released into the atmosphere. What are the controls? From the same source it can be read that the water consumption in m^3 per capita for 1980 ranges from 108 for Switzerland to 2306 for the U.S. For the same year it is reported that only 70% of the population was served by wastewater treatment plants in both countries.

The generation of hazardous waste alone in 1980 was approximately 250 million tons in the US and 100 million tons in Switzerland. In the meantime the only landfill for hazardous waste in Switzerland has been closed. This illustrates the problem dramatically.

A few of the major disasters in each decade, which we think have had a significant impact by forcing an increase in the efforts to control their negative effects, are also listed in figures 4-1–4-4.

RISK ASSESSMENT ISSUES: EXISTING WASTE SITES 107

Table 4-1. Cost for Prevention

Year	Incident	Compensation	Cost for Prevention
1976	Exploding reactor, Seveso	US $ 150 M	<US $ 10,000
1981	Collapse, Hyatt Regency Hotel. Kansas C.	US $ 90 M	<US $ 1,000
1984	Union Carbide incident, Bhopal	>US $ 200 M	<US $ 50,000
1986	Schweizerhalle fire	US $ 60 M	<US $ 100,000

Source Zurich Insurance Company (1987). Catastrophic losses — a problem for insurers only? Interlaken Symposium.

The situation until the 1950s can be described as follows (see figure 4-1).

- The water quality of lakes and rivers deteriorated rapidly in the 1950s. It was quite common that the color of certain rivers changed to red or yellow and that their beds were seriously polluted with heavy metals. The river beds acted as purification filters — at least in theory. The waste that could not get disposed of in a river or lake went into the normal household garbage that was dumped in every community. To fight air pollution from industrial emissions, stack heights were increased. In many areas the owner of the highest stack felt superior to the others. The governing principle was, "Dilution is the solution to pollution."

The description of the situation in the 1960s is as follows (see figure 4-2).

- The persistence of DDT in food and in Swiss cheese was discovered! In Japan, a large number of people either died or suffered permanent disability after consuming PCB-contaminated cooking oils. Massive birth defects occurred due to administering Thalidomide — daytime sedative and sleeping drug — to pregnant women. The controls were intensified with broader testing for potential negative health effects. The construction of garbage incinerators was initiated and large wastewater treatment plants were constructed. Rivers and lakes were no longer accepted as waste dumps. The oceans, however, kept this function.

The description of the situation in the 1970s is as follows (see figure 4-3).

108 INSURANCE AND RISK MANAGEMENT FOR HAZARDOUS WASTE

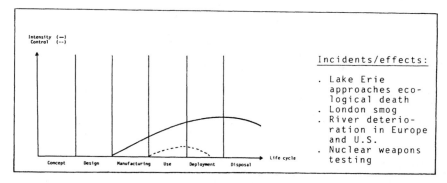

Figure 4-1. Pollution Intensity vs. Control until the 1950s.

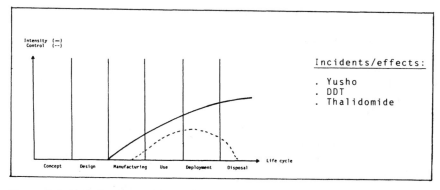

Figure 4-2. Pollution Intensity vs. Control in the 1960s.

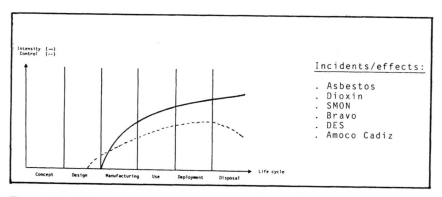

Figure 4-3. Pollution Intensity vs. Control in the 1970s.

RISK ASSESSMENT ISSUES: EXISTING WASTE SITES

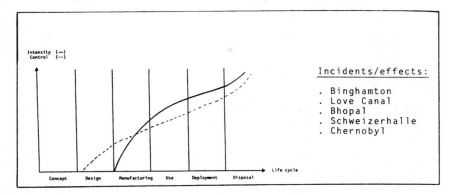

Figure 4-4. Pollution Intensity vs. Control in the 1980s.

- The WHO declared asbestos to be a proven human carcinogen. The Seveso accident happened, releasing approximately 2 kg of accidentally produced Dioxin into the environment. In Japan and elsewhere, 1000 people died (including suicides) and 10,000 were disabled to varying degrees by taking quinoform-type drugs such as clioquinol as treatment and prophylactic for diarrhea, which was often caused by the poor quality of drinking water.
- A blowout at the offshore oil platform *Bravo* in the North Sea spilled many millions of gallons of crude oil into the sea. DES, a hormone to control pregnancy, led to malignant tumors in women at the age of 20 after their mothers had taken the drug during pregnancy. DES is still used in animal feeds in some countries.
- A supertanker with only one rudder system crashed into the rocks of the north coast of France, spilling more than 100,000 tons of crude oil onto the shore, killing thousands of animals.
- After Flixborough, where a devastating vapor cloud explosion occurred, risk analysis methods in the chemical industry were introduced, especially the so-called HAZOP method. For the first time, the safety for systems was systematically addressed in the design phase. So the control methods for pollution effects started significantly earlier than the manufacturing process. Animal testing to detect negative side effects of harmful substances in humans was expanded, thus creating new resistance against a misuse of animals.

The description of the situation in the 1980s is as follows (see figure 4-4).

110 INSURANCE AND RISK MANAGEMENT FOR HAZARDOUS WASTE

- In Binghamton NY, a fire involving a PCB-filled transformer lead to a dioxin and dibenzofuran contamination of an entire 18-story office building. A hazardous waste dump that had been reclaimed leaked and threatened hundreds of people and their property. Bhopal was a human tragedy killing more than 2000 people and disabling an unknown number through a single industrial accident. Finally, Schweizerhalle near Basle did not injure or kill a single person, but resulted in the massive destruction of fish in the Rhine by pesticides in fire-extinguishing water.
- Environmental regulations forced preventive thinking. Risk assessment became widespread and as a consequence preventive pollution control moved further into the design phase of the product life cycle.

If we summarize the status of this development, we can state that the gap between emissions and control has become significantly smaller. However, effective control demands even more activities in the design phase.

4.3 The Concept

The question remains: how we can move the control–effort curve further toward the beginning of the life cycle? If we look at other industries, we find that one important industry has done this for decades — the aircraft industry — because of the immediate effects of a failure. They declared the curve illustrated in figure 4.5 as the curve to be followed if accidents are to be prevented and costs are to be minimized.

The safety efforts were governed by system safety, which actually developed as a result of the missile programs of 1950 and 1960. The liquid propellant missiles blew up frequently, unexpectedly, and devastatingly. In addition, the highly toxic and reactive propellants were sometimes more lethal than the poisonous gases used in World War I, more destructive than many explosives, and more corrosive than most materials in industrial processes. Also, the Air Force had many fatal accidents.

Until missiles were developed, aircraft losses were generally blamed on the pilots. Since there were no pilots aboard the missiles, no blame could be put on them. It became apparent that the causes of accidents were defective concepts, design, manufacture, maintenance, or other activity prior to use. The philosophy developed that safety programs had to be planned and initiated almost as soon as a new system was conceived, and

RISK ASSESSMENT ISSUES: EXISTING WASTE SITES   111

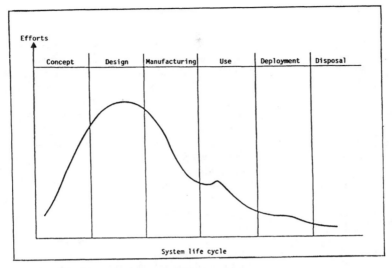

Figure 4-5. Safety Efforts Over the Life Cycle.
Source: Roland, H. E., and B. Moriarty (1983). *System Safety Engineering and Management.* J. Wiley & Sons, New York.

then carried on through the system's life cycle and up to the time when it is finally eliminated.

Initially, there was a lack of methodologies such as hazard analysis methods, safety program activities and control measures. But those who worked in this discipline were forced by necessity into organized approaches to any form of accident prevention. So system safety was born — a forward-looking approach to safety with a climax of the safety efforts in the design phase, as shown in figure 4-5, so that now all identified hazards can be engineered out before a system is built.

As we all know, the aircraft industry has done a marvelous job. Many times it was said that this was only possible with the tight regulations of this industry. But the accident analyses of other incidents prove that no catastrophe comes out of the blue. There are always early warnings, as can be shown in the case of asbestos and dioxin (tables 4-2 and 4-3).

The real question is: Given the early warnings available about asbestos and dioxin, why did the catastrophes associated with these substances occur? Well, there are two major problems: first, small incidents have to be recorded, and second, they have to be interpreted as an indication of a trend. The small incidents are normally qualified as an exception — this is much easier to do than an in-depth trend-finding exercise. Based on the

112   INSURANCE AND RISK MANAGEMENT FOR HAZARDOUS WASTE

Table 4-2.   The Hazardous properties of Asbestos

79 AD	Plinius describes the use of transparent leather skins as asbestos respiration protection for slaves
1897	A. Netolitzky, Vienna: Asbestos weavers and their families show respiration troubles as a result of inhaling asbestos
1906	The autopsy of an English asbestos worker reveals asbestosis
1918	Prudential Insurance, U.S.A.: No insurance for asbestos workers
1930	Mereweather and Price, U.K.: Examination of 363 asbestos workers shows a 26% asbestos rate. The rate increases to 81% with more than 20 years asbestos exposure
1947	English Factory Inspectorate: of 235 asbestos deaths, 13% had lung cancer. The respective rate for silicon deaths is only 1.3%
1974	WHO considers asbestos to be carcinogenic
1981	European manufacturers still question the hazards of asbestos
1983	Asbestos-free production begins (Switzerland)

Source: Castleman, B. I. (1984). *Asbestos — Medical and Legal Aspects.* Harcourt Brace Jovanovich, New York.

Table 4-3.   The Knowledge of Dioxin Incidents

Year	Country and Producer	Production
1949–1956	U.S.A, FRG, France: six incidents	
1962	Italy: ICM	TCP
1963	Netherlands: Philips–Duphar	TCP
1964	U.S.A.: Dow Chemicals	TCP
1966	France: Rhône–Poulenc	TCP
1968	UK: Coalite, Chemical Products	TCP
1970	Japan	PCP,2,4,5-T
1973	Austria: Linz Chemie	2,4,5-T
1973	*NZZ Article on Dioxin*	
1974	FRG: Bayer	2,4,5-T
1976	Italy: Icmesa	TCP

Sources: Hoffman–LaRoche (May 1986). Roche Magazine, Basle; and Neue Zürcher Zeitung (February 1973). Dioxine — Heimtuckische Gifte in unserer Umwelt, Zürich.

many negative experiences with these two substances, the catastrophes would have been preventable if the prior lessons had been learned.

The aircraft industry has learned its lessons. But another contributory factor was certainly that in 1947 it had already been declared that safety must be given the same attention as, e.g., aerodynamics or structural design. Once this decision was made, the necessary tools were developed and system-safety thinking was established. All the major risk analysis techniques in fact originated from the space and aircraft industry. With a long time delay, they are reaching the chemical industry and the other major industries that are part of the pollution problem.

4.4. Risk Assessment Issues

With the above facts in mind, we can comment on the assessment issues. But before stating the relevant facts, we have to choose one analysis methodology as a carrier of our comments.

There are two basic methods in risk analysis: inductive and deductive. The inductive methods go from the part to the whole and try to find out what can happen. Well-known methods of this type are, e.g., FMEA, FMECA, HAZOP, and PHA.

The deductive methods go from the whole — e.g., an undesirable event — to the part, and try to find out how it could happen. As a liability and property insurer being faced with a huge variety of products, and environmental and property risks, we had to develop our own methodology — the *"Zurich" Hazard Analysis*[2], which is now being introduced in major chemical and other companies. It is an inductive method, flexible and efficient. The method can be briefly described as follows.

A successful method calls for distinct, well-defined steps. For the "Zurich" Hazard Analysis, the first such step of prime importance is that of scope definition by identifying the system, systems, or part of a system to be analyzed. This is followed by collecting the needed information and, for maximum advantage, by forming an optimum analysis team. Next, the methodology calls for systematic hazard identification and consecutive listing of hazards in a hazard catalog. Once a hazard is identified and entered in the hazard catalog, together with its possible cause and its potential effects, each hazard can simultaneously be assessed on a comparative scale as to its probability of occurrence and its related severity. On the one hand, we thus wish to assess, with respect to the other hazards identified, where each particular hazard lies with respect to its effect — should an event be triggered. On the other hand, we wish to assess, still on a comparative

114   INSURANCE AND RISK MANAGEMENT FOR HAZARDOUS WASTE

basis, the relative likelihood that such an event may be triggered. This step of assessing the probability of a cause and the severity of an effect is called hazard assessment.

The use of a two-component definition for hazard assessment — probability and severity — allows a grid to be established. Entering the consecutively numbered and identified hazards in this grid reveals their relative position to each other. This is called the risk profile. It provides the desired visibility and forms an important management tool for proper risk reduction. With its help, the desired protection level can be established. Any risk above this protection level will have to be eliminated or reduced. Any risk within the protection level can be transferred or retained.

Based on this methodology, the risk assessment issues regarding existing hazardous waste facilities can be discussed systematically.

4.4.1 The Scope

The scope needs to be defined carefully and thoroughly. Full consideration has to be given to the information and time available and the matching results possible: with only layout information available, conceptual results are possible. If the system to be analyzed has human factors as major components — a situation prevailing in most facilities — a high degree of uncertainty comes in because human reliability is very low, as indicated by the Basic Error Rates (BER) in table 4-4. These data developed by the U.S. Navy provide a good overview of how often, out of a million tasks, simple tasks can be done wrongly.

Table 4-4.   Basic Error Rates (BER)

Task	BER
Read instructions resulting in procedural error	64,500
Install "O" rings improperly	66,700
Use wrong adjustment on mechanical linkage	16,700
Torqueing fluid lines incorrectly	104
Nuts and bolts improperly installed	500
Machined (drilling and tapping) a valve to the wrong size	2,083
Omitted parts in connector assembly	1,000

Source: Ferry, T. S. (1981). *Modern Accident Analysis, An Executive Guide.* J. Wiley & Sons, New York.

RISK ASSESSMENT ISSUES: EXISTING WASTE SITES

If the scope is defined for such a facility, we can say that in such a location an attempt is being made to dispose of an unlimited variety of compounds occurring in different physical forms. This daunting task, which has to be carried out in such an installation, already defines the high degree of uncertainty and the imprecise results possible. But who provides the data for all these substances if not the manufacturer? In the traditional waste business, the manufacturer is in most cases not involved, a fact that leads to a lack of the most important data — the substances to be disposed of.

So it can be concluded that the scope is easier to be defined if the facility has the following:

- Industrial, automated, and controlled processes rather than manual, vaguely defined procedures
- Known and reliable operating conditions, rather than dependence on the erratic behavior of a handful of manipulators of drums
- Well-known substances to treat, instead of many unknown compounds from unknown sources

Unfortunately, these criteria for a proper scope definition are very rarely met, so the first step of risk assessment is hardly possible.

4.4.2 The Team

No analysis is better than the knowledge of the persons doing it and the supporting documentation available. Where are the persons who have the intimate knowledge of the waste from products and manufacturing processes? Where are the persons who know what the safe levels of substances are for humans and for the environment? Where are the persons who know what the regulatory agencies will determine to be acceptable? Who represents the public in the team? This flood of questions leads to the following conclusions:

- Only a capable team will produce an accurate analysis
- Capability must also embrace the manufacturer of the substances or products
- Capability also means that those involved know the safe dose of a given substance to man and to environment (GRAS-levels)

These criteria can hardly be met, so also the second step of risk assessment can in most cases not be carried out satisfactorily.

4.4.3 Hazard Identification and Hazard Catalog

The more competently and precisely the pertinent hazards are identified by the team, the better the chances for a comprehensive hazard analysis. In this sense, a hazard is defined as a potential threat to persons and/or property and/or environment.

A systematic approach to hazard identification assures the high level of confidence needed in a professional job. The "Zurich" Hazard Analysis sequentially looks at five major aspects of a clearly defined system or process.

The first, and perhaps the most obvious, such aspect concerns the hazardous characteristics of the system or process as defined by the scope. It essentially addresses the hazards of materials used and the various forms of energy present in the design.

The second aspect, usually requiring more imagination, covers possible malfunctions of the same scope. Can the system or process become hazardous due to a malfunction? In this respect, failures under otherwise safe conditions are addressed and hardware as well as software is included.

The third aspect of hazard identification is that of the environmental influences on the defined scope. Can the various environments of the scope negatively influence its performance so as to render it hazardous? Any and all aspects that come from outside the defined scope are considered and thus include external causes affecting hardware and software within the scope.

The fourth point is that of the scope's intended and foreseeable use and operation. Is there a certain kind of use and/or operations that could be hazardous? Here emphasis is given to the man/machine aspects and interfaces, including ergonomics and foreseeable misuse.

Finally, as the fifth aspect, the life cycle is investigated. What are the potential changes over the entire life of the system, process, or facility as defined by the scope that might introduce hazards? This last point looks at hazards introduced by the time element and includes aging, changes in design and/or organization, occupancy, and use.

Each of these five major aspects has a listing of ticklers to stimulate, in a structured way, the thought process needed to become aware of hazards, their associated potential causes or triggers and their respective possible effects. However, the answers are only as complete as the ability of the team.

The consecutive listing of hazards, potential causes, and possible effects together with hazard assessment, makes up the hazard catalog (see figure 4-6).

RISK ASSESSMENT ISSUES: EXISTING WASTE SITES

	"ZURICH" RISK ENGINEERING	"Zurich" Hazard Analysis **HAZARD CATALOG** Company: Scope:		Page of By/Date: /

No	Hazard	Cause	Level	Effect	Cate-gory

Figure 4-6. The "Zurich" Hazard Analysis Hazard Catalog.

It can be concluded that a systematic approach to identification of hazards and their causes and effects can provide a high confidence level when the relevant knowledge and experience are available. However, a 100% identification is never possible. As defined in sections 4.4.1 and 4.4.2, the properties of the waste to be disposed of are not well known as long as the manufacturer is not in the analysis team. So by definition, the hazards of hazardous waste facilities that have to dispose of waste from different sources cannot be properly identified. Even for a known supplier, the properties of his waste may only be very poorly known, since the waste often consists of or includes unwanted byproducts. Consequently, the environmental effects based on this limited knowledge will be even less understood and therefore ill-defined. A limited level of reliability will only be possible if assured complete destruction at high temperatures can be achieved in suitable installations fitted with the necessary safeguards.

4.4.4 Hazard Assessment

In the hazard assessment process a rating is given to the effect of a hazard — the hazard effect category — and to the cause of a hazard — the probability of occurrence.

Absolute numbers, such as the ones used in statistics, cannot be com-

118 INSURANCE AND RISK MANAGEMENT FOR HAZARDOUS WASTE

pared without exact knowledge of the data base. The data bases, as already stated, are very limited. For this reason a relative assessment is used. The reliability of the assessment again highly depends on the available experts' knowledge in their fields.

For the relative severity of occurrence of a possible effect, the "Zurich" Hazard Analysis makes use of four categories. They can be defined as follows.

I.	Catastrophic	Death, total disability; loss of company image; detrimental financial loss; system loss; widespread environmental damage
II.	Critical	Severe injury with partial disability; severe loss of image; large financial loss; partial system loss; local environmental damage
III.	Marginal	Injury; transient loss of image; indirect financial loss; system damage; temporary local environmental impairment
IV.	Negligible	Minor injury; minor image or financial loss; minor system damage; little environmental disturbance

The latency periods of many substances emitted from a hazardous waste facility regarding the damage to human health and to the environment are certainly unknown in the low-dose area. This might pose considerable difficulties in allocating the right effect category. But in general, the better the destruction process is defined and the better the relevant process variables can be monitored, the easier it will be to assess the effect categories.

Since the information available to assess the probability of occurrence is also very limited, a relative scale of six levels is used. They can be described with respect to the life cycle as follows:

A.	Frequent	Often experienced or likely to occur frequently (= upper limit)
B.	Moderate	Experienced or occurring several times
C.	Occasional	Sometimes experienced or occurring
D.	Remote	May be experienced or may occur
E.	Unlikely	Unlikely to be experienced or to occur
F.	Impossible	Practically impossible (= lower limit)

If the effect cannot clearly be defined, the uncertainty in the probability of occurrence also remains very high.

In conclusion, we may state that the assessment of the effects is very

RISK ASSESSMENT ISSUES: EXISTING WASTE SITES

inaccurate for low-dosage exposure. Low dosages are emitted from ill-defined and controlled facilities. On the other hand, the probability of occurrence becomes very uncertain the more human operations are involved. However, even a highly automated waste treatment process needs a highly reliable maintenance program. Maintenance is again a human task with a high degree of uncertainty. It is known that the various degrees of maintenance generally decline with changing work ethics; if a need for more maintenance exists, it may increase the probability of failures significantly.

4.4.5 Risk Profile and Protection Level

By scaling the six hazard cause levels of the "Zurich" Hazard Analysis on the vertical and its four hazard effect categories on the horizontal, the risk profile grid is drawn (see figure 4-7).

Entering the consecutively numbered hazards in the appropriate profile location provides the much-desired visibility as to where each risk lies with respect to the others. Further, by using the established level and category definition for a particular analysis, the acceptable risk, named protection level, can be determined and drawn (see figure 4-8).

The acceptable protection level for the public should logically be at least that sought by a company risk policy. Disagreement or a differing sense of values between the risks acceptable to the user or neighbors and those acceptable to a company need to be resolved by an appropriate adjustment in the company risk policy.

Those risks from the hazard catalog that end up below and to the left of the line are within the desired protection level. The risks above and to the right of the line are outside the desired protection level and need to be further considered under the aspect of risk reduction.

If we conclude here, we can easily expect that the definition of the protection level or acceptable risk line will be the most difficult task. Who is willing to commit himself to the acceptable risks? The owner and operator? The supervising agency? The local authority? There have been cases where the local government took over the responsibility for the safe destruction of waste, such as in the case of the 41 waste drums from Seveso in the city of Basle. But in this particular case, they did not have to pay for the fastidious and technologically highly sophisticated process. And the total costs for the destruction of one drum exceeded $100,000!

But also in all other cases a protection level is needed, otherwise risk reduction to acceptable levels is not possible. Only if a zero-risk level is

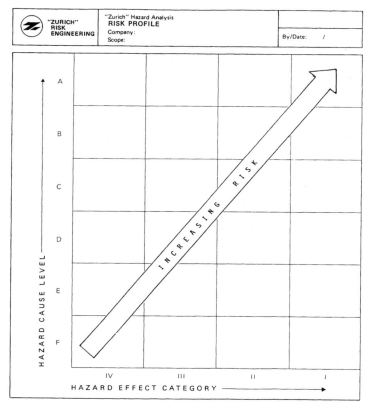

Figure 4-7. The "Zurich" Hazard Analysis risk profile grid.

assumed do we not have to ask, but in this case all activities in our society have to be immediately interrupted. So the discussion should continue to agree on an acceptable risk line that is certainly above zero in the effect categories II through IV. In category I, a zero level can be accepted as long as the effects of very low emissions — that means those within the permissible limits — are not assessed with a category I.

4.4.6 Risk Reduction

Any risk above the determined protection level as it is drawn into the risk profile can now be addressed for the purpose of risk reduction. Within the

RISK ASSESSMENT ISSUES: EXISTING WASTE SITES

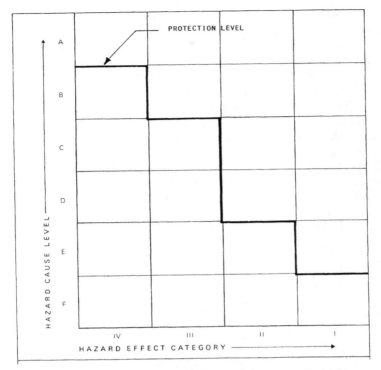

Figure 4-8. Protection level determined by the "Zurich" Hazard Analysis.

framework of cost–benefit considerations, an attempt should be made to eliminate the risks above the protection level altogether. If this is not reasonably possible or is outside acceptable cost–benefit considerations, the risks should at least be reduced.

Two components are available for risk reduction: that of reducing the hazard cause level for a given hazard effect category and that of reducing the hazard effect category for a given hazard cause level. With respect to the risk profile, the first one is on the vertical from the top down and the second on the horizontal from right to left (see figure 4-9).

The severity of the occurrence by itself provides the first priority. In this sense, the most severe hazard effect category, namely catastrophic I, is addressed first.

Following the priorities established by the risk profile, each risk above the protection level will now be addressed. For general considerations, the recommended sequence for risk elimination/reduction is as follows:

(1) The first step is that of preventing the existence of the unsafe condition or triggering of the unsafe event, thus eliminating the effect altogether. Any undesirable event that can be prevented should be; only the means vary.

(2) Should this first step not yield the desired results or not be reasonably possible, then, as a next step, an attempt should be made to protect against the potentially unsafe condition or event.

(3) If, due to the nature of the object or binding circumstances, this is again not successful, then, as a last step, the consequences an unsafe condition or event might generate must be minimized.

While the first step attempts to prevent or eliminate the cause and as a result the effect altogether, the second and third steps accept the fact that the unsafe condition or event might indeed exist.

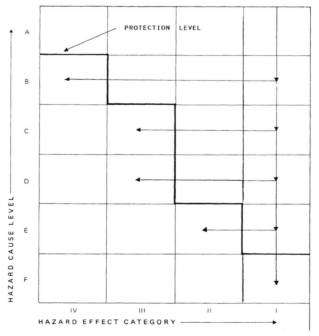

Figure 4-9. Risk reduction with reference to the protection level established by the "Zurich" Hazard Analysis.

RISK ASSESSMENT ISSUES: EXISTING WASTE SITES

The carrier of risk reduction is the safety organization. It has the charter to see to it that the weak links from a safety standpoint are recognized and that they are properly addressed. Such an organization also assures that the preventive, protective or minimization measures that have been taken are maintained and adhered to. In its own right, a safety organization can be a corrective action.

This sequence is in keeping with long-established system-safety considerations and has been widely adopted in Western countries.

In conclusion, the risk reduction process makes it clear that the identified and assessed hazards have to be treated by specific steps. The steps cannot be just organizational measures, but have to address engineering solutions to achieve an acceptable safety level as defined by the protection level. If risk reduction measures are initiated for an existing plant, they will affect a lot of hardware, which will generate high costs. It is our experience that this money will normally not be spent, thus leading to a nonacceptable situation.

It certainly cannot be our intention to insure risks above the protection level. It now becomes apparent that an economic risk reduction can only be achieved if it starts in the design phase. This is exactly the way the aircraft industry has approached its problems. But this implies that the manufacturer of the waste has to be involved, because he has the knowledge of the substances. Unfortunately, such is not normally the case. Unless these two conditions are met, the discussion on risk assessment and risk reduction will be purely theoretical.

4.5. Conclusions

The "Zurich" Hazard Analysis is a well-suited instrument for a thorough approach to the management of hazardous waste facilities. It allows for systematic hazard identification and assessment, and provides visibility for the identified and accepted risks: and finally, it allows risk reduction measures that follow clearly set priorities and an established sequence. Applying the methodology to operating hazardous waste sites reveals a number of important issues:

(1) The safety efforts should climax in the design phase of a facility. For an existing facility, a systematic and thorough hazard analysis should be conducted. It should be repeated periodically and performed afresh if changes are made or have to be made.

(2) A properly designed process with as few human interfaces as possible

124 INSURANCE AND RISK MANAGEMENT FOR HAZARDOUS WASTE

provides better and more reliable results. As much sophistication should be used as has been applied for the original production process.

(3) Since there is almost no other industrial process that has to cope with such a variety of substances, sometimes with unknown properties, and with a smaller budget, further problems have to be expected. The better known the substances to be treated are, the easier it will be to design such a facility.

(4) More knowledge of the nature of the waste has to be incorporated in the waste treatment process. This demands that the manufacturer be integrated in the analysis team and in the operation of a hazardous waste facility. This in turn leads to the demand that the manufacturer of the waste has to destroy it, because he alone has the intimate knowledge about his substances.

(5) Low-dose exposures are unknown. As long as this is the case, no risk analysis process will yield results for corrective actions within the permitted emission levels. Certainly, today's operations can be made much safer than they are, but the price for it will change considerably.

(6) All process variables need careful and exact monitoring. This implies well-defined and automated processes. Manual operations cannot be monitored sufficiently.

(7) More reliability in a waste facility demands simple and low maintenance processes. This can only be achieved by a before-the-fact-approach on the manufacturer's and operator's side. This means far less waste, the properties of which are better known, and far better treatment processes.

(8) An acceptable risk has to be defined for each hazardous waste facility. All risks above the protection level have to be eliminated and/or reduced in the proper priorities and sequence to below this level. The reduction step has to favor engineering solutions. The manual options do not prove to be reliable. Unfortunately, operators handling drummed waste of ill-defined composition still set the stage in the field of hazardous waste treatment.

The evaluation of the risk assessment issues clearly demonstrates that we have to do things right the first time — a solution that helped quality to become a positive factor in today's economy. For the safe treatment of hazardous waste, the problem must be studied carefully and the associated risk analyzed in depth. The use of sophisticated engineering solutions will increase the price of waste management dramatically, and this fact might help to provoke the development of better manufacturing processes to avoid the waste in the first place. The wheel will then have turned full circle.

RISK ASSESSMENT ISSUES: EXISTING WASTE SITES

Notes

1 OECD Environmental Data Compendium, 1987.
2 Zogg, H. (1987). "Zurich" Hazard Analysis, Zurich Insurance Group, Risk Engineering.

Discussant: John Amore

There are really only a few points that I can make in reviewing Peter Schroeder's paper. I would like to do so and then delve into the area of how these risk assessments interface with providing environmental insurance for currently operating facilities. I don't think we can take much exception to the types of analyses that Peter has made. These would work very well if we were starting up a new facility with no past environmental history and with access to substantial economic and human resources for side assessments. The only vagaries that would develop would be on undetermined toxicity levels of the substances being used within the facility and on undetermined changes to the environment that surround the facility as we move into the future.

However, it's been our experience that the environmental problems of today and of the future are not going to be created in pristine new facilities — they already exist. So one of the major objectives and one of the major differences of a risk assessment for environmental insurance is that it has to begin with establishing what type of pollution conditions exist at that site. What type of pollution loss potentials are already there? They have to be documented before any underwriters can begin evaluating loss potentials from ongoing operations.

Risk assessments also need to be designed for the users' purposes, and there are differences between insurance buyers and insurance providers. The buyer has a more severe commitment, in the sense that as a corporation, the liabilities cannot be limited to a specific dollar amount; they can therefore threaten the total assets of the corporation. Moreover, they cannot be limited to a time frame other than the life of the entity. The insurance provider can say, "We're interested in providing this insurance coverage, but will limit the amount of coverage to a specific amount and a specific period of time." Another major difference is that the insurance provider does not have the same level of specific expertise about the site and the operations conducted at the site as the buyer has. This is true even in the case of specialized underwriting facilities that pride themselves on their general expertise. Lastly, the demand for pollution insurance is elastic. It is sensitive not only to the premium to be paid, but also to the acquisition costs associated with gathering information for detailed submission requirements and expenses associated with risk assessments.

Because of these differences and the need to establish an environmental benchmark for the underwriter to evaluate existing pollution conditions at a point in time, the risk assessment model actually being used today has a more deductive than inductive approach. The core of the model is an analysis of pollution pathways (air, surface, and ground) that could interface with potential claimants and cause property damage, bodily injury, or cleanup costs. Risk assessments for other types of insurance coverage (e.g., fire, workers' compensation, etc.) generally cost less than $1000 per site. Environmental risk assessments under the current model have costs ranging from seven to 10 times greater than this. The inductive model, expanded to apply to sites already in operation, would substantially increase both the expense and the amount of time needed to obtain a risk assessment for underwriting purposes.

The risk assessment program should be designed to achieve a goal. The goal for an underwriting program is to allow the underwriter to select among the potential insureds in a manner that limits losses to an acceptable level. That level is directly related to the premium rate as it is expressed relative to the limits of liability being sold. If a program is offering $1,000,000 policy limits and if a premium equivalent to 2% of the limit is the average rate, then the program can fund one full-limit ($1,000,000) loss for every 50 policies sold. While this example oversimplifies, since any pricing program would also have to load the rate to cover sales commissions, underwriting expenses, partial losses, risk assessment expenses, taxes, and presumably a profit, it clearly demonstrates how critical a successful risk assessment is to the program. If the assessment program could select potential insureds so that there would be only one full-limit loss for every 100 policies or 1000 policies sold, the rate could be lowered substantially.

This is a very critical factor for the pollution underwriter, because the frequently observed impression that the demand for this product is inelastic is untrue. The success rate for obtaining an order versus the number of submissions actually quoted varies between 15% and 25%. There are many potential insureds who decline buying the coverage, even though they meet the underwriting and risk assessment criteria of the program being offered. Price (premium and cost of risk assessments) is one of the major explanations given for not purchasing the coverage.

Environmental risk assessments have not been very successful in helping the underwriter distinguish between low- and high-severity loss potentials.

RISK ASSESSMENT ISSUES: EXISTING WASTE SITES

Underwriters tend to evaluate most environmental loss potential as severe because of a judicial system that has been erratic, public sentiment that is frequently intolerant of even the lowest risk events, and a tendency to apply state-of-the-art information to past events.

Underwriters of pollution liability should be dedicated to a risk assessment technique if they are to have a long-term viable program. Further improvements in techniques, and an understanding of risk assessments such as Peter Schroeder's paper provides, are essential to broadening the availability and demand for pollution liability insurance. The ability to perform risk assessments that provide historical perspective and evaluation of operational expertise in a cost-effective manner are critical to addressing the liability needs of the operator who has relatively minor pollution exposures. The importance of accomplishing this is even more critical since much of the future economic growth in this country is forecasted as coming from small technical companies. Many of these companies will have minor but significant pollution liability exposures. Insurance for these types of companies would be helpful to their development.

Discussant: Martin Katzman

I can scarcely contain my enthusiasm for Peter Schroeder's paper. The reason is that he has pointed to a solution to two problems that have stumped me for several years. First, how can we craft a public policy to induce the consideration of these environmental impacts in the design phase? Second, how can we bring market considerations closer to the center of risk regulation? These questions have been particularly vexing in research I have been undertaking with the scrap metal recycling industry. I would like to share these insights with you.

The scrap recyclers are basically guys who wear white hats. They take everybody's junk, extract usable products, and put raw materials back into the economic stream. Some of these products contain toxic substances — perhaps lead, cadmium, or PCBs — that served a useful function during the product's service life. Since they cannot recycle every molecule, they leave behind a little bit of this toxic gunk that migrates into the environment. Although they serve the public purposes of waste reduction, of energy conservation, and of bolstering the competitiveness of the basic metals industry, about one third of the nation's scrap processors have been nailed under Superfund for clean-up liability.

The scrap metal recyclers note correctly that they did not put these toxicants into consumer products originally. These substances were embed-

128 INSURANCE AND RISK MANAGEMENT FOR HAZARDOUS WASTE

ded in manufactured products deliberately, by the manufacturers. Recyclers argue further that they should not be held liable for release of this toxic gunk into the environment. The policy question is this: How do you make manufacturers sensitive to long-term consequences or downstream pollution from the product life cycle, without imposing another irrational layer of regulation?

Some chemical companies, like DuPont and 3M, were sensitive to the impact of process design on their own waste streams 10 or 20 years ago. Perhaps they correctly anticipated that future regulations would render waste minimization more profitable than waste treatment, or perhaps they thought it inherently sound corporate citizenship. Today most chemical companies are attending to downstream impacts, perhaps to forestall the wrath of environmentalists, who are prone to froth at the mouth and to lobby for seemingly irrational legislation. We must admit that the crazies among the environmentalists have served a useful purpose in getting the attention of people who design chemical production systems. The message of environmentally sound process and product design, however, has not yet gotten to all the manufacturers of products that release hazardous chemicals upon disposal or recycling.

So far in this Conference, we have discussed the use and misuse of the regulatory mechanism. Another mechanism, the tort system, has also proven efficacious in grabbing the attention of firms producing and disposing of hazardous chemicals. A third alternative is the market mechanism.

I have written some on the theory of using insurance markets as an alternative to bureaucratic regulation. Much to the dismay of my socialist ancestors, this writing has attracted a coterie of libertarian fans and right-wing fanatics. The problem with my theory is that the pollution liability insurance market has totally broken down, for reasons exhaustively explained by Leslie Cheek. I am thoroughly persuaded by Les's argument about the impossibility of reviving pollution liability markets as long as the insurance pool is plundered to finance Superfund cleanups. Liability insurance, an instrument deterring future harm and compensating injured third parties, is ill suited to serve the social function of cleaning up the damage of the past. Experience on the Congressionally-mandated advisory board on the pollution insurance crisis has resulted in my total despair about the political feasibility of extricating the prospective and retrospective objectives in the United States.

Peter has examined the insurance market mechanism in the European context, where concerns of deterrence and of cleanup are more neatly separated. I was enlightened by his diagram that showed how environmental policy since the 1960s has become increasingly focused on the front-

RISK ASSESSMENT ISSUES: EXISTING WASTE SITES

end of the product cycle, especially on the design phase. Peter shows how the "Zurich" Hazard Analysis methodology can make the design phase much more sensitive to long-run impacts.

The Zurich methodology examines the risks attendant to all phases of product manufacture, use, and disposal. An insurance company can go to a manufacturing establishment and say, "Look, this is our way of thinking about your life cycle risks, based on the design of your products. This is a starting point. You can come back and refute us." The manufacturer may come back and say, "Well, no, there aren't any risks," but at least the framework for dialogue has been established.

If a manufacturer does not like the results of the Zurich methodology, then it can solicit analysis from a competing underwriter. The competing risk analyses by competing underwriters cannot help but avoid the biases of the regulatory system. Regulators have a tremendous incentive to be biased toward conservatism. If an underwriter is extremely conservative, it may require the manufacturer to overdesign the product as a condition of insurability. In a competitive market, this underwriter will lose the client. On the other hand, if the underwriter is sloppy or lax in its risk assessment, it is going to lose money.

The appeal of techniques like the "Zurich" Hazard Analysis methodology is that they force the attention to the entire life cycle and they are amenable to competitive implementation. By the use of risk-based premiums, market discipline can replace the cumbersome regulator. This dual function of the Zurich system of focusing attention on design and relying on the market is very exciting.

We know what does not work in the United States, and we would gain courage from the example of alternatives that worked elsewhere. We would like to know how the lessons of the Hazard Analysis methodology in a European context can be applied to the American context.

So much for the theory. What I would like to ask Peter Schroeder is this: Does the theoretical model work in fact? In Europe, is there competition among insurance companies in the use of such methods? What has been the response of manufacturers? Do underwriters and manufacturers negotiate the reassessment of a risk analysis? In other words, what is the track record of this system?

The Zurich method is not going to contribute to decisions about Superfund cleanups. It is not designed to do so. We must recognize that Superfund is mainly cleaning up synthetic organic toxicants that were discharged into the environment over the last 30 or 40 years. Even if we redesigned all products today, we will continue assaulting the environment with toxics as products embodying them are discarded over the next 10 or

20 years. What this model can do is to prevent the problem from recurring in the future. As presented by Peter Schroeder, the Zurich method provides an example of a flexible tool that is forward-thinking in attacking pollution hazards at the design phase. It can apply some of the best features of both regulation and the market to tackling the hazardous waste problem. The Zurich model provides a tool for a pro-active public policy. If it works, it may be the greatest invention since Saran Wrap.

5 INSURABILITY ISSUES ASSOCIATED WITH MANAGING EXISTING HAZARDOUS WASTE SITES

Baruch Berliner and Juerg Spuehler

5.1 Executive Summary

5.1.1 The Basic Situation

After introducing criteria of insurability and other tools for an insurability analysis of risks in general, endogenous and exogenous components of hazardous waste risks that impede insurability are examined. The analysis pinpoints properties of hazardous waste risks and of the social and legal environment that have driven these risks out of the domain of subjective insurability for many professional risk carriers. By doing so, the analysis not only uncovers what has been done wrongly but also shows how improvements could have been made and how they should be made in the future in order to make lost insurance capacity available again for at least part of the hazardous waste risks.

5.1.2 European Insurance Setups

In Europe, various specific schemes are in operation to cover pollution risks: the GARPOL in France, the Italian *Pool Inquinamento*, and the

132   INSURANCE AND RISK MANAGEMENT FOR HAZARDOUS WASTE

MAS-Pool in the Netherlands, as well as the existing concept of EIL covers in Sweden, which is to be supplemented by a public fund or an equivalent insurance scheme (both of a compulsory nature). All these approaches avoid the conflict between *sudden and accidental* and *gradual*, and abstain from applying the named-perils concept. However, most offer rather limited financial cover, all do not consider in a clear way the consequences either of the accumulation of legally allowed polluting activities or of unlawful behavior, and all leave restoration and cleanup costs outside the scope of coverage.

5.1.3 The Consequences

The question arises whether a scheme based on tested principles of insurance coupled with untraditional elements could be established to meet the expectations of the facility operators, namely to finance at any time (by disregarding legal developments and terms possibly being subject to litigation) first-party losses (restoration and cleanup costs) as well as third-party injury and damage. If a bridge could be built between these expectations and the general requirements of the insurance carriers (limitation of engagement in time and in amount), new frontiers would be set for hazardous waste facilities.

5.1.4 Outline of a Solution: The Liability Life Policy (LLP)

A possible solution to the problem faced may be found in the Liability Life Policy (LLP). The term *liability* signifies that LLP takes care of liabilities of any kind imposed on the facility operator; the term *life* demonstrates that the whole lifespan of the facility is considered; and the term *policy* is applied since the risk is transferred like insurance to a traditional insurer. The LLP takes care of the first- and third-party losses during the whole active and passive lifespan of the facility. It is financed by an initial down payment (as part of the whole investment for a waste site) and by periodical contributions incorporating incentives to the site operator to outlay and to run the facility as safely as possible.

The risk carrier guarantees from the inception of the scheme a prefixed maximum amount for balancing out the possible losses falling on the facility operator. At any time, gaps may arise between the accrued fund and the amount due for losses occurred. The contingent amount resulting therefrom constitutes the risk element to be borne by the risk carrier. His obligations also extend to assessing the facility, servicing and supervising it, rendering technical support, handling and adjusting the claims and investing the funds accruing in the course of time.

A. COMPONENTS IMPEDING INSURABILITY OF HAZARDOUS WASTE FACILITIES AND APPROACHES TO COMBAT THEM
Baruch Berliner

5A.1 Introduction

No community of people can live without restrictions called rules of laws. Restrictions determine the individual's degree of freedom. The restrictions characterizing social relationships include restrictions in every branch of economic activity; these show typical exogenous and endogenous initial and/or boundary conditions.

Sometimes, due to conflict of interest, such conditions may be changed to such a degree that all activity dies out. Such a dramatic development usually occurs unintentionally and unexpectedly because reactions of other interested parties are miscalculated.

In an insurance branch, for instance, activity dies out when insurance coverage becomes unavailable, i.e., when all risks in that branch become subjectively uninsurable for all professional risk carriers.

How available is insurance coverage for hazardous waste risks in the U.S.A.?

In a report to the Congress entitled *Hazardous Waste Issues Surrounding Insurance Availability*, the United States General Accounting Office (GAO) stated the following (Report to Congress by GAO, October, 1987):

> Pollution liability insurance continues to be generally unavailable. Although more than 100,000 companies generate, handle, or dispose of hazardous substances, few of them have insurance for pollution risks . . . [O]nly one insurance organization is actively marketing pollution insurance. A few hundred companies are insured under its policies. The maximum annual coverage that can be purchased is $12.5 million. Several other insurance organizations provide limited coverage pollution risk insurance.

We can conclude that insurance availability for hazardous waste risks is very restricted in the U.S.A. compared to cover needs. Hazardous waste risks are not very attractive to insurers. Such risks have drifted out of the subjective domain of insurability for most of those professional risk carriers

134 INSURANCE AND RISK MANAGEMENT FOR HAZARDOUS WASTE

who were once prepared to cover them, mainly due to changes in exogenous conditions that altered the hazardous waste risks substantially and badly influenced their insurability.

Before we turn to the characteristics of waste disposal risks that render insurability difficult and how these difficulties may be mitigated, let us turn to the criteria of insurability as a basis for the analysis.

5A.2 The Criteria of Insurability

An analysis of the insurability of risks in a certain branch should be based neither on the objects insured nor on the opinion of all risk carriers and risk owners, but only on the opinion of the professional risk carriers who insure risks in that branch (Berliner, 1982):

We follow a pragmatic approach by basing insurability on the opinion of professional risk carriers. A risk is subjectively insurable for a risk carrier if he is prepared to cover it. If he covers, for example, an underrated risk, the risk is subjectively insurable for him, even if he should not be covering it at all.

The following nine criteria must be evaluated by a professional risk carrier with respect to each risk (Berliner, 1982):

(1) Randomness (of the loss occurrence)
(2) Maximum possible loss
(3) Average loss amount upon occurrence
(4) Average period of time between two loss occurrences (i.e., loss frequency)
(5) Insurance premium
(6) Moral hazard
(7) Public policy
(8) Legal restriction
(9) Cover limits

These nine criteria form a concise and nearly complete system, i.e., in nearly all practical cases they determine unequivocally for each professional risk carrier whether a risk is subjectively insurable for him or not.

If a criterion confirms the insurability of a risk, we say that it is satisfied. The criteria are not independent of each other. Most of them contain subjective as well as objective aspects.

If at least one criterion is not satisfied for a risk carrier, then the respective risk is subjectively uninsurable.

INSURABILITY ISSUES: EXISTING WASTE SITES

For a professional risk carrier, the subjective domain of uninsurability consists of the set of all subjectively uninsurable risks in the lines of business which he is underwriting or would underwrite in principle.

The subjective domain of insurability that was mentioned in the introduction is complementary to the domain of subjective uninsurability and consists of the set of all subjectively insurable risks.

The objective domain of insurability is the intersection of all subjective domains of insurability, i.e., it consists of the set of all those risks that are subjectively insurable for all professional risk carriers. A risk lying in the objective domain of insurability is objectively insurable.

The intersection of all subjective domains of uninsurability is the objective domain of uninsurability. It is not complementary to the objective domain of insurability. Between the two domains lies an area of separation consisting of all risks that are for some professional risk carriers subjectively insurable and for others not.

The criteria of insurability can be quantified and interpreted in a geometric model as dimensions of insurability. To each risk a point can be assigned in a nine-dimensional Euclidean space.

The dimensions of insurability can be plotted in such a way that the objective domain of insurability is surrounded by the area of separation, which is in turn surrounded by the domain of uninsurability.

A degree of insurability can be defined that allows us to develop a strategy of selecting the best risks out of a set of offers for coverage of insurable risks (Berliner and Buhlman, 1986).

Keeping such a strategy in mind, it is important to make hazardous waste risks not only insurable again, but also attractive in comparison to risks from other branches that are competing for the available insurance capacities.

If a risk drifts out of the subjective domain of insurability for one professional risk carrier after another, it moves from the area of separation more and more towards the objective domain of uninsurability, becoming less and less attractive for insurance. For risks in the objective domain of uninsurability, no insurance availability exists.

Insurance availability can decrease for a risk in three ways:

(1) A risk may remain insurable in principle for a professional risk carrier, but other risks may become more attractive for him and fit better into his strategy. He may then reduce the capacity that he allocates for that risk.
(2) The boundary lines of a risk carrier's subjective domain of insurability change in time. Although the mapping risk point does not change if

136 INSURANCE AND RISK MANAGEMENT FOR HAZARDOUS WASTE

the risk itself remains unchanged, the same point may fall outside an increasing number of subjective domains of insurability in the course of time due to exogenous factors involving the risk directly or indirectly.

Example: Better information on a risk may cause professional risk carriers to avoid the coverage of that risk by changing their subjective domains of insurability.

(3) Endogenous or exogenous influence may change the risk itself and correspond to another point closer to or in the objective domain of insurability.

Insurance availability for hazardous waste risks originally decreased in the U.S.A. due to the third of these reasons, and the first step to move such risks in the direction of the objective domain of insurability must therefore be to change the risks themselves accordingly. How to make them more attractive for capacity disposal, once they are insurable in principle, is the next step.

5A.3 Endogenous Components of Hazardous Waste Risks that Impede Insurability and Possible Measures to Fight Them

Endogenous components are risk-inherent components. One risk-inherent component that characterizes every risk is an *element of unpredictability*. Every risk premium must therefore contain a *safety* or *fluctuation loading* (Berliner, 1982). A professional risk carrier must take into account *two types* of safety loadings, one for the *uncertainty in the estimation of the expected total loss*, and the other as a *compensation for covering the hazards inherent in the risk* (Berliner, 1982). Both components may heavily impede the insurability of hazardous waste risks.

The hazards inherent in waste risks are of a *catastrophic nature*, and the fluctuations are characterized by large standard deviations. Consequently, this component of the safety loading significantly raises the price for coverage of hazardous waste risks.

The other safety loading component, the uncertainty in the estimation of the expected total loss, originates from *lack of knowledge about the risk* and specifies the extent of *incalculability and guessing* about the risk's correct pure risk premium.

In case of claim, lack of knowledge can lead to uncertainty about *whether or not the insured or his insurers are liable for indemnification.*

INSURABILITY ISSUES: EXISTING WASTE SITES 137

The uncertainty about the liable party arises especially in cases of gradual pollution caused by waste deposits and includes in the first place the time element. It is usually very difficult:

(1) To pinpoint the moment when pollution started
(2) To determine for how long it went on
(3) To specify the intensity of pollution in different time intervals
(4) To specify (in the case of huge waste deposits used by different waste producers) whose waste actually polluted the water, soil and/or air respectively

Final uncertainty often exists on whether or not a specific pollution adversely affected the health or property of claimants. *The connection between cause and effect is not always clear.*

The two components of safety loading must take into consideration all the elements of risk and uncertainty described above and their potential influence on the technical results, i.e, the reserving and accounting of the professional risk carrier. Safety loading therefore makes the insurance coverage of waste risks very expensive, possibly too expensive for companies looking for insurance coverage.

The insurability criterion *insurance premium* is then not satisfied.

Additional entrants to a huge waste deposit increase the probability that the other companies will be considered liable in the case of a pollution event, due to uncertainty about whose waste triggered pollution. For the insurers of companies participating in a waste deposit, newcomers increase the *risk of contagion*, thus reducing the degree of randomness. This effect may make the criterion of *randomness* unsatisfied (Berliner, 1982).

The possibility that a waste producer and his insurers might be penalized for pollution caused by other waste producers contradicts the *insurance principle of fairness*, is *inconsistent with* the insurability criterion of *public policy* (Berliner, 1982), and should be eliminated.

Availability of coverage for hazardous waste risks could be immensely increased if insurers would in the future only cover companies that use their own waste deposits and impose on those companies the obligation to guard their deposits to avoid the storage of foreign waste in them. Such a measure should satisfy the *randomness* criterion and eliminate the above-mentioned component of unfairness that is inconsistent with the criterion of public policy. Moreover, the measure should reduce the uncertainty about the waste deposits and allow for a reduction of the insurance premium.

The criterion of *cover limits*, which includes clauses, restrictions, exclusions, and the like, is a key to changing risks to a level of insurability. This criterion can reduce risks to partial risks and move them into the domain of subjective insurability. Partial risks are defined as risks in themselves. Thus, different parts or percentages of a risk are considered to be different risks. We met the criterion of cover limits in a first approach by our advice to exclude waste deposits used by several waste producers from coverage in the future. It is important that the cover limits be worded very clearly, allowing no room for misinterpretation.

Uncertainty about the liable party, which is inherent in cases of gradual pollution, must be excluded from coverage as a first step back to insurability of hazardous waste facilities, as long as sufficient capacity has not been found for sudden and accidental pollution coverage.

Since a great need for coverage against gradual pollution exists, great efforts should be made to create insurance capacity for this risk.

No effort should be spared in implementing these further steps:

(1) To fix the time when gradual pollution began as precisely as possible
(2) To distribute liabilities among the insurers who cover the same facility as fairly as possible
(3) To prove the connection (and not only a vague possibility of a connection) between cause (gradual pollution) and effect (impaired health of persons claiming for indemnification
(4) To award affected persons a fair, but by no means an exaggerated, compensation
(5) To reduce additional costs to a minimum

We are trying to describe here how the confidence of commercial insurers could be improved by exogenous measures, so that they may be willing to cover risks lying at the limits of insurability. Moreover, such measures reduce the uncertainty about hazardous waste disposal risks and contribute to the possibility of reducing the risk premiums.

A great uncertainty exists about the maximum possible loss that may be caused by a hazardous waste risk. The criterion *maximum possible loss* would therefore probably not be satisfied unless the professional risk carrier introduced aggregate cover limits per client and per deposit. Moreover, it is sensible to give cover with more specific limitations in time, to introduce a high retention and to let the insured participate in the risk proportionately. All these measures inserted into the criterion *cover limits* reduce the risks of covering waste disposals for the professional risk carrier and may move them into his subjective domain of insurability. On the one hand, these limitations reduce the coverage available to the insured from the

INSURABILITY ISSUES: EXISTING WASTE SITES

private insurance industry below the optimal coverage that the insured might desire; on the other hand, they increase the degree of insurability of hazardous waste risks and make more and more insurance capacity available for such risks.

Such measures could even fill substantial capacity gaps if they were accompanied by measures like the following:

(1) Allowance for the waste producers to build tax-free reserves up to certain limits
(2) Setting up of hazardous waste risk pools
(3) Building up of increasingly self-insured retentions by individual companies, preferably with the help of professional risk carriers
(4) Making optimal use of the capacity available from a state fund
(5) Making optimal use of the capacity of professional reinsurers and similar facilities

5A.4 Exogenous Components of Hazardous Waste Risks that Impede Insurability and How to Combat Them

When discussing endogenous components inherent in hazardous waste risks, we already met with exogenous components, because the two categories are interwoven and it is difficult to draw a clear line between them. If, for example, as a consequence of the many years that gradually polluting waste disposals can be effective, it becomes difficult to determine the liable party or to distribute fairly the liabilities among involved parties, are these difficulties risk-inherent, i.e., endogenous, or are they exogenous since it is the task of courts or other exogenous factors to overcome these difficulties?

The following exogenous components that adversely affect the insurability of hazardous waste risks by letting them drift alongside critical dimensions of insurability out of the subjective domain of insurability for one professional risk carrier after another are in many cases a consequence of risk-inherent properties of waste disposal risks. We shall point out the critical criterion or criteria of insurability associated with an exogenous component that is impeding the insurability of hazardous waste risks and try to indicate how to combat or eliminate the component.

(1) It is a natural expression of sympathy that court awards to a private person whose health may be badly impaired are generous. Such generosity, however, not only increases the critical criterion of *insurance*

premium for insured waste producers, possibly to a level where insurance becomes too expensive for them, but it is also inconsistent with the critical criterion of *public policy*. It may leave later pollution victims with no compensation because of unavailability of insurance coverage at bearable terms, and because of the financial incapability of the accused waste producer to indemnify.

(2) A legal system that encourages private persons to prefer charges against companies for damages for which they are not liable renders the critical criterion *moral hazard* unsatisfied for most professional risk carriers. Such a system, which rewards dishonesty and pushes hazardous waste risks into the area of separation and even further towards the objective domain of uninsurability, should be changed no matter how good the intentions may have been upon which it was originally based. Undesirable developments like the one described above cannot always be foreseen. Cases like that of Jackson Township, where the financial recovery granted by the jury to residents of the municipality was based solely on worry, must be avoided in the future.

What a judicial system should aim at is that victims of pollution receive *correct* compensation from the *responsible* party (or its insurers). If the responsible party cannot be found, a state fund or funds to be set up by waste-generating industries should provide for compensation.

(3) As we have seen, the criteria *randomness, public policy*, and *insurance premium* may become unsatisfied due to the use of the same waste deposit by several waste producers. Due to the rule of *joint and several liability* imposed on generators, site owners, operators, and transporters in the U.S.A., "any party that handles or stores the waste is potentially liable for a substantial portion of the damage, no matter how careful the operation may be or how modern the equipment and methods" (Kunreuther, 1987). This exogenous rule not only increases the fear of insurers of being used as deep pockets and drives the waste disposal used by many waste generators still further towards, if not into, the objective domain of uninsurability; it also prevents serious cleanup operators, who are exposed to joint and several liability and can find no insurance coverage for their operations, from touching such hazardous waste sites.

Joint and several liability, which may have been set up with the good intention of securing compensation for every impaired person, actually turns out in the long run to be inconsistent with public policy. It not only decreases the degree of insurability of many sites, but also hampers cleanup operations and with them a better protection of the population from pollution.

INSURABILITY ISSUES: EXISTING WASTE SITES

In the U.S.A., "So far, only 23 of the 950 Superfund sites that the EPA says need immediate attention have been cleaned up" (Business Insurance, February 29, 1988).

(4) Misinterpretations of the wordings of insurance contracts that favor plaintiffs are *inconsistent with public policy* and *enhance moral hazard*. They impede insurability, which may leave persons whose health is impaired later without any compensation, and they should be avoided. The most significant case in the U.S.A. in recent years involved Jackson Township in New Jersey, where the court interpreted *sudden and accidental* as *neither expected nor intended* (Kunreuther, 1987).

(5) Increasing sensitivity of the population to environmental impairment is certainly very welcome. A legal system, however, should not stimulate *moral hazard*. Since the party who is responsible for damage should be liable for indemnification, and no one else, the *burden of proof* of connection between the cause (pollution) and the effect (harmful results to health) should be on the plaintiff's side. If no generator of the effect can be found, but an indemnification is very important, funds like the Superfund in the U.S.A. should be available.

(6) In case of litigation between a waste producer and his insurers, the burden of proof should likewise be on the side of the waste producer, to avoid a situation in which the wrong insurer reimburses under covers for which he should not be liable. Since, as a risk-inherent component, the timing of hazardous waste liability damages may involve a multiple trigger, it is of utmost importance for the *calculability of a hazardous waste risk premium* that courts distribute liabilities on several insurers of waste damage *as justly as possible*. They contribute in this way to satisfying the critical criterion *insurance premium* for some professional risk carriers. Furthermore, it is important that all courts follow the same criteria in the judgment of similar cases.

(7) It is often extremely difficult to relate carcinogenicity to chemical waste (Kunreuther, 1987). An interpretation that chemical waste is ultra-hazardous and an application of *strict liability* in accordance with the new toxic tort law, i.e., a determination that a chemical waste producer is liable for cancer of plaintiffs regardless of cause, would make the criterion of *loss frequency* critical and for most professional risk carriers even unsatisfied for chemical and other specific waste product risks.

An example of the danger of impeding insurability by application of strict liability is the case of *United States vs. Waste Industries Inc.* (Katzman, 1985).

142 INSURANCE AND RISK MANAGEMENT FOR HAZARDOUS WASTE

(8) The criterion of *maximum possible loss* must become unsatisfied if a policy covers, for example, a polluting leak in a waste site as an occurrence and defines an upper limit to it, but a court decides that each impaired or even only stressed individual who files a claim is to be treated as a claimant under a separate policy. In the Jackson Township case, over 300 persons were treated as claimants under separate policies. In this case, the insurers found the maximum possible loss to be more than 300 times as large as they had believed it to be (Kunreuther, 1987). This case illustrates the importance of a very clear contract wording and the *responsibility of a court to support and not undermine* the insurability of hazardous waste risks.

(9) Exogenous factors that could contribute to improve the insurability of hazardous waste risk are:

- Improved research of waste hazards
- Intensive risk management
- Better definition of contract terms
- Stimulation by tax laws and authorities for waste generators to increase their retention by self-insured retention programs and to participate in special pools established to increase capacity for hazardous waste risks
- Cooperation of authorities with waste-producing industries, insurers, and reinsurers to make enough capacity available for hazardous waste risks and to make optimal use of that capacity

B. MAJOR APPROACHES FOR POSSIBLE INSURANCE SOLUTIONS AND THEIR EVALUATION

Juerg Spuehler

5B.1 Introduction[1]

5B.1.1 Character of the Views Expressed

A suitable starting point for evaluating major approaches to possible insurance solutions for existing hazardous waste facilities is found in the present situation in Europe. Quite a few alternatives to traditional insurance covers have been established there in the past few years in order to solve, or at least to ease, the problems attached to pollution in general

INSURABILITY ISSUES: EXISTING WASTE SITES

and to waste facilities in particular. Although these alternatives have not fully proved to stand the test of time and have not met all the needs of industrial manufacturers and waste site operators, they represent quite a feasible platform for setting up a general framework for a realistic solution to today's problems as discussed below.

5B.1.2 The Principal Role of Insurance

To cope with risks means managing risks. For quite some time, very useful instruments have been developed that today reach a state of considerable complexity. They follow the basic line starting from *avoidance* of risks with technical and organizational means and measures, shifting to *suppression* or *reduction* of unavoidable risks with such means and measures, embracing the *transfer* of unavoidable risks that cannot be borne by the risk creator, and ending up with *self-financing* of residual risks.

This clearly reveals that *the function of insurance uniquely consists of transferring the negative consequences of risk from the risk creator to the professional risk carrier.* At the same time, the general line of development demonstrates — contrary to public opinion and public expectation — that *the existence of insurance does not guarantee a safe or safer state or a more acceptable condition at all.* Therefore, insurance must not be understood as the solution in itself to risk problems in general and to pollution problems in particular. And finally, it must be added to this basic qualification that insurance is not a social compensation system as such!

This paper analyzes how the transfer of risk consequences may be accomplished in a reasonable way with reference to pollution resulting from existing hazardous waste sites.

5B.2 Approaches to Balance Out Existing Deficiencies for Traditional Insurance (see appendix 5-1)

First the question arises if it is possible to balance out all or at least part of the deficiencies in the requirements in order to enable the transfer of risk attached to hazardous waste facilities by traditional insurance setups. The primary measures with such an objective that are within the range of direct influence by insurers are principally those of a technical nature that focus on the risk as such and on the shaping of the insurance cover-

144   INSURANCE AND RISK MANAGEMENT FOR HAZARDOUS WASTE

age in a specific way. What is the general situation with respect to these possibilities?

5B.2.1 Approaches from the Risk Engineering Side[2]

5B.2.1.1 Avoidance and Suppression or Reduction of Hazardous Waste. Negative effects on the environment originating from production and distribution processes and from the residues of goods and services after their consumption were, for a very long time, neither disclosed nor considered when making economic decisions. This situation abruptly changed when the negative effects, which tended more and more to outweigh the positive side of produced goods and services, were recognized on a large scale and became a primary topic in public discussions. The goal of avoiding negative consequences as such by avoiding the creation of hazardous waste at all is obviously unrealistic. We are forced to accept the fact that there will always be hazardous waste of substantial volume and intense toxicity. Therefore, we must look at constructional and organizational approaches with respect to waste sites.

5B.2.1.2 Constructional and Organizational Approaches. Existing facilities do not always adequately match constructional and organizational standards that protect against pollution of the air, the soil, and the water — be it for economic reasons (cost–benefit considerations) or due to non-availability of modern technical knowledge at the time of establishing the facility. In view of the volume and kind of inherent toxicity, most existing sites cannot fundamentally be turned ex post to meet the standards that would be required, at least not within the period of time necessary to combat the current, general situation of danger. There is no way left other than to intensively observe the development of the condition and behavior of such sites around the clock and to interfere in the best possible manner on an emergency basis if the potential risk becomes acute.

The situation just described applies to existing facilities. The situation is in general more favorable for *new* facilities, i.e., those that are designed, constructed, and operated according to the latest standards with respect to drainage, evaporations, the kind of depositing waste, recultivating procedures, and the like. Furthermore, the kind of deposit is of considerable importance: the more earthcrustlike it is, the safer the waste site is with reference to the extent of possible environmental impairment, and consequently, the more the site qualifies to be insured on a traditional basis. However, such deposit sites are few in number, so that the general situation

INSURABILITY ISSUES: EXISTING WASTE SITES

145

for insurance issue is not improved significantly by constructional and organizational approaches.

5B.2.1.3 Conclusions. Major approaches from the risk engineering side with respect to avoidance and suppression or reduction of the risk originating from hazardous waste facilities are not in a position to offset the deficiencies significantly. However, this does not mean that operators of waste sites should abstain from applying risk engineering principles. The extensive and prudent application of these principles represents the basic requirement for considering the transfer of risk by insurancelike or insurance-related means, as discussed in section 5B.4.

5B.2.2 Approaches from the Side of Available Insurance Techniques

5B.2.2.1 Gradual vs. Sudden and Accidental. The confusion with respect to the qualification *gradual* on the one hand and *sudden and accidental* on the other hand exists as ever before (even by disregarding the interpretation that *sudden* is to be understood as *unexpected and unintended* which results in affirming the coverage for *gradual* situations). There is no clear understanding of what situation necessitates that *gradual* be undeniably barred from insurance coverage.

Looking at this issue, coverage is to be given when the emission of polluting substances occurs in a sudden and accidental (nonrepetitious) way. It is of no importance if, later on, the harmful emission on persons or on property is of a gradual (nonsudden) nature[3]. The reason for this is that situations of a sudden and accidental nature are, as a rule, coupled with deviations from the planned and normal running of a manufacturing plant (breakdown of plant, machinery, and processes). Such a situation must be honored by underwriters, since it is unexpected and fully recognizable in due time. And therefore the underwriter is able to form the necessary loss reserves early enough.

The standard wordings for industrial liability policies in the vast majority of European countries correspond to the mechanism described. Reference is specifically made to the clear wording in the General Conditions for Liability Insurance in Switzerland (Art. 7 f AVB)[4] and in Austria (Art. 6, Pkt. 4.2.1 in connection with Art. 7, Pkt. 9 AHVB)[5]. In Germany, (4 , Ziff. I./5. AHB)[6] the intention is not plainly expressed, whereas in Denmark the interpretation (not the formulation itself!) of the present ruling

146 INSURANCE AND RISK MANAGEMENT FOR HAZARDOUS WASTE

implies a very strict cover: emissions as well as immissions must be of a sudden and accidental nature in order to secure coverage.[7]

In the case of hazardous waste facilities, it has to be recognized that such an approach in coverage is not suitable, because it does not meet the specific situation linked to it, and therefore it does not come up to the needs and expectations of facility operators. As a rule, emissions from waste facilities and the immissions resulting therefrom are of a gradual nature. Therefore, the standard insurance technique to handle *gradual* and *sudden and accidental* risk situations in an acceptable way fails with respect to the basic requirements to transfer the risk of hazardous waste facilities from the operator to the insurer by traditional insurance setups.

5B.2.2.2 Unlawful vs. lawful behavior. In all industrialized countries, the emissions linked with manufacturing processes and plants (and also emissions of a polluting nature) are allowed up to certain levels, as specified in laws and ordinances, and are not subject to penal sanctions. On the legal liability side, the question arises whether emissions that lead to harmful effects on persons and property due to their accumulation in the course of time are subject to compensation according to legal liability rules. In addition, there is the question whether a sudden (and strictly terminated) decrease in the levels of emissions, as requested by the authorities, potentially turns the former higher immissions therefrom into legal liability cases at the same time the new ruling takes effect. Manufacturers claim that what has been allowed by laws and governmental authorities can never create any liability to pay compensations to third parties. So far, this question has not been answered clearly.

In general, accumulations over longer periods of time that lead to pollution of the air, the soil, and/or the water, with negative consequences on persons and property, are subject to the legal liability rules. Therefore, the way out of the controversy regarding *gradual* and *sudden and accidental* cannot be found by applying instead the terms and qualifications *unlawful* and *lawful* behavior. Since all polluting emissions once falling under lawful behavior and the immissions linked to them must be kept away from traditional insurance coverage, the concept of breakdown of plant, machinery, and process comes to light again from these views.

At present, the existing wordings in industrial liability policies used in Switzerland and Germany have been rewritten with respect to the pollution coverage contained therein. In Switzerland, as of January 1, 1989, the concept of *breakdown* replaces (and in Germany will most probably replace) the former concept of *sudden and accidental* with respect to emissions falling under the coverage granted hitherto. The event linked with the breakdown must be *sudden, unexpected, of a single nature, and accidental.*

INSURABILITY ISSUES: EXISTING WASTE SITES 147

In the case of hazardous waste facilities, it is easy to recognize that such an approach does not suit the general purpose and does not meet the inherent situation there. Most causes of pollution are not found in a breakdown, in the sense of the term as mentioned. This fact is responsible for excluding waste sites from the new wordings referred to above; an acceptable solution has obviously not yet been found, since the wording for waste sites is left to the individual companies willing to underwrite such risks at all, without offering them guidelines from the National Insurance Associations advising their member companies how to proceed.

5B.2.2.3 The Named-Perils Approach. For a few years now, the pollution coverage in industrial liability policies has been on a case-by-case basis defined by specifying which perils causing polluting emissions are considered to fall within the scope of cover. The list applied originally contains the following perils: (1) unintended fire, lightning, explosion, or implosion, and (2) collision or overturning of road vehicles. This approach is strongly related to that of breakdown of plant, machinery, and processes. However, since the span of events triggering the coverage is rather limited, the kind of perils was occasionally extended to *bursting, rupture or explosion of valves, tubes, pipes, and the like*, thereby leaving the strict path of sudden and accidental events, and partly turning to gradual incidents, since such bursts were not defined and specified as those resulting from unique, unexpected, unintentional, and accidental mishandling of plants and processes or the like.

5B.2.2.4 Environmental Impairment as Target of Coverage. If the air, the soil, or the water as *free goods* are disturbed or changed in their natural state or condition, and if nature itself is not in a position to balance out or neutralize such an interference within a reasonable time, we are faced with environmental impairment.

Since the polluted free goods act as carriers, thereby spreading their impairment, damage to property as well as injury to persons may result in addition to the environmental impairment itself.

The question now arises whether the environmental impairment itself or only the consequences therefrom on third-party property and persons should be the subject of the insurance coverage. The answer to this very fundamental question cannot be given in an absolute manner. It may best be found in the setup of the legal liability rules in general or of those referring to environmental impairment in particular. In most countries, only specific liability rules with respect to water (including groundwater) exist today, mostly coupled with the legal obligation of the polluter to finance prevention as well as cleanup costs related thereto. In addition,

148 INSURANCE AND RISK MANAGEMENT FOR HAZARDOUS WASTE

such rules focus the compensation on impairment, and on reduction of and interference with individual rights and amenities protected by law.

From this situation it can be concluded that the subject of the insurance coverage cannot be found along the dividing line between *impairment of the free goods* and *damaging consequences to third parties resulting from such impairment*. The target of the insurance coverage must therefore be set within the area fixed by the legal liability rules. In other words, the scope of cover must, in principle, extend to the same impaired free goods that are subject to the legal liability rules. This guideline does not, however, quite correspond to the needs of operators of hazardous waste facilities and to the public interest attached to the potential danger resulting therefrom. Taking this into account, some distinction must be made: The protection granted by insurance should also extend to the prevention cost of any contingent damage or injury, even with respect to those who are not subject to compensation according to legal liability rules. It has to be considered that the public interest is focused on keeping the free goods unconditionally unimpaired and on taking all possible measures to reverse in the best possible way any impairment that may have occurred.

5.2.2.5 Conclusions. The realized major approaches by the insurance techniques are of such a nature that they are not in a position to outweigh the existing deficiencies regarding covers for hazardous waste facilities. An attempt to reword them would hardly be successful without substantially narrowing the protection expected and needed by facility operators.

Thus, the conclusion must be drawn that the application of traditional insurance techniques — even coupled with suitable risk engineering approaches — fails to cope adequately with the risk situation under review. Modified ways and means must be defined in order to meet the situation. The question arises whether some elements contributing to the accomplishment of this goal can be found in the setup of specific European insurance programs regarding pollution.

5B.3 Present Approaches Guiding a Modified Future Solution (see appendix 5.2)

5B.3.1 General Remarks

In 1977, the first significant specific insurance approach was started in Europe to cope with pollution problems. Three other programs have fol-

INSURABILITY ISSUES: EXISTING WASTE SITES   149

lowed since that time. With one exception, they do not specifically focus on hazardous waste facilities but on the risk of pollution in general.

5B.3.2 The GARPOL in France[8]

The cover provided by GARPOL extends, in principle, to those risks regarding environmental impairments that are excluded from the French Public Liability Policy. This is the reason why GARPOL cover refers not only to pollution of the soil, the air, and the water resulting from the site of industrial activities, but also to the negative effects on the neighborhood of such a site due to noise, vibration, odor, radiation, and temperature. There is no restriction to sudden and accidental, but the event causing the negative effects covered must be unintended and unforeseen by the insured party. As to the limitation in time, the coverage is given if the damage is manifested and the compensation resulting therefrom is claimed during the policy period. Furthermore, the events covered must be the consequence of facts unknown to the policyholder at the time of inception of the cover. If the policy is not renewed, the claims must be made within one year thereafter on the condition that they are the consequence of events that took place during the policy period. On the other hand, all consequences from the nonobservance of legal rules set out in the Special Conditions of the policy are excluded from the cover. It must, however, be pointed out that noncompliance with them only leads to the refusal of coverage if such noncompliance is the fault of the person responsible for the prevention of pollution.

Subject to a favorable inspection report as a basic requirement, coverage is provided for bodily injury and property damage, as well as for pure economic loss of third parties that has resulted from the limitation in making use of specific rights or from the impossibility to render services personally or by real estates and the like. Furthermore, expenses of the operator to reduce the covered loss, or to stop its aggravation, or to prevent its realization at all, fall within the scope of cover. This extension refers also to the neutralization, isolation, and elimination of polluting substances spreading into the atmosphere, the soil, or the water if such measures are ordered by law or ordinance.

There are specific provisions with respect to covering closed waste sites. Coverage is given for claims lodged within five years after closing of the site (extension in time may be possible).

The Pool capacity presently reaches FF 30 million (about $5 million) per any one event and in the aggregate per any one year, whereby the

150   INSURANCE AND RISK MANAGEMENT FOR HAZARDOUS WASTE

very same kind of pollution is considered as being one single event whether one or more third parties are involved (attempts are being made to increase the capacity to FF 100 million in the future). The limit for expenses covered extends to 20% of the said amount.

By way of qualification, GARPOL was the first facility in Europe to deal with pollution damage. It inspired the Italian Pool and influenced that setup extensively. The positive side of GARPOL is found in the fact that it does not distinguish between *sudden and accidental* and *gradual*, and that it abstains from applying the named-perils principle. The aspects of unlawful behavior are quite dominant, and constitute explicitly the Special Conditions of the policies by quoting therein numerous laws and ordinances. There is no clear stipulation regarding the consequences in cover for former but still undiscovered causations and occurrences when such laws and ordinances are changed, thereby restricting the extent of lawful behavior.

5B.3.3 The Italian Pool Inquinamento[9]

The Italian Pool, now supported by about 80 direct insurers, started its operations in 1979. It extends to the coverage of environmental impairment of the soil, the air, and the water, as well as the negative consequences therefrom on persons and property, regardless of whether the consequences are of a sudden and accidental or gradual nature. Therefore, latency risks fall under the scope of the protection provided. Furthermore, the coverage includes loss of use in the sense of interruption or suspension of industrial, commercial, or agricultural activities of third parties. In addition to pollution from the site of the policyholder, coverage is also available for the transportation of hazardous goods.

As to the definition in time, the claims-made basis has applied since 1988, in order to escape the difficulties in establishing the specific point in time when the pollution was caused or when it occurred.

The present capacity of the Pool is rather substantial, as compared with similar European facilities. It reaches Lit 16 billion (about $11 million), which, if needed, can be extended up to Lit 27 billion by readily available reinsurance. It is the declared objective of the Pool Management to increase the total capacity to Lit 50 billion within the next few years.

The specific Italian setup demonstrates that there is a real possibility to raise quite substantial financial support for pollution coverage in Europe. Linked with the positive setup of covering sudden and accidental as well as gradual pollution and of abstaining from the restriction of named perils,

INSURABILITY ISSUES: EXISTING WASTE SITES

this facility must be considered as quite progressive. Judging from the qualifications expressed in the present view, only the aspect of lawful and unlawful behavior in the course of time remains unsolved.

5B.3.4 The MAS-Pool in the Netherlands[10]

Since 1985, potential polluters can obtain cover for sudden and accidental as well as gradual pollution due to lawful behavior by liability insurers registered in the Netherlands through the intermediaries of the MAS-Pool, in which more than 50 insurance companies participate. Part C of the Dutch Public Liability Policy, which focuses specifically on environmental risk, is subject to the Pool scheme. It covers damage resulting from environmental impairment emanating from the insured's site. Insofar as the damage arises suddenly and unexpectedly, there is an overlap with the standard public liability cover of part B of the policy; in such a case, part C (and therefore the MAS-Pool) grants, in principle, primary cover.

According to the scheme, *environmental impairment* in the meaning of the Pool is understood to be the release of any liquid, solid, or gaseous material that has an irritating, contaminating, deteriorating, or polluting effect in or on the soil, the air, or the water (including groundwater).

The cover extends to bodily injury and material damage, the latter of which embraces property damage (including polluting or dirtying of objects), pollution or dirtying of water, and impairment of the economic value of tangible property other than by damage, destruction, or loss.

The financial capacity is quite restricted, namely to hfl 5 million (about $2.7 million) per claim. In order to cope with accumulation and the latency character, the indemnity for all claims related to one single impairment or imminent impairment of the environment is restricted to the same amount. And, furthermore, indemnity applies to the maximum coverage per policy within one and the same annual period. Thus the indemnity available is rather moderate, although the limitation in time is based on the claims-made principle, with a retroactive date set at the day of incepting the first policy extending to part C. In case of cancellation of the policy, there is an irrevocable offer by the insurer to cover the run-off risk for a period of one year.

The cover also extends to salvage costs, which include expenses for measures reasonably taken to avert imminent danger of third parties suffering damage for which the policyholder would be liable and for which the cover provides protection, as well as for measures to limit any damage being covered. Cleanup costs are understood to be covered to the extent

152 INSURANCE AND RISK MANAGEMENT FOR HAZARDOUS WASTE

that they have an effect as mentioned. However, there is no explicit coverage for cleanup costs as such at the policyholder's site.

The availability of the protection restricted by limits is linked to a favorable result of the inspection made before inception of the cover. This prerequisite is more and more frequently not accomplished so that no cover can be secured by the industrial entrepreneur requesting it.

All in all, the setup of the MAS-Pool provides only limited protection to industries. Its major deficiency is found in the very limited financial capacity compared with the dimensions attached to pollution. However, the following positive elements are present: coverage exists for sudden and accidental as well as gradual pollution, and there is no restriction to named perils.

5B.3.5 The Swedish Approach[11]

The Swedish approach falls into two parts: (1) the Environmental Damage Act (since 1986) represents the fundamental basis for specific EIL policies that are coordinated with the pollution coverage provided by ordinary Public Liability Policies; (2) in the near future, the Environmental Damage Compensation Fund Bill will most probably be enacted; it takes care of the consequences of pollution whose creator is insolvent or not traceable, as well as of pollution that is caused more than 10 years before claiming for compensation (so that claimants are barred from collecting by the statute of limitation). Both laws do not distinguish between sudden and accidental or gradual pollution, and there is no restriction on the basis of named perils.

The EIL policies provide coverage for gradual pollution up to a limit of about SEK 25 million (about $4.3 million), which constitutes, on average, a far lower amount than that available under Public Liability Policies covering the consequences of pollution of a sudden and accidental nature only. The specified EIL policy covers bodily injury, property damage, and economic loss, and excludes the consequences of unlawful behavior. Up to now, the demand for such cover has been rather restricted.

The proposed new Act aims at creating a specific fund to finance the cases falling under its provisions. Two alternatives are discussed. One system involves a fixed annual charge that is differentiated by taking into account the influence on the environment and the extent of the operations; it is imposed on most of the plants or activities for which a permit or notification is compulsory. The second aims at charges relating to the emissions being particularly dangerous for the environment and to the extent of the emission.

INSURABILITY ISSUES: EXISTING WASTE SITES

The Federation of Swedish Industries, in cooperation with the leading Swedish insurers, favors — as an alternative to the funding system proposed by the government — a collective insurance system. Such a system could take advantage of the insurers' vast experience in managing and preventing environmental losses; therefore it can come into force at short notice. Furthermore, an insurance program is more apt to set the demarcation line between pollution liability insurance of the traditional kind and that of other kinds of insurance referring to environmental damage; in addition, the insurance solution makes the existing risk capital of the carriers available (whereas the other solution would still have to build up such capital over a long time).

The insurance alternative seeks cover limits up to SEK 100 million (about $17.2 million) for bodily injury and SEK 50 million (about $8.6 million) for property damage, with a combined annual aggregate of SEK 200 million (about $34.5 million). The limitation in time is based on the principle that all losses are attributed to that year in which the insurer is notified of the loss or the first of the losses due to the same type of damaging effect. It is immaterial whether such loss results from sudden and accidental or gradual emissions or immissions. Furthermore, no exclusion is made for consequences of unlawful behavior of the polluter.

Looking at the perspective in Sweden, there will be an embracing program for balancing out the financial consequences of pollution. The interaction of the coverage under the Public Liability Policy, the EIL Policy, and the new proposed law (be it funding or insurance) is, however, rather complicated — apart from the fact that the limits of indemnity available differ considerably between the different parts.

5B.3.6 Conclusions

Although there are quite a few deficiencies in the present specifically pollution-oriented programs in some European countries, some vital conclusions can be drawn.

Compared with the extent of possible pollution by industrial activities the financial capacity of these programs is too small; they do not meet the situations to be expected in the near future, and, with a few exceptions, they do not take into account the specific needs of hazardous waste facilities still in operation. Furthermore, the accumulation of polluting activities within the existing legal boundaries as well as the negative consequences of unlawful behavior are not considered, especially given the fact that the line between lawful and unlawful is not uniform at a specific point in time

154 INSURANCE AND RISK MANAGEMENT FOR HAZARDOUS WASTE

within different geographical areas and not uniform over the period of time as well. Lastly, the coverage provided is strictly focused on third-party claims; therefore cleanup costs in the true sense of the term are completely left outside the scope, even though such costs may be considerable and thus not bearable by a possible polluter.

On the positive side, the avoidance of the conflict between sudden/ accidental and gradual must be mentioned, along with the fact that the named-perils concept has not been taken up by any of the specific pollution facilities.

5B.4 Outline for Setting New Frontiers

5B.4.1 The General Framework

New frontiers to solve the present problems attached to hazardous waste facilities can only be realized within a specific framework clearly defined and honored by all parties concerned in the course of time. Such a framework is characterized by the following elements: the kind of sites eligible (since the characteristics of eligible sites are material to the setting of new frontiers), the general needs and expectations of the facility operator, and the general requirements of the insurer as a risk-bearing institution.

5B.4.1.1 Eligible Sites. The line has to be drawn between *existing* sites and *new* sites. The former must fall outside the framework to be defined since they are attached to already ongoing polluting effects that could not be separated in a reasonable and strict way from polluting effects resulting from future activities. Even the unknown extent of the already set situation is to be placed outside the framework in order to establish a clear boundary line. New sites are equivalent to sites that in construction and operation match up-to-date standards, and do not just stand in line with minimum governmental requirements but rather take into account the techniques available today. Thereby the kind of waste deposited, as well as its chemical and physical condition, is of decisive importance. The closer the material is to the earthcrustlike condition, the further back in time a waste facility can be qualified as *new* in order to be eligible for the new frontier.

Eligible for the program outlined below are not only sites of commercially managed facility operators but also those of manufacturing companies.[12]

INSURABILITY ISSUES: EXISTING WASTE SITES

5B.4.1.2 Major Characteristics of Eligible Sites. The characteristics of eligible sites are material to the building up of the new frontier. The primary characteristic to be taken into account is the expected active lifespan (the period of time over which the material can be delivered and deposited) of the facility, whereby some differentiation must be made between the planned lifespan on the one hand and the possible real span of shorter duration in view of possible deficiencies that might crop up later on the other hand. Material to the setup of the new frontier must be the most probable worst situation that can be judged at the inception of the risk-bearing setup of the new frontier type.

Secondly, the postlife situation must be taken into account for the assessment of eligibility. It has to be realized that there is a substantial probability that sooner or later a situation may arise on the site itself or in the environment that could call for remedial action (for example, cleanup and reconstruction with respect to the site itself) or compensation payments regarding property damage or bodily injury of third parties. Depending on the construction, former operation, guarding after shutdown, and the kind of deposited material, the propensity to such actions and compensation is more or less severe.

Furthermore, the flows of income and of outgo as well as their incongruence as to amount and period of time are material for setting new frontiers. The higher and the earlier the income is with respect to the outgo, the more favorable the starting point for setting new frontiers.

5B.4.1.3 General Needs and Expectations of the Operator. The needs and expectations of the operator are very extensive. They extend to the transfer of financial consequences resulting from first-party damages, including restoration and cleanup costs at the site, as well as those resulting from third-party damages and injuries. Thereby the yardstick with respect to financing such consequences must be the actual as well as future legal rules of any kind and nature. The development of the applicable technical standards must be immaterial. Furthermore, legal liability rules should be left out as a single yardstick whether compensation is due or not. Just the fact that financial means are needed to outweigh first- or third-party expenditures should be the guideline for the setting up of new frontiers. By applying this general approach, the possible twisting in the interpretation of terms, legal rules, origins of impairments, and the like can be avoided. Thus, for example, it is immaterial whether the negative polluting effects are of a sudden and accidental or gradual nature. And furthermore, the criteria applied within the named-perils approach can be skipped. Lastly, the limitation of financial means on the operator's side must be taken into

156   INSURANCE AND RISK MANAGEMENT FOR HAZARDOUS WASTE

account, primarily with a view to the flows of income and outgo (in time as well as in amount).

5B.4.1.4 General Requirement of the Insurer   The role of the insurer as bearer of the risk transferred to him is, in general, limited by two requirements: there must be a limitation of his engagements in amount and in time. Otherwise the transfer and the mix of different origins of risk do not meet his own general position, which is characterized by a limitation of financial means, manpower, and know-how. Finally, a fair chance to secure an adequate surplus in the course of time must be safeguarded.

5B.4.1.5 Bridging the Positions of the Primary Parties Involved.   The facility operator as risk creator and the insurer as bearer of the transferred risk are the primary parties involved. Their needs, expectations, and requirements are, to a considerable extent, opposed to each other. The possibility of setting new frontiers therefore requires the bridging of the two standpoints. In the past such bridging was not accomplished since the economic risks on each side were not safeguarded in the course of time, primarily due to changed and twisted legal frameworks as well as to imperfect technical knowledge. Since in the future these threats cannot be eliminated, the new frontiers must as much as possible be made independent of them and must focus primarily on the financial needs resulting aperiodically from the facility operations.

5B.4.2 Outline of a Solution: First View

The general setup and the financial approach are outlined below in order to demonstrate the fundamental mechanism of the program of setting new frontiers.

5B.4.2.1 Setup of a Liability Life Policy (LLP) (see appendix 5-3)   When opening up a hazardous waste facility, one fact must be taken into account: most probably, first- or third-party losses will arise sooner or later, and compensation payments will become due. Therefore, there is the fundamental need to secure the financing of these payments by a specific program that adequately takes care of such a situation. In addition, there is the contingency with respect to the point in time when such losses or costs must be met. Furthermore, the length of time over which the adequate prefinancing can be accomplished is not known in advance at the time the waste facility is opened. It may well be that after a relatively short period of time — due, for example, to inadequate depositing activities or misin-

INSURABILITY ISSUES: EXISTING WASTE SITES

terpreted geological surveys — the facility must be closed long before the envisaged capacity has been used up. Then the facility operator has no income any longer, even though after that point in time substantial financial means must be at hand in order to cope with possible cleanup and environmental impairment.

The program to meet all the requirements resulting from this general situation may be found best in the form of a *Liability Life Policy*. The term *liability* signifies, on the one hand, that it takes care of third-party claims based on legal liability rules and, on the other hand, that the first-party expenses imposed on the facility operator by law and specific requirements established by the licensing authorities are taken into account. The term *life* is intended to demonstrate that the program aims at working during the whole lifespan of a hazardous waste facility, embracing also the time after its closing (even though in a more or less limited way). And the term *policy* is used since the risk transfer will be contracted by an insurancelike policy through a traditional insurance carrier. The capitals of the three terms used then give the acronym LLP.

5B.4.2.2 Financial Approach. In line with the opening of a new hazardous waste facility, fundamental investments are necessary. Part of them fall under the setup of LLP in the form of an initial lump-sum payment (down payment). Such payment constitutes the primary financing of compensations accruing in the course of the time for first- and third-party claims. The initial lump-sum payment is increased by periodic (as a rule annual) payments, the extent of which must stand in line with the degree of risk set by the deposits of the same period.

In the course of time, the down payment and the periodic payments — both reduced by the risk bearer's own cost and profit margin, and the risk part counterbalancing payments for possible premature first- and third-party claims — add up to a fund that is increased by the net interest earned on it. The risk bearer guarantees at any time during the period for which the program runs an agreed maximum amount for the payment of first- and third-party compensations falling due. As long and as far as the guaranteed limit exceeds the fund accrued, the risk bearer has to pay the difference out of his own assets, primarily financed by the risk part of the initial and periodic payments under LLPs.

In principle, the program aims at raising the necessary financial means over the period of time itself. However, at any time gaps may exist and — depending on the extent and frequency of payments for damage, injury, cleanup costs, and the like — may change their extent. The contingent amounts resulting from such gaps constitute the risk part (contrary to the financial part) of the scheme, and thus represent the insurance-related element borne by the risk carrier. Furthermore, the activity of the risk carrier consists of the assessment of the facility at the beginning of the program, permanent servicing and supervising of the facility during its whole lifespan, rendering technical support to the facility operator, taking care of claims made against the facility operator, and investing the funds accrued. Though the basic idea originates from the mechanism of a mixed life insurance policy, the contractual services of the risk carrier go far beyond those attached to such a policy.

From the point in time when the facility is closed, there will be no income to the facility operator any longer. Consequently, there are no further periodic payments to be made under the program. The only financing sources left are those from net interest resulting from the fund accrued.

The duration of the program is a question to be discussed; however, the program is focused on the long term. Various alternatives are at hand. There may be a prefixed period of time (including part of the time after the planned closedown of the facility), or an agreement according to which the program is to run until the unknown contingent date of effectively closing the facility (with a subalternative that extends the expiration date of the program by a definite preagreed number of years after the unknown contingent closing date). In order to balance out the risk and the chances of both parties involved in the LLP, the program should, as a principle, be unredeemable. Or, as a possible alternative, the facility operator should be firmly obliged to immediately pay the discounted total of all periodic payments still outstanding if he withdraws from the program prior to the ordinary termination. Such a barrier is necessary to make cancellation unattractive, above all in those cases where at a specific moment there is a considerable gap between the total of the payments already made and the compensation for first- and/or third-party losses being due. It is thus possible to allow for cancellation by either of the two parties. In such a case, the cancellation terms must be of such a kind that they are to the disadvantage of the withdrawing party.

The final question concerns the ownership of any remainder of the amount accrued, once the program expires. In principle, this amount belongs to the facility operator, since he financed it primarily. On the other hand, the carrier to whom the contingent risks are transferred should, as a super-margin, participate partly on such an accrued amount since there is a substantial possibility that at the end of the period during which the program was in force there will be a deficit, which falls completely to the insurer.

The LLP corresponds to the general requirements of the insurer. There is a limitation of engagement in amount, namely the guaranteed maximum of total compensations to be paid. And there is a limitation of engagement

INSURABILITY ISSUES: EXISTING WASTE SITES

in time, though not always stipulated by a specific predetermined number of years; but there is definitely no open-endedness in time as to the obligation of the insurer to pay compensations within the agreed scope of the scheme.

5B.4.3 Outline of a Solution: Second View

The first view taken on the outline of a solution must now be defined more precisely regarding necessary differentiations in determining the contributions from the facility operator, and regarding the kind of possible incentives for his benefit.

5B.4.3.1 Differentiation of Contribution. Obviously the contributions from the facility operator (i.e., amount of the down payment and periodic payments) depend on the period for which the program is concluded in relation to the total lifespan of the facility as planned and expected according to the assessment done. These two contributory components can be calculated in combination with the guaranteed maximum of total compensations from the insurer. In order to escape the possible burden resulting from substantial changes in the rate of interest over long periods of time, the institution of zero bonds may be used; by approaching on these grounds, the calculation of accruing interest is eased.

However, in relation to hazardous waste facilities a very dominant aspect must be observed that may outrule all the considerations mentioned above regarding the financing. These are the incentives for the risk creator to be incorporated in the program.

5B.4.3.2 Kinds of Incentives and Their Evaluation. The incentives that qualify for application must be of such a nature that they substantially contribute to making the hazardous waste facility and the operational activities resulting therefrom safer. Therefore, the incentives must already be attached to the very first decision taken by an operator, namely the determination of the site's location and the layout of all the installations and operational procedures. The risk carrier is to be involved from the beginning in order to bring in all his past experience with handling and depositing hazardous waste. The earlier and the more intensive his involvement, the more substantial the incentives offered. Such incentives are reflected in the amount of the guaranteed limit, the relation of that amount to the amount of the down payment, the minimum periodic amounts to be paid, and the period of time during which the program is to run. The more favorable the early involvement, the higher the amount of the guar-

160   INSURANCE AND RISK MANAGEMENT FOR HAZARDOUS WASTE

anteed limit, the smaller the down payment in relative terms, the smaller the minimum periodic amounts to be paid, and the longer the period during which the program is to run.

The same effects may originate from the kind of waste to be deposited as well as its state and condition at the time of deposition. The more the waste deposited is of an earthcrustlike state and condition, the more favorable the setup of the program can be.

Finally, it must be mentioned that each and every payment for first- and third-party losses increases, according to a predetermined schedule, the future periodic payments to be made by the facility operator, whereas claims-free periods of 12 months initiate a rebate, in line with the schedule mentioned (bonus/malus system according to a prefixed schedule).

These aspects demonstrate clearly that the overall assessment by experts is the fundamental prerequisite for realizing the solution outlined. Furthermore, strong incentives must be built in. They assure that the facility operator and the risk carrier are partners and not opponents.

5B.4.4 Outline of a Solution: Third View

Finally, the outline of a solution must be defined more precisely with respect to the qualification *new site* and the determination of the earliest inception of coverage under the LLP program.

Those sites are also qualified for the scheme that are already in operation, provided that they fulfill the requirements for being new as referred to earlier. In all these cases retroactive cover should be obtainable from the original start of the facility operations. Such cover must, however, be tied to the exclusion of all those facts known at the time of inception of the retroactive cover that presumably might give rise to compensations later on. Furthermore, to establish such cover, a lump sum must be paid that amounts to the total of those contributions that otherwise would have accrued since the opening of that facility.

In this way the extent of application of the LLP can be appropriately expanded.

5B.5 Final Conclusions

The presented outline for a solution setting new frontiers can only be realized when the playing rules embodied in the original objectives and

INSURABILITY ISSUES: EXISTING WASTE SITES 161

mutual agreements between the parties are honored over the course of time. Let us hope that, by recognizing the urgent need to solve the problems now and speedily with respect to *new* hazardous waste facilities, this will be the case, so that the present generation can meet its far-reaching obligation for the benefit of the future generations where hazardous waste facilities are concerned.

Appendix 5-1. Approaches to Balance Out Existing Deficiencies for Traditional Insurance

Source	Ways and Means	Suitability for Hazardous Waste Facilities
(1) Risk management	(1.1) Avoidance/suppression of waste at all	Not present: unrealistic as contrary to economic activities as such
	(1.2) Reduction of waste	Partly present: may be technically, but not economically feasible
	(1.3) Constructional/organizational approaches	• For existing facilities not present: not suitable as being contrary to technical and/or economic acceptability or possibility • For new facilities partly present: earthcrustlike deposits are most favorable
(2) Insurance techniques	(2.1) Gradual vs. sudden/accidental	Not present: judicial interpretation contrary to insurers' intention; sudden/accidental not corresponding to existing risk situation
	(2.2) Unlawful vs. lawful behavior	Not present: abrupt changes of legal regulations and allowed emission limits; accumulation of hazardousness in the course of time possible
	(2.3) Named perils	Not present: not corresponding to existing risk situation

Appendix 5-2. Major European Approaches on Pollution Coverage.

Country/Facility	Capacity	Principal Coverage
(1) France: GARPOL (since 1977)	FF 28 million (maximum) (= about $5 million)	Sudden/accidental as well as gradual pollution due to lawful behavior; no restriction to named perils
(2) Italy: Pool Inquinamento (since 1979)	Lit 16 billion (= about $11 million) (including reinsurance capacity: up to Lit 27 billion) per any one event	Sudden/accidental as well as gradual pollution due to lawful behavior; no restriction to named perils
(3) Netherlands: MAS-Pool (since 1985)	hfl 5 million (= about $2.7 million) per claim/year with aggregate per any one emission	Sudden/accidental as well as gradual pollution due to lawful behavior (however, excluding claims originating from accumulation); no restriction to named perils
(4) Sweden: (4.1) Environmental Damage Act (since 1986)	EIL Policy (individual, non-compulsory)	Gradual pollution only (sudden/ accidental covered by Public Liability Policy) due to lawful behavior; no restriction to named perils
Statute of repose: 10 years	SEK 25 million (maximum) (= about $4.3 million)	
(4.2) Environmental Damage Compensation Fund Bill (draft, not yet enacted) Compensation for damage/ injury caused by • Insolvent polluters	(1) Government proposal: funding collectively by the industrial enterprises (compulsory) by either fixed annual charge or emission-related charge	Sudden/accidental as well as gradual pollution due to lawful behavior; no restriction to named perils

Appendix 5-2 (continued)

Country/Facility	Capacity	Principal Coverage
(Sweden: continued) • Unknown polluters	(2) Collective insurance scheme as alternative to government's proposal by leading Swedish insurers	Sudden/accidental as well as gradual pollution due to any behavior; no restriction to named perils
as well as compensation for • Claims made more than 10 years after closing the environmental disturbance (as not falling any longer under the Environmental Damage Act)	SEK 100 million (= about $17.2 million) per disturbance of the same kind for bodily injury and SEK 50 million (= about $8.6 million) per disturbance of the same kind for property damage In all SEK 150 million (= about $25.8 million) per any one year for bodily injury and property damage combined Premium charged according to the nature and extent of activities of insured plant	

Appendix 5–3. Mechanism of the Liability Life Policy (LLP)

Maximum of total compensation to be paid ▶

Lump sum payment at inception as part of the total starting investment for the facility ▶

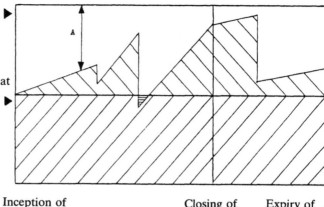

Inception of the scheme

Closing of the facility

Expiry of the scheme

A *Amount at risk* for the carrier at that point in time.

 Lump sum (down payment) at inception.

 Accumulated sum resulting (1) from the net interest on the unused capital amount (surplus) and (2) from periodical payments by the operator.

 Possible situation where the accumulated sum is exceeded by the compensation payments.

Closing of the facility — Starting from this point in time the net interest on the unused capital amount (surplus) is the only source of income.

Notes

1) The views expressed in this paper must be understood as a contribution to the discussion about new frontiers in managing hazardous waste facilities. They do not represent a commitment by Swiss Reinsurance Company to offer appropriate capacity as such and unconditionally with respect to the outline for setting new frontiers (section 5B.4)

2) McCann, P. G. (1987). Waste minimization and recycling options. *Hazardous Materials and Waste Management*

3) Example: Due to a handling error, the pressure in a tube containing a toxic substance is increased to such an extent that the tube bursts (qualification for *sudden and accidental*). The toxic substance is pressed high in the air where it evaporates to a considerable degree and, in turn, spreads gradually over a vast area around the location of the production plant.

INSURABILITY ISSUES: EXISTING WASTE SITES 165

In the course of time, the evaporated toxic substance falls upon persons and property in that area, causing step-by-step (this is *gradually*) injury and damage to them.

4) Current edition of the General Conditions, which will be altered by January 1, 1989: "Von der Versicherung ausgeschlossen sind: . . . , die Haftpflicht fr Schden an Sachen durch allgemeine Einwirkung von Witterung, Temperatur, Rauch, Staub, Russ, Gasen, Dmpfen oder Erschtterungen, ausser wenn die allmhliche Einwirkung auf ein pltzlich eingetretenes, unvorhergesehenes Ereignis zurckzufhren ist."

5) "Versicherungsschutz fr Schden durch Verunreinigung von Erdreich und Gewssern: Der Versicherungsschutz bezieht sich auf Sach- und reine Vermgensschden - einschliesslich des Schadens an Erdreich und an Gewssern -, die Folge einer vom ordnungsgemssen, strungs-freien Betriebsgeschehen abweichenden, pltzlichen Ursache sind, auch wenn diese Schden allmhlich eintreten. Insoweit ist Art. 7, Pkt. 9 nicht anzuwenden."

6) Ausschlsse (current edition): "Haftungsansprche aus Sachschaden, welcher entsteht durch allmhliche Einwirkung der Temperatur, von Gasen, Dmpfen oder Feuchtigkeit, von Niederschlgen (Rauch, Russ, Staub, u. dgl.), ferner . . ."

7) "The insurance shall only cover liability for claims arising out of pollution or contamination to or through air, soil or water in consequence of the operation of the business of the insured, its waste dumps or its removal or waste production, surplus production or scrap production as well as any injury to persons or damage to property occurring in the course thereof, insofar as any such injury or damage occurs unexpectedly, unintentionally and accidentally . . ."

8) Deprimoz, J. (1987). Les garanties Garpol et les perspectives de couverture en Europe. L'assurance Francaise: 908–912.

9) Bernardi, Alfredo (1988). Il mercato assicurativo italiano di fronte ai rischi da inquinamento, Convegno "Il Danno Ambientale, Regolamentazione, Prevenzione, Sicurezza", ANIA, 30.11.1987.Piace a pochi il grande Pool *Giornale delle assicurazioni* 87.

10) Wansink, John H. The New Environmental Impairment Liability Insurance. Undated manuscript.

11) Swedish Government Official Report, SOU 1987:15. In the middle of June, 1988, the Swedish Parliament approved the insurance solution taking effect on July 1, 1989.

12) According to recent statements made by experts in current discussions, 96% of the total waste is deposited at sites belonging to manufacturing companies and only 4% at commercially operated facilities.

References

Berliner, B. (1982). *Limits of Insurability of Risks*. Prentice Hall, Inc., Englewood Cliffs, NJ.

Berliner, B. (1983). Gedanken zur Versicherbarkeit und zur Schiedsgerichtsklausel, Mannheimer Vortrage zur Versicherungswissenschaft, Heft 28, Verlag Versicherungswirtschaft, e.v., Karlsruhe.

Berliner, B. and N. Buhlman (1986). Subjective determination of limits of 05555insurability on the grounds of strategic planning. *The Geneva Papers on Risk Insurance* 11 (39).

General Accounting Office (1987). *Hazardous Waste — Issues Surrounding Insurance Availability*. Report to Congress, October.

Katzman, M. (1985). Critical issues in tort law reform: A search for principles. *Journal of Legal Studies,* December, 1985.

166   INSURANCE AND RISK MANAGEMENT FOR HAZARDOUS WASTE

Kunreuther, H. (1987). Problems and issues of environmental liability insurance. *The Geneva Papers on Risk and Insurance* 12 (44).

Pollution / Insurer's proposal could cut EIL costs. *Business Insurance*, February 29, 1988.

Discussant: Dennis Connolly

This morning we had a discussion about worst-case scenarios multiplied out. Sometimes those things can work out rather nicely. We've also had a discussion of some of the numbers involved in environmental liability, particularly under Superfund, and I'm pleased to tell you that the numbers you were giving this morning were all too low. If you go back to the CERCLA 301(a) study, the actual number of potential sites was 670,000. In addition, some experts and even some people at EPA after SARA was passed said that the average cleanup might be between $30–50 million per site. I took those two improbable, but possible, numbers and I multiplied them together. And then I went to our management and said, "I have a solution for these liabilities that we can sell to our clients, a way to settle these cases. It is cost-effective and socially beneficial." I took the grand total and told our management that if we reduced our rates for providing this service, we would still produce between $600 billion and $1.5 trillion worth of revenue for Johnson and Higgins. So they gave me a nice office and promised me a bonus. We're still working on the program and getting closer to success.

I want to start with the premise that insurers' fears of environmental liability are probably justified; this is even without considering the bodily injury aspect. One of the things that is interesting about environmental liability is how new it is. People act as if an insurer who was underwriting in 1940 was expected to know exactly what was going to happen in 1988. Well, in Gene Lucero's recent article there's an interesting description of the early attitudes of the EPA. Even they didn't really know whether the liability system just enacted in 1980 was going to be joint, whether it was going to be strict, and whether it was going to be retroactive.

We've now learned that it is each of those. We're even learning day by day how extreme the liability system can be. There is recent legislation showing how broad retroactivity can be. The Justice Department is after the Pennsylvania Railroad, which went into bankruptcy and was discharged in 1978. Justice is actually trying to go after them for their pre-Superfund exposures and activities. It gives you an idea of some of the retroactive problems that an underwriter getting into this field has to think about.

INSURABILITY ISSUES: EXISTING WASTE SITES

I think, nevertheless, that underwriters have in general been a little bit short on imagination. I think that the Life Liability Policy proposed by Zurich is a step in the right direction. It's imaginative. It's a way of thinking about how the insurance industry might provide some useful coverage for the future. I'm not entirely sure that this is the answer. I'm not entirely sure that it works, and I have some problems with how one would create an insurance relationship that might have to last 60 years, running from the opening of the site to the very end, when the variables that one would have to deal with would be changing science and changing law. In the discussion this morning, we noted that now we know exactly what the problem is, but 10 years ago we didn't. And one suspects that maybe 10 years from now we may have another view. Knowledgeable people disagree about what things are harmful.

As an underwriter you've got to know, because it's your money that may be on the line if you stay with the risk for 60 years. You also have to predict social attitudes. Most of the speakers this morning seemed to indicate that we have a system that doesn't make any sense, that spends money in the wrong place, that may be totally unjustifiable, and that may have all the wrong incentives. But we have the system because of the social attitudes. Social attitudes can change, which means that the system could get worse in the next 60 years.

How?

Well, I can think of ways. You may not remember H. R. 4813, a bill introduced in 1984. It was a federal cause-of-action for bodily injuries. If you want to think about something that could make it worse, try that one. It would have created presumptions of causation and expanded the definition of injury.

For insurance, we've got to remember the coverage litigation. We've seen how that went. In 1982, every case was in favor of the insured. Nowadays it's a little different. I wrote an article called *Litigation Schizophrenia: Insurance Coverage and the Superfund*, in which I posited a proposition that for each and every coverage determination on environmental exposures, there's an equal and opposite decision somewhere else. And I think that's still true.

Another problem we have with any solution we've talked about today in the area of risk assessment is the comparative ability of the big corporate entity versus the small entity. The fact of the matter is that we don't have a solution in any of these systems. Insurance, Superfund, RCRA, nothing works for the small companies, counties, and municipalities.

We have reports that there is one possible solution.

There's a case in Florida of a garage owner who took his waste oil and dumped it into a hole at the back of the garage. Florida has come up with a solution to that kind of illegal activity. They fined him $37.5 million. I suspect that that really taught him a lesson and that he'll never do that again. I wanted to write to them that they really ought to impose punitive damages, because I found out that his hydrologist was only a master of science, and not a Ph.D.

One of the predicates of the Zurich system was an honorable playing field — I think that was the phrase that they used — and they meant a level playing field where the system would be predictable. What I'd like to suggest is that if we think we can get an honorable playing field, then we really shouldn't strive for such a modest improvement in the system. We really ought to take the system, revamp it, and make it work. We should evolve the system as a social system that says that if there are social responsibilities that need to be taken care of, then they ought to be spread broadly. Unfortunately, I don't think even collecting data and showing that the system doesn't work now is going to change the current system. I think it's unfortunate, but it is my prediction that the system will not change unless there is an unfortunate economic collapse of some sort.

I started off by saying that I didn't think the insurers were imaginative enough; I think they've fallen short. I think that there are targets of insurability, and the AIG has hit upon some of them. There are opportunities to pick out varieties of activity where the liabilities are limited as to their exposure, limited in terms of time, and limited in terms of the types of events, so that insurers really can craft coverage for those targets. One of the better examples is that there are some states where there is special legislation for cleanup contractors. That legislation isn't perfect, but it can be improved.

The other opportunity is one that may fall outside the ambit of the insurance industry, and that is the opportunity for insureds to form their own groups. Such groups would take advantage of multiple disciplines. They would work out a system that coordinates legal and risk assessment and insurance underwriting. A lot of the policy interpretation problems might be avoided because these groups would be drafting their own policies. For example, when you have a risk retention group that has drafted its own policy, the contract-of-adhesion argument doesn't have its normal impact. In addition, there are opportunities for creative use of deductibles to make some of these risks insurable.

So, I understand the insurers' fear of the past, I am concerned about the fear of the future, but I think that the insurance industry must be more imaginative in the way it handles the opportunities that it does have. I commend our European friends for their effort to build such a solution.

6 RISK MANAGEMENT ISSUES ASSOCIATED WITH MANAGING HAZARDOUS WASTE SITES

Richard Fortuna

6.1 Overview

This paper will examine the evolution of the nation's basic authority that governs the daily management of hazardous waste, the Resource Conservation and Recovery Act (RCRA), its emphasis on preventive hazardous waste management, and its role in establishing a market for technologies that protect public health. The paper will examine the lessons learned from the implementation of RCRA's regulatory programs since its enactment in 1976, as well as other mechanisms to ensure proper waste management, including a strict liability standard, direct and indirect economic incentives, and corporate conscience. In addition, the role and ruse of risk assessment in a preventive program for existing hazardous waste management facilities will be examined.

The paper observes that both a formal regulatory system of structured Agency discretion to limit inherent regulatory uncertainties and a strict liability standard are essential to establishing a functioning marketplace of economic incentives to properly manage hazardous wastes. The success and significance of direct economic incentives and corporate conscience are predicated on the existence of a legislative and liability system that defines the market.

170   INSURANCE AND RISK MANAGEMENT FOR HAZARDOUS WASTE

The risk assessment process is important to the establishment of national minimum performance standards for existing waste generators and/or treatment storage and disposal facilities. However, the risk assessment process is distinctly limited if not destructive when used to establish facility standards and/or definitions of hazardous waste that vary on a site-specific basis. Moreover, caution must be used to ensure that the risk assessment process reveals, rather than disguises, the value and assumptions inherent in the process. Facility-specific risk assessments may be useful to provide an additional level of protection when they are imposed by private insurers or by the facility itself, but not as a basis for establishing facility standards on a site-specific basis.

The uncertainties and complexities of introducing facility-specific risk assessment into the RCRA program cannot be justified. Members of the industry would rather face the unpleasant than the uncertain. That is, they would rather engage in a greater level of control than necessary, in exchange for the certainty of knowing what standard must be met to ensure compliance and protection.

6.2 Historical Perspective on Controlling Current Waste Management Practices

Because of the central role of the Resource Conservation and Recovery Act (RCRA) and its regulatory programs as the risk management strategy established by Congress for existing hazardous waste facilities, sections 6.2 and 6.3 of this paper examine the historical evolution of the national hazardous waste program and RCRA's role as a cornerstone in establishing a market for preventive hazardous waste management, respectively.

6.2.1 Humble Beginnings (1976-1981)

Fall 1988 will mark the thirteenth anniversary of the enactment of the Resource Conservation and Recovery Act (RCRA), the nation's basic authority for the preventive management of hazardous wastes. In 1976, the Congress gave EPA 18 months to promulgate all the necessary regulations to protect human health and the environment from exposure to hazardous wastes. The findings of the 1976 Act reflected a concern about the land disposal of hazardous wastes, but also emphasized a pending shortage of land-disposal capacity as a major reason to develop alternatives to land disposal.

RISK MANAGEMENT ISSUES: EXISTING WASTE SITES

In 1980, RCRA was amended, and May of that year saw the first major set of regulations issued, establishing interim status standards for treatment, storage, and disposal facilities. The expectation at that time on the part of many in the treatment industry, the Congress, and the general public was that these standards would bring about better ways of managing hazardous wastes through the increased use of treatment, given that they were the first national standards for any hazardous waste facility. While it seemed like a good idea at the time, from the standpoint of treatment the promise of previous legislative endeavors has proved illusory and elusive. The regulations merely gave legal sanction to many loopholes or previously unregulated activities. They proved to be far less than stringent controls, and moreover proved the axiom that a statute present in name only is worse than no statute at all.

In 1982, the final regulations for land disposal facilities were issued along with final facility standards for surface impoundments. While being hailed as the remedy for all previous ills, the land disposal regulatory medicine may be worse than the disease, and provides more questions than answers.

From these brave but humble beginnings, the national program of preventive hazardous waste management began to take shape. In retrospect, it may be said that we in the Congress underestimated the scope and significance of the problem. At the time nobody, least of all those that were the crafters of the original legislation, believed that reversing the inertia of past land disposal practices and establishing a preventive program of hazardous waste management — one that did not propagate future Superfund sites — would turn out to be the environmental equivalent of balancing the federal budget.

6.2.2 Origins of the Transition (1982-1984)

By early 1982, the Gorsuch Team was firmly established at EPA. A fledgling and flawed program was coming under intense scrutiny, and if anything it was being strangled in the crib. Even some of the most fundamental and basic requirements, such as restricting the placement of containerized liquids into landfills, was being challenged as inappropriate, unnecessary, and too costly to industry. This might be because of the absence of any requirements for explicit cost consideration under RCRA.

It was against this background that Congress began the reauthorization process of 1982, which led to the most significant changes in any environmental law to date — a true watershed change to RCRA as well as to Congress's approach to environmental legislation in general. That era was

172   INSURANCE AND RISK MANAGEMENT FOR HAZARDOUS WASTE

filled with discoveries of other sites like Love Canal. A true dump-of-the-month club was being assembled at an alarming pace, as well as repeated incidence of illegal dumping. It was this discovery of a new and seemingly endless stream of Superfund sites, many of which appear to be attributable to weaknesses in the program itself, that so accentuated the laissez-faire policies of the Gorsuch era. The casual observer might assume that a Congressional reaction such as was reflected in the 1984 RCRA Reauthorization (i.e., the Hazardous and Solid Waste Amendments of 1984 (HSWA)) was attributable solely to political arrogance and a misguided philosophy.

For some this was no doubt the case. However, for those of us involved in rewriting the statute, the principle reaction to the Gorsuch era was the prompting of perhaps the most detailed and exhaustive examination of any environmental program to date. We took the program by its heels and shook it loose, expecting to find nothing but political contraband falling to the floor. What we discovered is that just as many dentures hit the floor as did daggers. That is, the statute lacked many of the teeth necessary under any administration or political regime to promulgate a purposeful program aimed at prevention.

In short, the Act itself had to share the blame along with the appointees. It was not so much *illegal dumping* that was creating new Superfund sites, but rather *legal dumping* occurring under the full sanction of the existing RCRA regulations. More waste generators and facilities were exempt from RCRA controls than were subject to it. The hole was quite literally bigger than the doughnut. Many wastes, including dioxins, were not listed as hazardous. Many recycling practices, such as placing wastes on roads to control dust (i.e., Times Beach, Missouri) were considered exempt recycling practices. During permitting, firms were increasingly engaged in *plume gerrymandering* and attempting to permit only the nonleaking units at their facilities while shifting the cleanup for leaking units to an already overburdened Superfund program. The gaps across the media coverage between water acts and the Clean Air Act were massive, with inadequate controls on sewer disposal and/or evaporative emissions.

And last, but by no means least, there were no restrictions whatsoever on land disposal. Generators and disposers were free to land-dispose any wastes they wished to.

Beyond the specific provisions and the land-disposal bans and other requirements of the 1984 Amendments, the most fundamental change brought about by this enactment is to the decision making process itself. The 1984 RCRA amendments accomplish change in four specific ways, by

RISK MANAGEMENT ISSUES: EXISTING WASTE SITES

(1) establishing a national policy that states that land disposal is the method of last resort in the management of hazardous waste (RCRA Section 1004(b), 42 U.S.C. 6982(b)); (2) instituting presumptive prohibitions against land disposal of all untreated hazardous wastes, (RCRA Section 3004(d), (e), (f), (g)); (3) avoiding a matching of specific waste management techniques to specific wastes and instead allowing competing technologies to achieve a performance standard; and (4) supporting the presumptive prohibitions with the statutory set of minimum controls and self-implementing prohibitions (or hammers) in the event that EPA does not act to override or modify them by certain dates.

6.2.3 Lessons Learned

The intervening years from 1976 to the present have been more instructive than constructive. They have yielded many valuable lessons on what will work and what won't in constructing a preventive hazardous waste management program. While this period was ostensibly dedicated to solving and/or preventing future hazardous wastes problems, in reality we have learned a great deal more about the nature and scope of the problem than we have implemented solutions to them. We have also learned about how the hazardous waste management industry functions, and moreover what makes it dysfunctional. By the end of 1984 we had a clear sense of the proper way in which to structure incentives/requirements to ensure both protective management of hazardous wastes, and to stimulate the emergence of a more responsible technology-based industry. In short, 13 years of attempting to implement a preventive program through a system of statutory controls have provided us with the most meaningful possible lessons on the role of legislation in establishing demand for proper management, and on the need to structure discretion to deal with inherent uncertainties of decision making in this field.

Agency discretion has become its worst enemy: too much of it is just as bad as too little. In the case of the hazardous waste program, there had been too much discretion for too long a period of time. In a program that sorely needed leadership and appeared unable to choose among potential policy options, Congress was forced to substitute its judgment on a wide range of hazardous waste issues.

No administration, even one that is well intentioned, can cope with open-ended discretion. An unlimited number of regulatory options leads to paralysis and maintenance of the status quo.

174 INSURANCE AND RISK MANAGEMENT FOR HAZARDOUS WASTE

The 1980 regulations, rather than increasing treatment, confidence, and certainty, merely redefined the loopholes due to their discretionary nature. The regulations themselves were proving to be the leading causes of our future Superfund sites, with the firms furthest out on the limb being those that had invested in the best methods of management well in advance of the regulations.

Legal dumping was a more prominent threat to public health and to the certainty of investment and proper treatment than some of its more notorious counterpart, illegal dumping.

Methods exist to treat every waste that is currently being generated (*National Academy of Sciences*, Committee on Disposal of Industrial Hazardous Waste, February, 1983). However, treatment methods will not be employed as long as there are so many legal ways to dispose of wastes that are not subject to control, when certain facilities are not required to meet public health standards, and when there are no qualitative or quantitative restrictions on land disposal to ensure that it is protective (*Wall Street Journal*, 8/15/83, p.19).

No matter how heroic the engineering effort, there is a wide range of wastes that cannot be contained by land-disposal facilities. As such, technical containment standards alone are a necessary though insufficient means of controlling the release of hazardous wastes from land-disposal facilities. Public health policies must approach this problem from both directions: stringent facility containment standards, and waste treatment requirements or outright prohibitions on those hazardous wastes that cannot be contained by land-disposal facilities. Improper management has significant impacts beyond public health and natural resource concerns, including economic development, property values, and property transactions.

6.3 The Role of Legislation Regulations in Creating Market Demand for Preventive and Proper Management.

There are few who would challenge the need for increased use of high-technology hazardous waste treatment. Moreover, based on the implementation of the 1984 RCRA Amendments, few could question whether such a transition is under way. The premise of this movement is the belief that treatment, while not being magical in and of itself, when properly conducted provides certainty in two key respects: certainty in knowing what was done to the wastes, and certainty in knowing that future generations will not be exposed to their hazards. While these are compelling and self-evident benefits of treatment, there are equally telling reasons why treatment has not emerged to any significant degree.

The specific reasons for low-cost unprotective land disposal are numerous: little capital is required up front; there are no inherent or technical limitations on what can be physically placed into a land disposal facility; many preventive measures such as dual liners and groundwater monitoring are either not required or avoided due to weak regulations and grandfathering; ultimate liability for the facility after closure may be shifted to governmental entities; and protection of public health is predicated solely upon physical barriers that cannot contain many wastes that are so disposed.

Just as the regulations in effect in 1982 and the inherent nature of the practice placed no restrictions on land disposal, a treatment facility invests the majority of its capital before a single load of waste is ever received. In addition, most forms of treatment operate under stringent standards governing process efficiency and duration, and are specific to certain waste streams. Unlike landfills, there is no current treatment process that can manage all types of wastes (i.e., incinerators must limit the concentration of metals in the waste feed; metal stabilization processes cannot accept high levels of solvents). Most importantly, it will always cost more to permanently render a waste nonhazardous at the time of generation under controlled conditions, than it will to simply bury and hope. The desire for increased certainty in the protection of public health cannot be separated from the inevitability of increased costs for proper hazardous waste treatment.

While treatment delivers greater certainty that wastes will be prevented from causing future threats to public health and the environment, certainty is a two-way street. The regulations and policies must provide greater certainty that there will be a market for something other than unrestricted land disposal. Without an explicit policy that requires treatment as the primary method of waste management, the envisioned transition is little more than a pipe dream. As an article in the *Wall Street Journal* (August 15, 1983) observed, "There is a lot of risk involved in designing and siting a waste treatment plant. Why should a company take a technological risk along with other risks?" One might also question why a society that prides itself on high-tech innovations has had a no-tech approach to hazardous waste management.

In surveying the scope and expanse of the changes to the RCRA and Superfund hazardous waste management statutes to reverse this trend, one cannot help but be struck by the proscriptive nature of these new legislative vehicles. Many in industry regard these changes as exceedingly unpleasant.

176 INSURANCE AND RISK MANAGEMENT FOR HAZARDOUS WASTE

However, it is important to place these concerns into a broader context in order to understand why a legislative/regulatory program is now and always will be the cornerstone of a preventive waste management program and the basis for establishing a market for protection.

Hazardous wastes are like water running downhill; they will always be disposed of along the path of least regulatory control and least cost. Market forces alone are an unreliable and indifferent broker when it comes to ensuring protective management of hazardous wastes in a cost-competitive environment. The forces and facilities that establish the lowest marketplace cost also underwrite the methods that provide the least protection for public health and the environment when practiced without restriction.

The public, federal, and state regulators do not view the chemical industry or the hazardous waste industry as just another business. They do not see the industry as being in the business of simply managing hazardous waste, but rather as being in the business of environmental protection.

Prior to 1984, the hazardous waste regulations were structured as if the two fundamental facts of life did not apply to hazardous waste management: you *get* what you pay for, and there is no free lunch — doing something always cost more than doing nothing.

Aside from these perceptions, there is no avoiding the fact that management of hazardous waste is a pay-me-now, pay-me-later proposition. The only thing that changes is the form and amount of currency.

To a large extent, these proscriptive changes were inevitable. Many states already were beginning to implement land-disposal bans. It is in the interest of all industries to have a nationally consistent scheme, rather than one that varies widely from one state to the next.

In addition, as the recipient of a disproportionate share of federal and state enforcement actions, the commercial hazardous waste management industry would rather face the unpleasant than the uncertain. It would rather face the unpleasant task of undergoing scheduled changes of facility operations, knowing that all competitors are doing the same, than have to undergo such changes sporadically based upon the individual state and federal agencies.

The treatment industry needs a level playing field. It could not allow a situation to persist where the land disposal was left largely unregulated, and where treatment methods such as incineration operated under stringent standards.

And last but not least, the commercial industry felt very strongly that it needed a national series of strict, minimal standards in order to close down a number of the fly-by-night operators that have given this industry such a bad name, and that have caused untold liabilities to waste generators.

RISK MANAGEMENT ISSUES: EXISTING WASTE SITES 177

6.4 Other Approaches to Risk Control: Strict Liability, Economic Incentives, Corporate Conscience
6.4.1 Introduction

In addition to the central role of legislation and regulations that establish the ground rules for preventive management at existing waste facilities, several other approaches have succeeded and are a necessary complement to ensure the viability of a legislative and regulatory system. These include a system of strict liability for the cleanup of hazardous releases, other programs driven by cost and corporate value considerations (i.e., waste minimization, recovery, and assets and liability evaluations), and corporate conscience. Waste managers and generators will control waste releases because they have to, want to, are afraid not to, or believe they can save money doing so.

The have-to part of this risk-control equation was discussed above: regulations and legislation requiring national minimum standards, specific prohibitions on certain forms of waste management, and sanctions for failing to comply. However, therein lies the weakness of regulations and legislation alone: they require enforcement. Enforcement is a discretionary activity that is only slightly less susceptible to changing political winds than a box kite. The incorporation of a joint, strict, and several liability standard into the Superfund waste cleanup law has provided a necessary adjunct to the pre-1984 glacial pace of the regulatory programs and the unpredictable whims of the Agency enforcement. Firms now know that merely complying with a regulation may by itself not be a sufficient defense in the event that those regulations later prove to be inadequate. Nor is there safety in numbers. Waste generators and managers alike know that they alone may be singled out as the firm responsible for the cleanup of an entire site, respective of their proportional contribution to the site.

These two approaches to risk management taken together — regulations and a free-standing joint, strict, and several liability standard — are the indispensible elements that create the market for protective management. Either one alone is a necessary but not a sufficient means to ensure that decisions made in managing hazardous waste are also consistent with both short-term and long-term protection of public health and the environment. The other market incentive emerging from the implementation of this national scheme will be discussed below in further detail.

Partly as a result of RCRA's regulations and CERCLA's liability standard, and partly due to a desire to do the right thing, corporations are also moving toward risk prevention through frequent auditing systems in order to both anticipate and prevent releases of hazardous wastes. Lastly, perhaps

178 INSURANCE AND RISK MANAGEMENT FOR HAZARDOUS WASTE

the least quantifiable and most intangible of all the incentives for risk control are those that arise from the individual rather than the corporation. Included in this latter category are fear of what improper management can do to one's career, reputation, or personal assets.

6.4.2 Joint, Strict, And Several Liability Standard

In 1980 Congress enacted the Comprehensive Environmental Response, Compensation and Liability Act (CERCLA), commonly referred to as Superfund. Superfund has fundamentally reformed corporate liability for environmental releases by holding a company potentially liable for past activities, and by holding each party as potentially liable for the entire cost of cleanup — joint, strict, and several liability. Under Superfund, any person who arranged for the disposal of hazardous substances, any person who at the time of disposal owned or operated the facility, any person who selected a site and transported a hazardous substance there, and the present owner or operator of the site may be liable for all the cost of removal and long-term remedial action at the site. Thus all parties in the chain of waste management either may bear some or individually may bear all of the responsibility for cleanup, including nonnegligent past generator or current site operators even if there was no knowledge or fault with respect to past waste disposal.

Prior to the enactment of the 1984 Amendments, the "evolving" regulatory system, rather than creating demand for proper management, really gave comfort to the status quo. Those firms that had invested in permanent protective treatment technologies based on the expectations created by the 1980 RCRA regulations found themselves furthest out on a limb without a market to speak of. They continued to compete against low-cost (i.e., short-term), unprotective land-disposal methods. The impact of a regulatory system that was underwriting continued land disposal had its most dramatic impact on hazardous waste incineration. For example, until 1986/1987 the only appreciable market that existed for commercial incineration consisted of polychlorinated biphenyls, which require incineration under TSCA regulation; the relatively few companies that believe that incineration was the proper way to manage wastes; and a significant segment of companies that had unfortunately become involved as a potential responsible party in a Superfund action and sought to ensure that past practices were not repeated.

In short, the joint, strict, and several liability standard under Superfund is now and will remain a leading incentive for changing corporate behavior

RISK MANAGEMENT ISSUES: EXISTING WASTE SITES 179

and accounting for the liabilities of improper management through the use of preventive and protective waste management technologies. To the extent that this approach to environmental liability has been counterproductive in prompting cleanup of the sites themselves, the solution lies not with the liability standard, but with its execution. The seemingly endless negotiations and transactions at every phase of the cleanup process, once parties have come forward, has tended to overshadow the significance of the strict liability standard in achieving timely cleanups.

6.4.3 Economic Incentives

The 1984 RCRA Amendments, coupled with CERCLA's joint, strict and several liability standard, have created a market for protecting public health and the environment. RCRA accomplished this end not by matching individual waste streams to specific technologies, but rather by eliminating the environmentally unacceptable practices that created Superfund sites in the first place, and by creating presumptions against ongoing land disposal and other unprotective practices.

Unless waste management choices are governed by something other than simply lowest cost, unprotective practices that yielded short-term cost savings to the generator but that are potential long-term threats to the environment would not have changed. People who sought to invest in capital-intensive operations needed to have a sense of certainty that the marketplace would not continue to simply perpetuate the status quo. With this program in place, coupled with the liability standards of Superfund, there is for the first time a reason for waste generators and managers to look at waste as something other than merely a discarded commodity, but rather as a resource and/or a social cost. However, unless and until these two cornerstones are in place, no market could ever exist, except for those rare situations where a waste stream was either so valuable or concentrated as to warrant recovery.

The economic underpinnings established by RCRA and Superfund have caused several private market mechanisms to account for past and potential environmental liabilities in the context of current assets and property transactions. A formal linkage between property transaction and accounting for past environmental liabilities has been incorporated into New Jersey law by the Environmental Cleanup and Responsibility Act (ECRA) (NJSA 13:1K-6 et seq. (1981), effective January 6, 1982). This law not only requires that at the time of a property transaction the owner reveal environmental liability associated with the site, but in addition engage in necessary cleanup

180 INSURANCE AND RISK MANAGEMENT FOR HAZARDOUS WASTE

actions to make it suitable for sale and protective of public health and the environment. In addition, the financial community in general recognizes that accounting for environmental liabilities in the evaluation of assets, liabilities, corporate worth, or stock value may be the single most important facet of a transaction.

Another emerging market response to control environmental liabilities is witnessed in the growth of waste minimization efforts. Not until generators are faced with the true cost of waste management do they begin to look in earnest for ways to minimize the volume of waste generated and to turn waste into a resource, rather than a societal liability.

6.4.4 Corporate Conscience

Another practice that can be directly linked to the growing stringency of RCRA and CERCLA's liability standard is the move by companies to conduct scheduled audits of waste operations to monitor practices and prevent releases and associated liabilities. Virtually every major company has an ongoing program of periodic audits of all of their waste management facilities to prevent waste releases and a future entanglement with Superfund liability or RCRA violation.

Beyond the national realm of regulatory agencies, federal and state governments, and corporate behavior is the individual. Prior to 1980, it was rare to see any individual that was held personally liable for the activities of the corporation. As the result of recent case law, this is changing. Personal accountability and fear for one's future career are becoming increasingly vital concerns. Many corporations have helped to affect and accelerate this level of consciousness by raising the level of decision making regarding the waste management to the level of the Board of Directors, rather than the sanitation department, as has been traditional in many companies both large and small.

6.5 The Role and Ruse of Risk Assessment In a Preventive Program

The risk assessment process is one of the tools used in establishing control strategies for existing hazardous waste facilities. However, there are distinct limitations on the use of risk assessment in a preventive program. The RCRA program relies on implementability, consistency, timely decision making, and prudent allocation of uncertainties. Risk assessment is being

RISK MANAGEMENT ISSUES: EXISTING WASTE SITES

so overprescribed in the standard setting and in waste-definition processes that a potentially useful antidote for uncertainty is being transformed into scientific snake oil.

The risk assessment process plays its most important role in establishing national minimum performance standards for wastes, facilities, and generators under RCRA. Wastes are identified or listed as hazardous and facility standards are developed based upon reasonable worse-case scenarios; that is, what will happen if the waste/facility is not subject to the controls. On a national basis, the process is useful for assembling and arraying the assumptions, facts, and the judgments inherent in virtually every aspect of regulatory development: level of acceptable risk, type of statistic used, confidence interval used, fate and transport assumptions, exposure assumptions, etc. This process then yields a national minimum standard that has been subject to national scrutiny and that reflects the harms likely to result from mismanagement. There never will be universal agreement with any number or standard yielded by a regulatory process. Risk assessment can be used to establish national, uniform, minimum standards that clearly reveal the factors used to develop a standard, which then can be consistently applied.

Beyond these limited uses, the risk assessment can undermine the certainty that is necessary in the system: certainty in knowing what wastes are in the system, and what the standards are that must be met. In recent months, the Agency has increasingly advocated the use of risk assessment to allow variances from the national minimum standards on a facilities-specific and/or site-specific basis. These variances are usually based upon assumptions about how different a facility may be from the national worst-case scenario (i.e., population density, future land use, groundwater quality, and attenuative properties of soil), and they create significant environmental inequities, demand excessive use of resources, and are an implementation nightmare.

By allowing generators and/or facility operators to either redefine what is hazardous for their situation or to alter facility standards based upon site-specific considerations, we are establishing a policy that allows the following to occur: areas of lower population density are afforded less protection due to the alleged lower probability for exposure; areas of high contamination could be redefined as non-hazardous based upon assumptions about the unlikelihood of future land use, attenuative properties of the soil, or poor quality of the existing groundwater; and the site-specific factors that justify the variances are assumed to be somehow immutable, never subject to human intervention, and impervious to breaches of personal commitments.

182　INSURANCE AND RISK MANAGEMENT FOR HAZARDOUS WASTE

This approach also turns the purpose of RCRA on its head by focusing on exposure, rather than prevention of releases. Establishing different standards for each facility based upon site-specific assumptions changes and perverts RCRA's approach to waste management regulation from one that emphasizes preventing releases to one that maximizes the opportunity for exposures that are not assumed to be harmful. The sordid history of this field has shown that wherever we have erred in the past, it has always been on the side of inadequate controls, where the burdens and the uncertainties associated with such assumptions have fallen on those least capable of dealing with them.

From a broader perspective, the risk assessment process does not reveal *the answer*, but rather is a subjective process that can yield many answers. The risk assessment process is being posited as the means to identify objective truth, when indeed it is in many cases being abused in a manner that conceals, rather than reveals, the value judgments underlying such objectives. Assumptions are masquerading as surrogate facts. As the site-specific variance example shows, the risk assessment process can be used to give a patina of credibility to public policies that are composed of little more than a particle board. We cannot avoid the fact that values and judgments are involved in these decisions, not just data. The data rarely, if ever, speak for themselves, because which major public health decisions have ever been arbitrated by scientific data alone? All involve elements of uncertainty, controversy, and judgment.

Rarely in this field do we ever act on pure scientific certainties; in fact, in its pure sense environmental legislation is merely a means to allocate uncertainties and structure the discretion under which these uncertainties are allocated. Who bears the burden of proof? Who pays the price if the assumptions are wrong? To begin establishing standards for each facility/waste/location combination not only misconstrues RCRA's statutory charge, but is an affront to practicality.

For current hazardous waste management facilities, the question is not "What standard do we need to establish." Rather the question is always "What type of mistake do you want to make and on what side do you want to err?" What better argument for erring on the side of doing more to prevent mismanagement under the RCRA program do we have than Superfund? Once the wastes have been mismanaged and have escaped in the environment, the panoply of imponderables never stops: Who contributed? How much? Whose contribution is more toxic? Who should pay? What should the cleanup level be? How should it be cleaned up? How are damages apportioned? The only way to prevent a continuing repetition of these questions is to put a maximum emphasis on preventing those situations

RISK MANAGEMENT ISSUES: EXISTING WASTE SITES 183

that caused mismanagement and created Superfund sites, even if that means doing more than a given generator/facility believes necessary for their situation. Otherwise, rather than encouraging investment in prevention, we will stimulate a foray for assumptions to redefine a waste /facility out of the system and to shift the emphasis of the program from prevention to a self-serving redefinition of pollution.

As the EPA itself observed in its now historic first series of hazardous waste regulations issued on May 19, 1980:

> The system may not work perfectly for every waste, however. It may overregulate in some instances and underregulate in others. This is an unavoidable consequence of attempting to develop a national hazardous waste management program which has to regulate thousands of wastes into hundreds and thousands of individual transportation, treatment, storage, and disposal situations. To develop a program which would provide precisely the right degree of environmental and health protection in each management situation would require regulations which would be either so vague that they would offer limited guidance to the regulated community and would be largely enforceable or so extensive and so encumbered with provisions for case-by-case variance that they would be an administrative nightmare for both EPA and the hundreds and thousands of persons and facilities which are potentially subject to them.
>
> Fed. Reg. 45: 33088 (May 19, 1980)

Unless restrained, the risk assessment process, rather than providing a noticeable improvement to regulatory standard setting, will become the nightmare the Agency foresaw over eight years ago: a way to encourage people to redefine and dissemble, rather than prevent and comply.

References

Fortuna, R., and D. Lennett (1987). *Hazardous Waste Regulations: The New Era*, McGraw Hill, New York.

Hall, R., and D. Case (1987). *All About Environmental Auditing*. Federal Publications Inc., Washington, DC.

Current Practices in Environmental Auditing. Report to the U.S. Environmental Protection Agency, Arthur D. Little Inc., February, 1984.

Discussant: Nicholas Ashford

Richard Fortuna's paper is splendid. I commend it to you. He makes two basic points. I want to comment on those and then make some remarks of my own.

First, he deals with the existing RCRA standards and how important those standards are for moving the market away from land disposal towards waste treatment as a preferred approach to hazardous waste. He argues emphatically for a strong and enforced regulatory system, and for a strict liability standard. He is in the minority arguing for that, and I join him. The joint and several liability standard has made it impossible for people to dispose on land, and that is the idea. The idea is that we should not dispose on land — we should make it difficult, make it impossible to do so, and define a market away from land disposal. At the same time, we should increase the uncertainty of liability to the extent that the only thing that really is certain is that you should not dispose on land. He makes a very strong case for that.

Parenthetically, I want to remark that his strategy is fine for what we want to do from now on — that is, we do not want to dispose on land. However, a special problem is presented for contractors who are working today to clean up past sites. We might want to reconsider applying a negligence standard for those contractors. Leave everyone else subject to strict liability; but no contractor is going to want to work to clean up a site if he is strictly liable. There is no inconsistency in applying strict liability for the owners and the dumpers, but applying a negligence standard for the contractors so we can clean the business up. But that is not the focus of this session.

The second thing that Richard talks about is the risk assessment issue and how it is very important for ranking and general waste-minimization activities other than land disposal. Regarding shifting the cleanup activity in hazardous waste, i.e., from land disposal to treatment, I would argue that this does not go far enough. Why not take this principle seriously and talk about things that have not been mentioned at this conference and that are not a subject matter of this session, i.e., *source reduction*? From the perspective of the future, source reduction has a lot going for it. First of all, one law that one can never avoid is the second law of thermodynamics. When the genie gets out of the bottle, it stays out of the bottle. Dispersed material cannot be reclaimed. The closer one gets to the front end of the production process, the easier it is from a technological perspective not only to control it, but also to perform a risk assessment and a comparative risk assessment on the alternative technologies that one should have. I believe that we have had too many environmental engineers and end-of-pipe treatment people dealing with hazardous waste, and not enough attention being paid to input substitution, product reformulation, and process redesign. Peter Schroeder's scheme of asking firms to think about what they do is very relevant here. That is, the risk assessment ought not to be exercised only on the basis of the best way to dispose of hazardous waste.

RISK MANAGEMENT ISSUES: EXISTING WASTE SITES   185

It is nice to lobby for environmental regulation by saying that there are 50,000 chemicals in commercial use, 1000 new ones every year, and that we need to employ toxicologists, economists, and lawyers. But the truth of the matter is that there are probably no more than about 1500 chemicals that are causing 95% of the problems with regard to human exposure to toxins. Those 1500 chemicals are known. Some of them substitute for each other. Some are less toxic than others. Some are not toxic under certain uses. Within a decade, I believe, from a technological perspective, we can really effectuate change in input substitution, product reformulation, and process redesign. We need, however, not only environmental engineers and toxicologists; we also need process engineers and material scientists, and these people have not been brought into the environmental field. Insurance and risk management are interlocked. An insurance company would have a better feeling for insuring a company with whom it had a dialogue about changing what it produced than a company that is simply going to talk about how to get rid of the nasty junk.

If this is a time for readjustment, I suggest we leapfrog, not just simply move one step back from disposal to incineration or waste treatment as the preferred solution to this problem. I suggest that we seriously begin to consider, as companies are doing themselves without the prodding of insurance or regulation, how to reformulate what it is they make and how they make it. We need to form a partnership between the producers, the chemical companies, the materials people, the insurance industry, and the environmentally oriented activists to get this all-win situation. Incinerators are a problem. Aside from the dioxin and other problems that exist, they have a community receptivity problem; they have an additional energy pollution problem with regard to the energy that is utilized. So many people, in undertaking risk assessments, leave out the energy that is used to run the incinerator. They simply talk about end-of-pipe emissions from waste treatment.

I would argue that we should discourage incineration. It is going to produce toxins. Alternative technologies are moving forward, but we have to change what it is we are doing. The insurance mechanism has a large role to play here. I think that whether it is reinsurance or insurance pools or a new line of insurance that really begins to deal with the manufacturing process, we have to back up to the front end of the production process.

Discussant: Marcia Williams

I think it is time we stop looking for magic bullets as solutions to all these waste problems. My own feeling is that there's a role for treatment —

186 INSURANCE AND RISK MANAGEMENT FOR HAZARDOUS WASTE

a really important role. There's a role for waste or source reduction, and there's going be a role, frankly, for land management, at least of treatment residuals. There's an integrated role for all these things, and that will be the situation for a long time. All these things can be done well, and as it was pointed out earlier, some of these can be done poorly. Treatment can be done poorly, and it can cause lots of problems. Recycling can be done poorly, and it can cause lots of problems. I think we've got to be able to learn to look at an integrated system and to be able to differentiate when things are being done well and when they're being done poorly.

I'd like to go through a number of points using the extremely useful structure presented in Richard Fortuna's paper. I will add my own viewpoint on some of those items.

With regard to the lessons learned from the RCRA program to date, I would add perhaps some others to the ones Richard mentioned. The first one I will mention is actually a point of disagreement. I think removing Agency discretion leads to a system that quickly becomes obsolete over time and that can cause very serious problems. I think that this is what's going on right now in the RCRA system, despite the best intentions of the people who wrote the statute in 1984. I'll give two examples. One is voluntary cleanup in this country by manufacturers who want to go in today and do cleanup. Voluntary cleanup is virtually stopped by a legislative system that was put in place to take away Agency discretion. That wasn't why it was put there. No one assumed that that's what the system was going to do. However, that's what it *is* going to do.

The second example is this: Richard Fortuna has rightfully come into the EPA saying that the mobile treatment unit provisions are being implemented right now in a way that stops all kinds of good things. But why is that happening? EPA didn't decide that they were going to try to stop the use of mobile treatment units to clean things up. However, the corrective action requirements and the public participation requirements of the statute, written for other situations, apply to this situation as well.

I feel very strongly that we have to continue to look to legislative frameworks that do not bind the Agency into what seems like the best idea at the time the legislation is written.

A second point in terms of lessons learned is, I think, that a black-and-white hazardous/nonhazardous dichotomization of the whole waste system is incredibly simplistic. It does not allow for the range of control practices that need to be put in place.

The third lesson learned is that complexity leads to more complexity. The tax code is a great example of this. RCRA is probably the next best example you can find of that particular problem, and it is really hard to fix.

The problem with such complexity is noncompliance, which happens for two reasons. First of all, because people don't understand what they're supposed to do, they cannot comply even if they want to do so. Secondly, you get noncompliance because the enforcement itself becomes very resource-intensive when the system is extremely complicated.

So those are some of my lessons, in addition to the ones that Richard raised.

Shifting over to the role of legislation and regulation in creating market demand for proper management, I'd again like to lay out for you some examples of what I see as major problems in the current legislative/regulatory framework in terms of getting the right management there. The first example is the inability to site any new facilities. Of major concern is the fact that it doesn't matter whether these facilities are *good* facilities or *bad* facilities. They cannot get sited today.

The result is that we've got inadequate old facilities that are going to stay in place for a long time because they're the only ones there. The problem also encourages extremely high prices for treatment and disposal. What all of us have said over time is that the cost of something should be internalized so that there's a legitimate high cost to do it if that's what it really costs. But raising the cost to ridiculous levels because there isn't a free market is not the intent. Frankly, I think this will ultimately lead to lots of noncompliance again. Noncompliance and illegal activities will victimize the environment.

What's the solution? I think there is a solution. I think SARA framed the solution when it required capacity planning through its new requirements. The states had to do planning, and do an adequate job of planning so that each state had capacity or needed arrangements to manage its own waste. The solution is good. The same thing ought to be done for solid waste and enforced by EPA to make sure that adequate capacity is there.

A second problem, to my mind, is creating a market demand for proper management. The system that's in place right now, be it Superfund or RCRA, does not differentiate between management of new waste and old waste. The very same system is in place if you want to manage newly created waste versus cleanup waste. That doesn't make any sense to me.

I agree with Richard Fortuna and many others who say you want to increase the cost of disposal of new waste so that ultimately people look for waste minimization, source reduction and so on. This perspective is fine for new waste, but it breaks down completely for cleanup waste. If you increase the cost of cleanup to a high enough level, all you're going to ensure is that a lot of it doesn't get cleaned up. You can't minimize it. It's already in place.

Moreover, technology is forcing standards like the land-disposal ban

188   INSURANCE AND RISK MANAGEMENT FOR HAZARDOUS WASTE

rule and saying that every amount of soil dug up will have to be treated to best-technology levels. This is going to ensure that lots of people will not do anything as long as they're not required to. As long as they don't have to clean it up, they won't clean it up, because they aren't allowed to clean up 75% of the problem. The regulations require a 99% perfect job or nothing at all.

The solution in my mind is a doable solution. It requires recrafting the RCRA program so that the subtitle C hazardous waste requirements do not become applicable across the board to all cleanup waste. There are better methods of defining requirements for cleanup waste, including the use of comparative exposure assessment.

The third problem with the current system, and this is not a statutory problem but rather a regulatory problem, is that the system is gridlocked right now. Whether we're talking about cleanup permits, closures, or de-listings, it's just plain gridlocked. Both federal and state people are very much in a "get more information" mode, and there are good reasons for this. The decisions that society asks for are hard, and the regulators are being crucified by the public and press every time they make a decision. So there's a natural tendency on the regulators' part not to want to make decisions. We're in a mode where many cases are getting information for information's sake because this delays decision making. In many cases, the information is not going to result in a better decision. I think the solution to a lot of the gridlock problems is to go back to some of the early concepts. These include permits by rule, class permits, and other kinds of ways to get facilities upgraded more quickly than they will be the way the system is working today. I would also put in a strong plug for the concepts of negotiated permitting (very much like alternative dispute resolution).

The last point I would like to make is that the system we have today persists in looking at wastes as if they are totally separate from the products that generate waste. Whether you look at the product as a target of source reduction (which I think is a good idea) or not, you also need to look at the context of whether the waste generated in making the product is acceptable in light of the benefits that occur from having the product. From a Congressional standpoint, from a regulatory standpoint, the Toxic Substances Control Act and the RCRA Act really ought to be brought together through an integrated kind of thinking so that more sensible decisions can be made.

As to other approaches to risk control, I would just like to add a couple to the ones Richard mentioned. These are more regulatory than statutory.

I think we ought to look at some risk-based financial assurance requirements, where the Agency makes the decision up front. I understand there is a lot of reluctance in the insurance industry to play the role of the most knowledgeable party in terms of which things are risky and which aren't, but the Agency is more comfortable with that and could, in theory, put out risk-based financial assurance requirements.

Secondly, we ought to look at compliance like financial assurance requirements, where there are rewards for facilities that have very good compliance records. Such rewards wouldn't be imposed by the insurance industry; they'd be crafted by the Agency in the way they structure the financial assurance requirements in the first place.

With regard to joint and several liability, my sense is that there is an important place for this concept. I am not one who thinks that we ought to get rid of the concept; I think there has to be some prosecutorial discretion laid on top of the joint and several schemes. People need to have an understanding that some parties are higher priority to go after than other parties. My own premise is that parties who basically follow the law at the time should not be the ones that are worrying about whether or not they're going to get caught later in the Superfund system. This concept is in place today with pesticide applications. I don't really see the difference between pesticides and the company that operates a facility, whether it's a waste facility or any kind of a manufacturing facility, as long as they're building and operating at standards that are approved and permitted at the time.

Finally, a few quick comments on risk assessment. I think a number of these points have already been made. There is an important role for risk assessment, and I see it in a similar fashion to that which other people have laid out. I divide RCRA into four pieces: identification of what ought to be in the program in the first place (what is a hazardous waste); prevention strategies; detection strategies for problems; and cleanup.

With regard to identifying what should be in the system, clearly I think risk assessment has an important place in bringing the worst waste into the system first. The key is not worrying about the exact cutoff point. Should we bring it in at 10 parts per million or 12 parts per million? Let's just make sure the stuff that's the absolute worst stuff across all of the waste universe is brought in before we perfect the exact level for each waste stream.

Secondly, I think risk assessment needs to be used in the waste identification phase to get stuff out of the system quickly. We are today wasting a tremendous amount of time worrying about whether stuff should be let out of the system, and that's the wrong thing for anybody to be spending their effort and their attention on.

With respect to prevention strategies, I think risk assessment should be used to decide what standards and what activities really need to be looked at first. For example, when the HSWA Amendments were passed, there were 70 requirements, all kicking in in a matter of four years, and I think it would have been really good to look at this list to figure out which of those 70 had the biggest payoff and set the deadline for EPA's requirements. That wasn't done. Some of the most important pieces, like location standards, didn't have any kick-in days at all. Some of the least important — from a risk effect perspective — had early kick-in dates. I would also use risk assessment to look at interactions between different prevention requirements. In other words, if there are treated wastes that are going to be put into a landfill, is it still necessary to have the same location standards for that landfill that would apply for untreated waste? Would the same kind of monitoring system be required? I think there are tradeoffs between these things, and I think risk assessment can be helpful in doing that kind of analysis.

And finally, with regard to the prevention side, I would disagree with Richard on the concept of site-specific adjustment of standards. I don't think there's any question that it's not just the type of waste that's the problem. It's the type of waste, the type of management, and the location that determine whether or not we've got a risk problem. We can write a set of national standards that are incredibly complicated to try to cover all these various kinds of situations, but that probably doesn't make very much sense. What makes more sense to me is to write a set of national standards and to allow, on a site-specific basis, the addition of extra requirements if that's necessary, or the deletion of certain requirements if that's appropriate. Then we end up with equal protection at all sites, which strikes me as what we're really trying to achieve.

With regard to detection, I don't really see a significant role for risk assessment here. With regard to cleanup, I think we've already touched on that. Essentially we need to prioritize which sites ought to get looked at first and make decisions on what levels of cleanup are appropriate. Risk assessment has a major role to play in cleanup decisions.

III THE SMALL GROUP DISCUSSIONS

A principal purpose of the conference was to enable participants from different interested parties to interact more informally in a small group setting in designing elements of an integrated waste management program. Six different perspectives on an integrated waste management program were explored through background papers prepared in advance of the conference. Each participant was assigned to one of these small groups. He or she also received the relevant background paper in advance of the meeting and was asked to prepare a written response to the following question:

> What is the highest priority action that should be taken to address the problems or challenges identified in the paper and what is the most significant obstacle you see to achieving it?

Responses to each of these questions were collected at noon of the first day. The groups met in small discussion rooms after dinner of the first day and all members were provided with the set of action items at the start of their session. Leaders of each of the small groups were asked in advance to prepare a summary of the key recom-

191

192 INSURANCE AND RISK MANAGEMENT FOR HAZARDOUS WASTE

mendations. These are presented in the concluding chapter of the book.

Part III comprises papers on six different perspectives. We have also included a summary of each paper, prepared by the author, and the small group chairperson's comments that were presented at the concluding plenary session.

Risk Communication Perspective. Paul Slovic's paper highlights the importance of improving communication about risks associated with hazardous waste, given the great concern the public has on this issue. Paradoxically, it appears that as society's scientific knowledge increases, so does its perception of risk. Slovic points out that this fact may be due to an unfamiliarity with new technologies, the extensive media coverage given to catastrophic mishaps, the lack of control we have over risks such as hazardous waste, and the loss of credibility of experts, since there is considerable disagreement between them. Slovic provides guidelines for improving risk communication, noting that information needs to be presented so as to increase trust between laypersons and experts.

Environmental Perspective. Sheldon Novick traces the development of environmental laws beginning in the 1960s and culminating with RCRA and CERCLA (Superfund). He highlights some of the difficulties EPA has had in implementing these laws, noting that a land-disposal ban emerged after the public recognized that such disposals would eventually leak. Novick feels that risk assessments should be used cautiously in setting priorities and that it may be useful to establish scientific *de minimis* standards that will eventually represent society's level of tolerance for environmental hazards.

Industry Perspective. Isadore Rosenthal and Lynn Johnson describe the programs used by one chemical company in managing risks from current waste streams, minimizing risks from future waste streams, and remediating the injury caused by past waste-handling practices. They detail the program of *prudent action* at their company as it relates to reducing risk and informing people about it. They also describe the different programs in the company related to waste control. In the concluding portion of the paper, they suggest changes that government could make that would help industry. These include setting performance standards across the country and establishing a coordinated management effort to tackle the problems of waste at the individual and industrial levels.

Insurance Perspective. Neil Doherty, Paul Kleindorfer, and Howard Kunreuther delineate the roles that insurance has been expected to play in the hazardous waste management process. These roles include policeman, regulator, and deep-pocket guarantor of financial responsibility, in addition to its traditional role as a risk-spreading mechanism. After delineating the criteria of insurability, the authors question whether insurance can play the role that Congress expected it to fulfill when passing RCRA and CERCLA. The paper examines different types of risks associated with inactive and active waste sites and concludes that traditional insurance mechanisms are limited in providing coverage except for vertically integrated firms that are concerned about their future (as opposed to past) liabilities. The authors propose a three-tiered insurance program consisting of self-insurance, private coverage (possibly through insurance pools), and government protection for losses above some prespecified limit.

Legislative Perspective. Jack Clough offers an insider's perspective on the political process as it relates to hazardous waste management. The theme of his paper is that legislation is an ongoing social process that is designed to protect individual rights and to resolve social conflict. Given these functions of legislation, Clough concludes that legislative changes can only occur if there is a broad base of public support. In order to make a case to the legislator, information needs to be framed to persuade. Congressmen themselves will utilize these data so that they do not set up losing propositions for themselves; hence legislation will continue to pursue the fictional concept of zero risk rather than explicitly addressing questions of acceptable risk.

Legal Perspective. Michael Baram discusses the implications of recent legal developments on industry behavior. Costly tort litigation, new federal and state *right-to-know* laws, and the unavailability of pollution insurance have led corporate managers to reduce their hazardous waste risks. He outlines an approach for an integrated waste management program that involves reducing offset risks and hence liability consequences. The paper concludes by discussing the types of state and federal policies that would encourage firms to pursue an integrated waste management strategy. These include authorization of waste reduction programs, enacting performance standards and reporting requirements, and providing small firms with training and technology transfer programs.

7 A RISK COMMUNICATION PERSPECTIVE ON AN INTEGRATED WASTE MANAGEMENT STRATEGY

Paul Slovic

Discussions, debates, and decision making pertaining to the storage and disposal of hazardous wastes invariably take place in a highly charged, confrontational atmosphere. It is essential to understand public perceptions of risk in order to design an environment in which diverse parties can cooperate in solving common problems of waste management.

Cooperative problem solving will not come easily. Perceived risk can best be characterized as a battleground marked by strong and conflicting views about the nature and seriousness of the risks of modern life. The paradox for those of us who study risk perception is that, as we have become healthier and safer on average, we have become more — rather than less — concerned about risk, and we feel more and more vulnerable to the risks of modern life. Studies of risk perception attempt to understand this paradox and to understand why it is that our perceptions are so often at variance with what the experts say we should be concerned about. We see, for example, that people have very great concerns about nuclear power and chemical risks (which most experts consider acceptably safe) and rather little concern about dams, alcohol, indoor radon, and motor vehicles (which experts consider to be risky).

As noted above, one of the important aspects of risk perception is that

196   INSURANCE AND RISK MANAGEMENT FOR HAZARDOUS WASTE

the American public is becoming more and more concerned about risk, particularly when it comes from nuclear and chemical technologies. Public opinion surveys have shown a marked increase in concern about the risks from nuclear power since 1979 (Rankin, Nealey, and Melber, 1984) and a similar increase in concern about chemical wastes and other environmental problems (Dunlap, 1985). In 1980, Louis Harris and Associates surveyed the risk attitudes and perceptions of a representative sample of Americans. One question asked people whether or not they agreed with the statement that society has "perceived only the tip of the iceberg with regard to the risks associated with modern technology." More than 60% of the public agreed with that statement. Another question asked people whether or not they agreed with the statement that people are subject to more risks today from technology than they were 20 years ago, less risk today, or about the same amount of risk as 20 years ago. Here, almost 80% of the respondents said that people are subject to more risk today than 20 years ago. When asked to project 20 years into the future, 55% said they thought technological activities would pose greater risks than they do today, 32% thought the risks would be the same, and only 18% expected the risks to be lower.

A particularly informative survey on the subject of hazardous waste was conducted for EPA in 1984 by the Roper Company. This survey indicated that

- Safe manufacture and storage of toxic substances is a major concern of the American public
- The public desires greater regulation of the chemical, drug, food, and electric-power industries
- Seven out of 10 people view contamination from chemical-waste disposal as "one of the most serious environmental problems"
- About as many people say that they follow chemical-waste problems closely in the media as follow media reports about U.S.–Soviet relations
- Eight out of ten people consider it industry's "definite responsibility" to clean up its air and water pollution
- Only four out of 10 people believe that industry is meeting that responsibility adequately. Industry gets lower marks on this responsibility than on any of 13 other responsibilities (e.g., producing quality products, hiring minorities, advertising honestly, paying fair share of taxes, etc.)
- Only 17% of the public have confidence in company assurances that their plant safety systems ensure against industrial accidents. 76% have doubts about such statements and don't believe one can trust them.

A RISK COMMUNICATION PERSPECTIVE

Perceptions of risk appear to exert a strong influence upon the regulatory agenda of government agencies. In 1987, an EPA task force of 75 experts ranked the seriousness of risk for 31 environmental problems. Allen (1987) observed that (1) EPA's actual priorities differed in many ways from this ranking, and (2) EPA's priorities were much closer to the public's concerns than to the experts' risk assessments. In particular, hazardous waste disposal was the highest priority item on EPA's agenda and the area of greatest concern for the public, yet this problem was judged only moderate in risk by the experts.

It is important to understand why the American public is so greatly concerned today about risks from technology and its waste products. The answer is unclear, but this author has several hypotheses about factors that might contribute to the perceptions that such risks are high and increasing. One hypothesis is that we have greater ability than ever before to detect minute levels of toxic substances. We can detect parts per billion or trillion or even smaller amounts of chemicals in water and air and in our own bodies. At the same time, we have considerable difficulty understanding the health implications of this new knowledge. Second, we have an increasing reliance on powerful new technologies that can have serious consequences if something goes wrong. When we lack familiarity with a technology, it is natural to be suspicious of it and cautious in accepting its risks. Third, in recent years we have experienced a number of spectacular and catastrophic mishaps, such as Three Mile Island, Chernobyl, Bhopal, the Challenger accident, and the chemical contamination at Love Canal. These get extensive media coverage that highlights the failure of supposedly fail-safe systems. Fourth, we have an immense amount of litigation over risk problems, which brings these problems to public attention and pits expert against expert — leading to loss of credibility on all sides. Fifth, the benefits from technology are often taken for granted. When we fail to perceive significant benefit from an activity, we are intolerant of any degree of risk. Sixth, we are now being told that we have the ability to control many elements of risk, for example by wearing seatbelts, changing our diets, getting more exercise, etc. Perhaps the increased awareness that we have control over many risks makes us more frustrated and angered by those risks that we are not able to control, when exposures are imposed upon us involuntarily (e.g., air and water pollution). Seventh, psychological studies indicate that when people are wealthier and have more to lose, they become more cautious in their decision making. Perhaps this holds true with regard to health as well as wealth. Finally, there may be real changes in the nature of today's risks. For example, there may be greater

198 INSURANCE AND RISK MANAGEMENT FOR HAZARDOUS WASTE

potential for catastrophe than there was in the past, due to the complexity, potency, and interconnectedness of technological systems (Perrow, 1984).

7.1 Psychometric Studies

Public opinion polls have been supplemented by more quantitative studies of risk perception that examine the judgments people make when they are asked to characterize and evaluate hazardous activities and technologies. One broad strategy for studying perceived risk is to develop a taxonomy for hazards that can be used to understand and predict responses to their risks. The most common approach to this goal has employed the *psychometric paradigm* (Slovic, 1986, 1987; Slovic, Fischhoff, and Lichtenstein, 1985), which produces quantitative representations or *cognitive maps* of risk attitudes and perceptions. Within the psychometric paradigm, people make quantitative judgments about the current and desired riskiness of various hazards. These judgments are then related to judgments of other properties, such as the hazard's status on characteristics that have been hypothesized to account for risk perceptions (e.g., voluntariness, dread, catastrophic potential, controllability). These characteristics of risk tend to be highly correlated with each other across the domain of hazards. For example, hazards judged to be catastrophic also tend to be seen as uncontrollable and involuntary. Investigation of these relationships by means of factor analysis has shown that the broad domain of risk characteristics can be reduced to a small set of higher-order characteristics or *factors*.

The factor spaces shown in figures 7-1 and 7-2 have been replicated often. Factor 1, labeled *Dread Risk*, is defined at its high (right-hand) end by perceived lack of control, dread, catastrophic potential, and fatal consequences. Factor 2, labeled *Unknown Risk*, is defined at its high end by hazards perceived as unknown, unobservable, new, and delayed in their manifestation of harm. Nuclear power stands out in this (and many other) studies as uniquely unknown and dreaded, with great potential for catastrophe. Nuclear waste tends to be perceived in a similar way. Chemical hazards such as pesticides and PCBs are not too distant from nuclear hazards in the upper-right hand quadrant of the space.

Research has shown that lay people's perceptions of risk are closely related to these factor spaces. In particular, the further to the right that a hazard appears in the space, the higher its perceived risk, the more people want to see its current risks reduced, and the more people want to see strict regulation employed to achieve the desired reduction in risk (Slovic, Fischhoff, and Lichtenstein, 1985). In contrast, experts' perceptions of risk

A RISK COMMUNICATION PERSPECTIVE

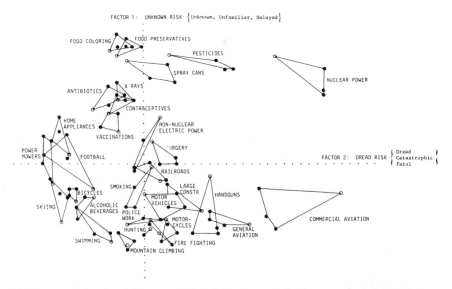

Figure 7-1. Location of 30 hazards within the two-factor space obtained from League of Women Voters, student, Active Club, and expert groups. Connected lines join or enclose the loci of four group points for each hazard. Open circles represent data from the expert group. Unattached points represent groups that fall within the triangle created by the other three groups. Source: Slovic, Fischhoff, and Lichtenstein (1985).

are not closely related to any of the various risk characteristics or factors derived from these characteristics. Instead, experts appear to see riskiness as synonymous with expected annual mortality. As a result, conflicts over risk may result from experts and laypeople having different definitions of the concept.

7.2 Social Amplification of Risk

Risk analysis typically models the impacts of an unfortunate event (such as an accident, a discovery of pollution, sabotage, product tampering) in terms of direct harm to victims — deaths, injuries, or damages. The impacts of such events, however, sometimes extend far beyond these direct harms, and may include significant indirect costs (both monetary and nonmonetary) to the responsible government agency or private company that far exceed direct costs. In some cases, all companies in an industry are

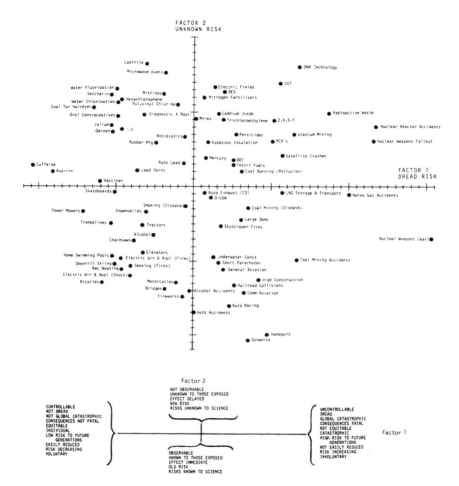

Figure 7-2. Location of 81 hazards on Factors 1 and 2 derived from the relationships among 18 risk characteristics. Each factor is made up of a combination of characteristics, as indicated by the lower diagram. Source: Slovic, Fischhoff, and Lichtenstein (1985).

affected, regardless of which company was responsible for the mishap. In extreme cases, the indirect costs of a mishap may extend past industry boundaries, affecting companies, industries, and agencies whose business is minimally related to the initial event. Thus, an unfortunate event can

A RISK COMMUNICATION PERSPECTIVE

be thought of as analogous to a stone dropped in a pond. The ripples spread outward, encompassing first the directly affected victims, then the responsible company or agency, and, in extreme cases, reaching other companies, agencies, and industries. In the case of a region stigmatized by environmental pollution, the indirect impacts might well include adverse effects on tourism, population, and economic development. This proliferation of impacts is an important element of the phenomenon that has been termed *the social amplification of risk* (Kasperson, Renn and Slovic, in press).

Some unfortunate events make only small ripples; others make larger ones. One important challenge for research is to discover characteristics associated with an event and the way that it is managed, that can predict the breadth and seriousness of those impacts. Early theories equated the magnitude of impact to the number of people killed or injured, or to the amount of property damaged. However, the accident at the Three Mile Island (TMI) nuclear reactor in 1979 provided a dramatic demonstration that factors besides injury, death, and property damage impose serious costs. Despite the fact that not a single person died at TMI, and few if any latent cancer fatalities are expected, no other accident in our history has produced such costly societal impacts. The accident at TMI devastated the utility that owned and operated the plant. It also imposed enormous costs (estimated at $500 billion by one source) on the nuclear industry and on society, through stricter regulation (resulting in increased construction and operation costs), reduced operation of reactors worldwide, greater public opposition to nuclear power, and reliance on more expensive energy sources (EPRI Journal, 1980; Evans and Hope, 1984; Heising and George, 1986). It may even have led to a more hostile view of other complex technologies, such as chemical manufacturing and genetic engineering. The point is that traditional economic and risk analyses tend to neglect these higher-order impacts, and hence they greatly underestimate the costs associated with certain kinds of events.

Although the TMI accident is extreme, it is by no means unique. Other recent events resulting in enormous higher-order impacts include the chemical manufacturing accident at Bhopal, India; the pollution at Love Canal, New York, and Times Beach, Missouri; the disastrous launch of the space shuttle Challenger; and the meltdown of the nuclear reactor at Chernobyl. Following these extreme events are a myriad of mishaps varying in the breadth and size of their impacts.

It appears likely that multiple mechanisms contribute to the phenomenon of social amplification. First, a large amount of media coverage of an event could contribute to heightened perceptions of risk, even more so

202 INSURANCE AND RISK MANAGEMENT FOR HAZARDOUS WASTE

if the information reported is exaggerated or distorted (Mazur, 1984). Second, a particular risk or risk event may enter into the agenda of social groups, or what Mazur (1981) terms *the partisans,* within the community or nation (see, e.g., Kunreuther and Linnerooth, 1983). This may occur either because a particular group has objectives pertaining to this risk issue or simply because political advantage is to be had by exploiting this particular risk. To the extent that risk becomes a central issue in a political campaign or conflict among social groups, it will be vigorously brought to public attention, often coupled with an ideological interpretation of the risk management process. Polarization of views and escalation of rhetoric by the partisans are frequent results. These social alignments tend to become anchors for subsequent interpretation and evaluation of risk management and may become quite firm in the face of conflicting information.

A third mechanism of amplification arises out of the interpretation of unfortunate events as clues or signals regarding the magnitude of the risk and adequacy of the risk management process (Slovic, Lichtenstein, and Fischhoff, 1984). The informativeness or *signal potential* of a mishap, and thus its potential social impact, appears to be systematically related to the characteristics of risk and the factor spaces shown in figures 7-1 and 7-2. An accident that takes many lives may produce relatively little social disturbance (beyond that caused the victims' families and friends) if it occurs as part of a familiar and well-understood system (e.g., a train wreck). However, a small accident in an unfamiliar system (or one perceived as poorly understood), such as a nuclear reactor or a recombinant DNA laboratory, may have immense social consequences if it is perceived as a harbinger of future and possibly catastrophic mishaps.

Adverse events involving technologies in the upper-right-hand quadrant of figures 7-1 and 7-2 (which includes nuclear power and chemicals) appear particularly likely to produce large ripples, because of the (perceived) unknown, dread, and catastrophic qualities of these hazards. As a result, risk-impact analyses and decisions involving hazards in this quadrant need to give serious consideration to possible higher-order impacts.

7.3 Environmental Stigma

Stigma is a concept that originated with the adverse characterization of people and is increasingly being applied to environments. Goffman (1963) noted that the word stigma was used by the ancient Greeks to refer to bodily marks or brands that were designed to expose infamy or disgrace — for example, that the bearer was a slave or criminal. As it is used today,

A RISK COMMUNICATION PERSPECTIVE

the word denotes a victim "marked" as deviant, flawed, limited, spoiled, or generally undesirable in the view of some observer. When the stigmatizing characteristic is observed, perception of the victim changes in a negative way. Prime targets for stigmatization are members of minority groups, the aged, persons afflicted with physical or mental disabilities and deformities, and behavioral deviants such as criminals, drug addicts, homosexuals, and alcoholics. Individuals in these categories have attributes that do not accord with prevailing standards of the normal and the good. They are denigrated and avoided.

Jones et al. (1984) attempted to characterize the key dimensions of social stigma. The six dimensions or factors they proposed were as follows:

(1) *Concealability*. Is the condition hidden or obvious? To what extent is its visibility controllable?
(2) *Course*. What pattern of change over time is usually shown by the condition? What is its ultimate outcome?
(3) *Disruptiveness*. Does the condition block or hamper interaction and communication?
(4) *Aesthetic qualities*. To what extent does the mark make the possessor repellent, ugly, or upsetting?
(5) *Origin*. Under what circumstances did the condition originate? Was anyone responsible for it, and what was he or she trying to do?
(6) *Peril*. What kind of danger is posed by the mark and how imminent and serious is it?

Although the sociological and psychological discussion of stigma typically pertains to contexts far removed from that of hazardous waste disposal, it seems evident that stigma can be generalized from persons to environments. Chemical waste disposal sites may be perceived as repellent, ugly, and upsetting (dimension 4) to the extent that they become visible (dimension 1). A waste site may also be perceived as disruptive (dimension 3) and dangerous (dimension 6).

A stigma resulting from pollution by a toxic substance is described by Edelstein (1986), who analyzed a case in which a dairy's cows became contaminated with PCBs for a short period of time. Once this contamination became known (a visible mark) the reputation of the dairy was discredited and its products became undesirable, even though the level of PCBs was never sufficiently high to prohibit sale of those products. Edelstein shows, step by step, how this incident meets the various criteria of stigmatization put forth by Jones et al.

Although Edelstein's case of stigma involved dairy products, only a short

204 INSURANCE AND RISK MANAGEMENT FOR HAZARDOUS WASTE

leap is required to extend the concept to environments that have been contaminated by toxic substances. Times Beach, Missouri, and Love Canal, New York come quickly to mind. Other well-known examples are the dioxin contamination at Seveso, Italy, which appears to have produced local economic disruptions in excess of $100 million, and the Amoco-Cadiz oil spill, which severely damaged the sea economy and tourist trade of the French Riviera (Lagadec, 1982).

We have been exploring ways to measure environmental stigma. One method that seems quite promising uses a word-association technique to evoke the imagery, knowledge, attitudes, beliefs, and affective states associated with specified environments. Word associations have a long history in psychology, going back to Galton (1880), Wundt (1883), and Freud (1924). More recently, Szalay and Deese (1978) have employed the *method of continued associations* to assess people's subjective representation systems for a wide range of concepts. This method requires the subject to make repeated associations to the same stimulus, for example,

war: soldier

war: fight

war: killing

war: fear

war: enemy

war: etc.

Szalay and Deese argue that this association technique is an easy and efficient way of determining the contents and representational systems of human minds without requiring those contents to be expressed in the full discursive structure of language. In fact, we may reveal ourselves through associations in ways we might find difficult to do if we were required to spell out the full propositions behind these associations through answers to questions.

We conducted a pilot study in which we asked University of Oregon students to produce continued associations to four states: New Jersey, Nevada, Colorado, and California. The format and instructions for the task are shown in figure 7-3. Subjects answered anonymously and without time pressure, to allow full expression of their thoughts.

Scoring of the associations was relatively straightforward. All responses were tallied according to their order in the hierarchy. A first response was

A RISK COMMUNICATION PERSPECTIVE

INSTRUCTIONS

This experiment is part of a study in verbal behavior, and this particular task involves word associations.

The task is easy and simple. You will find four words printed below. Reading each stimulus word will make you think of other associated words (objects, ideas, issues, etc.). You are asked to write as many separate responses as you can think of in the time allotted. Try to think of one-word responses and avoid long phrases or sentences.

It is important that in giving your responses you always take the given stimulus word into consideration. For example, if the stimulus word was <u>table</u> and your answer was <u>writing</u>, in giving the subsequent responses you must refer back to <u>table</u> and avoid "chain" responses (<u>writing</u>, <u>pen</u>, <u>ink</u>, etc.).

These are group experiments, and your responses will not be evaluated individually but collectively for your group. Your responses are completely anonymous. There are no bad or wrong answers, so do not select your responses but put them down spontaneously in the order that they occur to you.

Please work without hurrying, but do your best to give us as many answers as possible. Do not work longer than one minute on any word. The entire task should take you about 4 minutes to complete.

New Jersey _____	Nevada _____	Colorado _____	California _____
New Jersey _____	Nevada _____	Colorado _____	California _____
New Jersey _____	Nevada _____	Colorado _____	California _____
New Jersey _____	Nevada _____	Colorado _____	California _____
New Jersey _____	Nevada _____	Colorado _____	California _____
New Jersey _____	Nevada _____	Colorado _____	California _____
New Jersey _____	Nevada _____	Colorado _____	California _____
New Jersey _____	Nevada _____	Colorado _____	California _____
New Jersey _____	Nevada _____	Colorado _____	California _____
New Jersey _____	Nevada _____	Colorado _____	California _____
New Jersey _____	Nevada _____	Colorado _____	California _____
New Jersey _____	Nevada _____	Colorado _____	California _____
New Jersey _____	Nevada _____	Colorado _____	California _____
New Jersey _____	Nevada _____	Colorado _____	California _____

Figure 7-3. Instructions for the continued associations task.

worth six points, second was worth five points, third four points, fourth through seventh three points, and eighth two points, according to a system proposed by Szalay and Deese based on empirical studies of the reliability of associations as a function of order in the hierarchy. In cases where different associations appeared to represent similar content (e.g., ocean and beaches), we combined them into one category and added their scores.

The results of this pretest, shown in table 7-1, clearly document environmental stigmatization in the minds of our sample of Oregon students. We see the image of New Jersey dominated by pollution including toxic waste, filth, and garbage as members of that first associative category. It is also seen as overcrowded, ugly, and associated with crime. Its farms and

206 INSURANCE AND RISK MANAGEMENT FOR HAZARDOUS WASTE

Table 7-1. Hierarchy of Associates to Four States

Response	New Jersey	Nevada	Colorado	California
1.	Pollution (toxic waste)	Desert	Mountains, high elevation	Sun
2.	Overcrowded	Gambling	Skiing, outdoor recreation	Beach
3.	Ocean, beaches	Prostitution	Cold	Entertainment
4.	Accent	Entertainment	Clean, fresh	Sex
5.	Farms	Nuclear (test site), radiation	Rivers, water	Nature
6.	Small	Scenic	Scenic	Crowded cities
7.	Crime	Horses, ranches	Nature	Smog, pollution
8.	Ugly	Outdoor recreation		Agriculture
9.	Cold	Flat		
10.	Sports			
11.	Industry			

beaches are only secondary images. Nevada's imagery is dominated by its desert landscape along with entertainment, gambling, and prostitution. It is also becoming associated with things nuclear (fifth category). Its scenic beauty, outdoor recreation, and ranches are secondary to these other images. Images of Colorado and California are far more positive than images of New Jersey and Nevada. The predominant thoughts about Colorado are of its fresh and unspoiled natural beauty, which affords many opportunities for outdoor recreation. California's image is also dominated by outdoor fun and entertainment, but its crowded cities and pollution come through as secondary images.

More recently, we have found that people's preferences for places to vacation, site a new business, or relocate (to take a job or retire) can be predicted quite accurately from knowledge of the positive and negative qualities of their images of these places. Thus, a bad image can have substantial social and economic consequences through effects on tourism, economic development, and migration (Slovic, et al., 1989).

The stigmatization of environments has two important implications for hazardous waste management. First it implies that, whatever the health

A RISK COMMUNICATION PERSPECTIVE

risks associated with waste products, there are likely to be significant social and economic impacts upon regions perceived as polluted, or as dumps. Second, it also gives additional importance to managing wastes effectively so that stigmatizing incidents (even ones without significant health consequences) will not occur.

7.4 Risk Communication

Analysis of risk perception leads very naturally to the topic of risk communication. There is a great need to help people put risk into perspective. We think better if we have effective ways of comparing risks. Yet communication must not be merely one-way (expert to public). It needs to be a two-way process, based on trust and mutual respect. Each side has something to learn from the other. The experts' models and assessment methods are very sophisticated, but they tend to focus on the more easily quantified elements of risk. Lay people have a much broader, more qualitative view of risk. One can't say that they are wrong to be concerned about risk to future generations, catastrophic potential, issues of voluntary vs. involuntary exposure, inequitable distributions of risk and benefit, and so on. Underlying these concerns are values that need to be respected and integrated appropriately into risk management decisions.

7.5 Risk Comparisons

One frequently advocated approach to broadening people's perspectives is to present quantitative risk estimates for a variety of hazards, expressed in some unidimensional index of death or disability, such as risk per hour of exposure, annual probability of death, or reduction in life expectancy. Even though such comparisons have no logically necessary implications for acceptability of risk, one might still hope that they would help improve people's intuitions about the magnitude of risks. Risk-perception research suggests, however, that these sorts of comparisons may not be very satisfactory even for this purpose. People's perceptions and attitudes are determined not only by the sort of unidimensional statistics used in such tables but also by the variety of quantitative and qualitative characteristics reflected in figures 7-1 and 7-2. To many people, statements such as, "The annual risk from living near a nuclear power plant is equivalent to the risk of riding an extra three miles in an automobile" give inadequate consid-

208 INSURANCE AND RISK MANAGEMENT FOR HAZARDOUS WASTE

eration to the important differences in the nature of the risks from these two technologies.

However, a comparison among very diverse hazards is only one of many types of comparisons that can be employed. Covello, Sandman, and Slovic (1988) have recently produced a guide for using risk comparisons in communicating about chemical risks. This work, sponsored by the Chemical Manufacturer's Association, was designed to aid chemical-plant personnel when they meet with the public in the vicinity of their plants. However, it should be equally relevant for the purposes of communicating about hazardous waste management in general. Covello et al. describe a dozen types of comparisons that may be more useful than the comparison of unrelated risks. For example, one can compare

- Risk at time 1 vs. risk at time 2
- Risk vs. a standard
- Risk from doing X vs. risk from not doing X
- Risks from alternative solutions to the same problem
- Average vs. peak risk
- Risk from one cause of a health effect vs. risk from all causes of that effect
- Other sources of the same risk (e.g., one radiation risk vs. other types of radiation risks)
- Risk vs. cost
- Risk vs. benefit

Although brief guidelines are provided for employing these comparisons, Covello et al. stress that additional research is needed to determine how they can be used most effectively.

Much has been written about the need for risk communication and the difficulties of doing it (Covello, von Winterfeldt, and Slovic, 1986; Slovic, 1986; Slovic, Fischhoff, and Lichtenstein, 1981.). Covello and Allen (1987) condense this voluminous literature to seven key rules, which are shown in Table 7-2. The rules seem simple and easy, but are surprisingly often violated in practice. I have taken the liberty of adding a rule of my own to the list (#8): Don't trade dollars against lives (or health effects). It seems clear that the public has little sympathy with the view that resource limits justify exposing people to risk against their wishes. Kunreuther, Desvousges, and Slovic (1988) also have found that people are unwilling to consider compensation for accepting waste sites if they are not confident of the safety of such sites.

I would like to conclude with two rules that do not appear in table

A RISK COMMUNICATION PERSPECTIVE   209

Table 7-2.   Rules for Risk Communication (Covello & Allen; 1987)

Rule 1. ACCEPT AND INVOLVE THE PUBLIC AS A LEGITIMATE PARTNER
- They have a right to participate; involve them early
- They have a right to be informed
- Include all parties with an interest or a stake

Rule 2. PLAN CAREFULLY AND EVALUATE PERFORMANCE
- Know your objectives
- Consider your audience
- Select and train spokespersons carefully
- Pretest messages
- Evaluate your efforts

Rule 3. LISTEN TO YOUR AUDIENCE
- Find out what concerns them
- Recognize hidden agendas, symbolic meanings, and broader political considerations

Rule 4. BE HONEST, FRANK, AND OPEN
- Credibility is essential
- Don't expect to be trusted initially
- Admit uncertainties and work to reduce them
- Admit past mistakes
- Disclose risk information at the earliest possible time
- Develop good relations with the community

Rule 5. COORDINATE AND COLLABORATE WITH OTHER CREDIBLE SOURCES
- Avoid conflicts and public disagreements
- Build bridges
- Communicate jointly with trustworthy sources

Rule 6. MEET THE NEEDS OF THE MEDIA
- Be open and accessible to reporters
- Provide background material on risk issues
- Praise or criticize reporting as warranted
- Establish long-term relationships

Rule 7. SPEAK CLEARLY AND WITH COMPASSION
- Simple, nontechnical language
- Use examples and anecdotes
- Be sensitive to emotions: fear, anxiety, anger, helplessness
- Indicate actions you are taking
- Indicate what you cannot do
- Avoid distant, abstract, unfeeling language about deaths, injuries, and

210 INSURANCE AND RISK MANAGEMENT FOR HAZARDOUS WASTE

Table 7-2. Rules for Risk Communication (*continued*)

illnesses. Acknowledge and say that any avoidable illness, injury, or death is a tragedy.

Rule 8 (Slovic). DON'T TRADE DOLLARS VS. LIVES
• Don't use limited resources or cost-benefit arguments to justify exposing people to risks they cannot avoid

7-2. These additional rules are closely related in that both are designed to develop and maintain trust:

• Educate people about the nature and strengths of the risk management process. Without a trustworthy process, no form of communication will be satisfactory to the public.
• Create a decision-making process that involves the public early and in a meaningful way — and gives them control over risks to which they may be exposed.

Given the extreme negative perceptions of the public regarding chemical wastes, there is no easy path to ensuring constructive, cooperative problem solving. However, attention to the implementation of these various rules may significantly improve the current, unsatisfactory state of hazardous waste management.

Acknowledgements

The writing of this paper was supported by Grant SES 8796182 from the National Science Foundation to Decision Research. Any opinions, findings, conclusions, or recommendations expressed in this paper are those of the author and do not necessarily reflect the views of the foundation. I wish to thank Howard Kunreuther for his valuable comments on an earlier draft of the manuscript.

References

Allen, F. W. (1987, November). The situation: What the public believes: How the experts see it. *EPA J.*: 9-12.
Covello, V. T., and F. W. Allen (1988). *Seven Cardinal Rules of Risk Commu-*

A RISK COMMUNICATION PERSPECTIVE

nication. Unpublished manuscript, Columbia University, Center for Risk Communication.

Covello, V. T., P. M. Sandman, and P. Slovic (1988). *Risk Communication, Risk Statistics, and Risk Comparisons: A Manual for Plant Managers*. Chemical Manufacturers Association, Washington, DC.

Covello, V. T., D. von Winterfeldt, and P. Slovic (1986). Risk communication: A review of the literature. *Risk Abstr.* 3: 171-182.

Dunlap, R. E. (1985, July/August). Public opinion: Behind the transformation. *EPA J*: 15-17.

Edelstein, M. (1986). *Stigmatizing Effects of Toxic Pollution*. Unpublished manuscript, Ramapo College, Department of Psychology, Ramapo, NJ.

Electric Power Research Institute, Palo Alto, CA. *EPRI J*. (1980). Assessment: The impact and influence of TMI. 5(5): 24-33.

Evans, N., and C. Hope (1984). *Nuclear Power: Futures, Costs, and Benefits*. Cambridge University Press, Cambridge.

Freud, S. (1924). *Collected Papers*. Hogarth, London.

Galton, F. (1880). Psychometric experiments. *Brain* 2: 149-162.

Goffman, E. (1963). *Stigma*. Prentice-Hall, Englewood Cliffs, NJ.

Heising, C. D., and V. P. George (1986). Nuclear financial risk: Economy-wide costs of reactor accidents. *Energy Pol.* 14: 45-51.

Jones, E. E., et al. (1984). *Social Stigma: The Psychology of Marked Relationships*. W. H. Freeman, New York.

Kasperson, R. E., O. Renn, and P. Slovic et al., 1988. The social amplification of risk: A conceptual framework. *Risk Anal.* 8:177–187.

Kunreuther, H., W. Desvousges, and P. Slovic (1988). Nevada's predicament: Public perceptions of risk from the proposed nuclear waste repository. *Environment*, 30:16.

Kunreuther, H., C., and J. Linnerooth (1983). *Risk Analysis and Decision Processes*. Springer-Verlag, Berlin.

Lagadec, P. (1982). *Major Technological Risk*. Pergamon, Oxford.

Mazur, A. (1981). The Dynamics of Technical Controversy. Communications Press, Washington, DC.

Mazur, A. (1984). The journalists and technology: Reporting about Love Canal and Three Mile Island. *Minerva* 22: 45-88.

Perrow, C. (1984). *Normal accidents*. Basic Books, New York.

Rankin, W. L., S. M. Nealey, and B. D. Melber (1984). Overview of national attitudes towards nuclear energy: A longitudinal analysis. In W. R. Freudenburg and E. A. Rosa (Eds.), *Public Reactions to Nuclear Power*. Westview, Boulder, pp. 41-67.

Slovic, P. (1986). Informing and educating the public about risk. *Risk Anal.* 4: 403-415.

Slovic, P. (1987). Perception of risk. *Science* 236: 280-285.

Slovic, P., B. Fischhoff, and S. Lichtenstein (1981). Informing the public about risks of ionizing radiation. *Health Physics* 41: 589-598.

212   INSURANCE AND RISK MANAGEMENT FOR HAZARDOUS WASTE

Slovic, P., B. Fischhoff, and S. Lichtenstein (1985). Characterizing perceived risk. In R. W. Kates, C.Hohenemser, and J. X. Kasperson (Eds.), *Perilous Progress: Technology as Hazard* (pp. 91-123). Westview, Boulder.

Slovic, P., S. Lichtenstein, and B. Fischhoff, (1984). Modeling the societal impact of fatal accidents. *Management Science* 30: 464-474.

Slovic, et al., 1989. Perceived risk, stigma, and potential economic impacts of a high-level nuclear waste repository in Nevada. Carson City, NV: State of Nevada, Nuclear Waste Project Office.

Szalay, L. B., and J. Deese (1978). *Subjective Meaning and Culture: An Assessment through Word Associations*. Erlbaum, Hillsdale, NJ.

Wundt, W. (1883). Uber psychologische Methoden. *Philosophische Studien* 1: 1-38.

Summary: Paul Slovic

Communication is often thought of as a panacea for all the craziness that we see going on around us with regard to risk. Perception of risk is perhaps the driving force behind a lot of the phenomena discussed yesterday, and the view is that if only we could communicate with the public and educate them because they're ignorant, all these problems would go away and things would become rational and orderly.

My paper tries to dispel the notion that things are as simple as this. It starts from a discussion of the perception of risk. The assumption is that in order to communicate, you have to know how people view the situation, what's going on in their minds, what's driving their concerns, what their concerns really are, and so on. Some of what I discuss is a summary of work on public opinion polling, which I think shows that as a nation we all are environmentalists and, whether or not we openly identify ourselves as such, we are thinking that way more and more. There is tremendous concern about chemical risks. This is very clearly indicated in the polling data. There is evidence that most people believe that all use of chemicals must be risk-free. That's a rather strong statement, but we see in Sheldon Novick's discussion that this attitude works its way into policy as well. Industry is distrusted and is seen as not doing its job in cleaning up the environment. We have a social/political system for managing risks that is highly adversarial. And all of this goes on in a very open climate where the media is very intrusive and keeps bringing risk problems to the attention of the public.

All of this feeds on itself and leads to heightened perceptions of risk. Thus, despite the massive efforts and amounts of money that we're putting into managing risks and reducing risks, and despite the gains in health and longevity and reduction in accidents and so forth, people in this society see themselves as increasingly vulnerable, more at risk now than ever before, and likely to be more at risk in the future than today. So the harder we work to reduce risks, the more concerned and less satisfied the public gets. This is a situation in which some sort of unsatisfactory confrontation is inevitable.

All of this then sets the stage for research on perception, which tries to understand this phenomenon. Risk perception research shows that the lay public has a very different conception of risk than do the experts. The experts have a quantitative view, which is guided by their view of risk as determined by the probability and magnitude of consequences of various events — expected losses, and so forth. The public has a view that is much broader and more complex, and that takes into account considerations such as the potential for catastrophe, equity, the risk to future generations, and issues of power and control. All of these influences on public perceptions make for a situation where, when the technical community presents quantitative risk estimates to the public, the public may give them very little weight, because their view of risk is really something quite different.

There have been a number of attempts to develop better ways to communicate risk. We know that simple comparisons of nuclear power versus driving a car anger people rather than help them gain perspective. I discuss some recent work that attempts to show that these sorts of comparisons really can be broadened to include a dozen or more different types of comparisons that may be more effective in helping people put risk in perspective. But I finally conclude that no matter how much tinkering you do with the types of comparisons and with the language that you use to present risk information, what really matters is the social/political and procedural environment in which risk issues are discussed and decided. We need to somehow construct contexts in which people are respected for their views, where we have two-way communication, and where we give the public meaningful control.

Without this kind of climate of trust and participation, it doesn't matter how we attempt to communicate; the effort will not be successful.

Chairperson's Remarks: Frederick Allen

We had an interesting discussion last night. As I was trying to think of how to summarize it and give you a flavor for it, I recalled a story about Robert Benchley, which appropriately enough took place in an academic setting. He was faced with that ultimate nightmare for all students — what

214   INSURANCE AND RISK MANAGEMENT FOR HAZARDOUS WASTE

happens when you get to the exam and you really don't know what to say to the question. In this case, the question was to describe the origins and to discuss the implications of a certain North Atlantic fishing treaty, and he didn't know anything about it. He said, as he sat down to write, "Some people will talk about this treaty from the point of view of the North Americans, and other people will talk about this treaty from the point of view of the Europeans. I, however, propose to talk about this treaty from the point of view of the fish."

Now, as for our discussion, we didn't really have a lot of very good answers, but we had a lot of good points of view. I will talk about the many wishes that were expressed and a few of the proposals that were discussed.

There was no shortage of problem statements, and Paul Slovic laid out the problems very well. There were two things, I think, on which we did agree. One was the dilemma that we face in a democracy where the public is ultimately in charge of setting priorities, but where their estimates of risk may be very different from what experts are telling us — thus the choice of which estimates we follow.

Another thing we did agree on was an objective for risk communication — namely, to get a better understanding of risk so that real communication can take place on issues of values, allocations of costs, and benefits and risks. It is not necessary for risk communication to solve all those problems but just to facilitate that kind of discussion.

There were four categories of problems and solutions that we discussed:

(1) We began with the whole issue of public understanding of risks. The first suggestion that we had was that we could just wait for things to get worse and then something would happen. None of us were very pleased with that, so we went on to talk about how improving public understanding of risk issues ought to be a much higher priority nationally. The need for framing the discussion in terms of choices and priorities is often a productive approach. We discussed working towards a broader understanding of the many different aspects of risk, including the benefits associated with the activities producing those risks and the relationship between production and risk. We talked about the need for constructive as opposed to adversarial discussion.

(2) Another category of problems and solutions can be referred to as the NIMBY syndrome, the whole issue of *not in my back yard*. We talked about how to site particular facilities, and about issues of incentives and compensation, especially about their pros and cons, but I won't try to summarize all this now. We talked about issues of picking sites by lot. We talked about upgrading and expanding existing sites as a way of increasing

A RISK COMMUNICATION PERSPECTIVE   215

capacity, the idea being that those sites were already familiar to the people living near them; perhaps there are some opportunities there.

(3) Another category of problems and solutions that we discussed was message. We talked about working with the media and providing them with appropriate data. We had some disagreement about whether the media was really the problem in this case or not. We talked about going right past the media to public meetings and trying to have useful dialogue on all aspects of the issues at hand. We talked about focusing on different segments of the audience and the importance of deciding who it is that you're trying to reach and why. We talked about watching to see what happens over the next couple of years with SARA Title III — as companies release data about their emissions and effluents, seeing how they try to explain the risk and seeing if there are some useful approaches that come out of that. We talked about the desirability of finding a way to combine the qualitative and quantitative aspects of risk, but didn't come up with much. Finally in this category we talked about several new manuals that have been coming out. Paul Slovic discussed some work that he's participated in with the Chemical Manufacturers Association. The State of New Jersey has recently put out a manual on risk. There are indications that there are several others that are coming out as well.

(4) The fourth subject was the issue of attitudes. We talked about the importance of involving the public in many decisions because they will get involved even if they're not involved initially. And then there was the whole issue of the safety-first attitude as an approach to the problems that we were discussing.

One of the things that Paul Slovic mentioned in his paper was a recent effort by several of us to put down seven cardinal rules of risk communication. These were developed from a lot of the research and case study material that has been put together over the past several years. The interesting thing about them is how simple they are and yet how frequently they are disregarded.

Why don't people follow these rules if they are so simple? There are probably three reasons, each of which suggests some possible actions for companies, agencies, and others involved in these public discussions.

The first is the issue of skills. Simply put, most engineers and others do not get a great deal of training, if at all any, at a professional level, in communication. So one thing that needs to be discussed is the curricula of the different professional schools that have graduates going on to participate in many of these debates. Similarly, there is scope within our own institutions for providing training for people who will need it.

Second is the issue of attitudes. On that list of seven cardinal rules, it

216   INSURANCE AND RISK MANAGEMENT FOR HAZARDOUS WASTE

will be seen that many of them have to do with attitudes. There needs to be more discussion within businesses and within agencies, perhaps about particular cases that lead people to understand the importance of attitudes in this whole area.

Finally, there is a series of institutional barriers to good risk communication. For instance, there is the way that we set up management control systems in our organizations and the pressures that these put on our staffs. We're also talking about conflicting goals at different levels of organizations. One of the earlier speakers this morning talked about how it is important to have a firm goal throughout an organization to enable certain things to happen. Oftentimes, in a government setting, this means that there might be somebody at the staff level who thinks, "Here's the right way to do it" but for some other level, perhaps from the very top level, for political reasons, that is not possible. There are a number of other institutional barriers. In the short run, many of them are very difficult to change, but in the long run it is the leaders of these institutions that do have the opportunity to change them, and they should focus on them.

I would like to close by going back to Benchley and what he was saying in talking about the fish. We've really looked extensively at how the public understands risk, and at how the public react to different kinds of risk messages. In contrast, we've looked very little at the problems faced by risk communicators in companies and in government agencies and what they need to do a better job. Why is it that they don't do a better job? Perhaps some of the suggestions that we just made are things that should be followed up in future research.

8 AN ENVIRONMENTAL PERSPECTIVE ON AN INTEGRATED WASTE MANAGEMENT STRATEGY

Sheldon Novick

At the moment, formal risk assessment plays only a small role in hazardous waste management: Superfund remedial sites are ranked in accordance with risk assessments, for instance. Before the Superfund Amendments of 1986, the present Administration made some efforts to use formal risk assessment and management techniques to set the boundaries of cleanup at Superfund sites. As is well known, these efforts were abandoned as impolitic and impractical, and Congress closed this door by adding Section 121 to Superfund, setting a target of national environmental quality standards for cleanup, with only a narrowly drawn variance procedure for risk assessment to creep back in.[1] That action seemed to end the last important use of risk assessment in the hazardous waste program. (I speak only of risk assessment for regulatory purposes, and not the assessment of claims exposure that may be needed for insurance purposes.)

Like most environmentalists, I was glad to see Section 121 adopted. Risk assessment is not very useful for setting goals or general policies in environmental programs, for reasons I will try to explain. Once goals and policies have been set, however, risk assessment may play a modest role, in setting priorities and ensuring consistency.

To some people, however, especially to business people, it seems ir-

218   INSURANCE AND RISK MANAGEMENT FOR HAZARDOUS WASTE

rational for the government to set goals and policies without some formal assessment of the evil to be remedied, and the cost of the cure.

We continue to talk at cross-purposes about this, although it is not a new debate. Early in the nineteenth century, the followers of Jeremy Bentham thought English law was senseless and irrational. They proposed to replace the whole confused mass of common law with orderly, rational legislative codes. They wanted to weigh the costs and benefits to society of every act of government. The movement spread to the United States but it never accomplished much in a practical way, either in Great Britain or here.

One of its critics was Justice Oliver Wendell Holmes, who had two objections. The first was that we rarely know enough to assess costs and benefits very well. But the second objection was more fundamental:

The life of the law has not been logic; it has been experience.[2]

Many of my fellow environmentalists object to risk assessment because it cannot be done very well in the present state of knowledge. But that is not really the fundamental objection; it is only the easier one to express. Speaking just for myself, I would like to go over some recent and familiar history with you briefly.

Hazardous waste management policy, at least at the level of the federal government, is the confluence of three streams: the regulation of pollutants diverted from other media; the regulation of trash disposal; and the cleanup of abandoned chemicals. These three sources of policy have each contributed to a swelling national determination to do away with hazardous waste disposal entirely.

8.1 An Environmentalist's Perspective

8.1.1 Mountains of Trash

Trash disposal, once regulated — if at all — by county or municipal governments, is a surprising topic for federal law. Yet trash disposal has become a federal matter. There has been a federal program of assistance to state and local entities since the 1950s[3] and trash disposal is now nationally regulated along with air and water pollution.[4] The origin of trash regulation is quite different from some other environmental protection programs, however.

Progressives at the turn of the last century made trash disposal a national question. Cities were corrupt and dirty; the Progressives meant to clean

AN ENVIRONMENTAL PERSPECTIVE 219

them up, literally and figuratively. Edward Bok, the young editor of the Ladies Home Journal, was one of the more curious figures in the Progressive movement; urban tidiness became one of his magazine's crusades, along with the Boy Scouts, hygiene, and general uplift:

> Bok began to note the disreputably untidy spots which various municipalities allowed in closest proximity to the center of their business life, in the most desirable residential sections, and often adjacent to the most important municipal buildings and parks.[5]

Bok published photographs of the dozen worst municipal eyesores in *Ladies Home Journal* at the time Lincoln Steffens was writing "The Shame of the Cities" in *McClure's*. The Bok series may have had a more permanent effect than Steffens's. In his autobiography, Bok wrote:

> [L]ocal pride was aroused, and as a result not only were the advertised "dirty spots" cleaned up, but the municipal authorities went out and hunted around for other spots in the city, not knowing what other photographs Bok might have taken. . . . Cities throughout the country now began to look around to see whether they had dirty spots within their limits, not knowing when the . . . photographers might visit them.[6]

Bok's Progressive tidiness included a campaign against billboards, the forerunner of the Johnson Administration's highway beautification program; Bok took credit for stopping advertising signs on the top of the Grand Canyon and on the scenic bluffs of Niagara Falls.[7]

At about the same time, the Sierra Club was organized, and the conservation movement turned to preserving natural areas for recreation.[8] Municipalities began to set up trash disposal monopolies under city ownership, and the city or county dump was born, the child of a national reform movement. Forty years after Bok's crusade, the county dump had become an eyesore and a health hazard; litter, once an urban problem, was spreading over the countryside. Conservationists, therefore, extended Bok's urban tidiness campaign to the country. Hiking groups organized volunteers to pick up trash; Keep America Beautiful, Inc. (KAB), an organization of disposable packaging manufacturers, advertisers, and publishers, ran antilitter campaigns; and publishers and broadcasters donated time and space to the effort. Some conservation groups continue to join with KAB in voluntary efforts to reduce trash and litter; recycling campaigns are now common suburban phenomena.

In the 1960s, a new tributary began to flow. Environmentalist organizations, which were quite different in their attitudes and concerns from the older conservationist groups, began to search for the root causes of pollution.[9] A widespread theory traced the origin of pollution problems to

220 INSURANCE AND RISK MANAGEMENT FOR HAZARDOUS WASTE

the careless use of resources; the most common version of this theory held that the *external* environmental costs of disposal were not included in the costs of production.[10] Throw-away packaging accordingly became the symbolic enemy of the new environmental movement, which resembled its Puritan and Progressive forebears in its opposition to waste, untidiness, and dirt.

A less popular theory was that the waste of natural resources revealed a flaw in capitalism that could only be cured by radical change.[11] Curiously, the conservative and radical arguments supported a common cause against disposable packaging.

By the 1970s, the antilitter movement had pretty well spent itself, at least as a movement to support federal laws; the movement's ethic has been widely accepted, and campaigns to clean up city streets or ban rural litter continue; but the effort to enact federal antilitter legislation was defeated in 1976[12] and will not be renewed in the foreseeable future. KAB and its adversaries now fight their battles in state houses and municipalities. Litter control, the original motive for governmental regulation of trash disposal, is returning to the marketplace. The antilitter campaigns have left their fossil traces in the law, however; archaic language in federal statutes is a reminder of their origins; the term for a waste disposal facility that pollutes groundwater is, incongruously, an open dump.[13]

In sum, trash disposal became a proper concern of the federal government during the Progressive era, when national attention was needed to reform municipal corruption. In the 1970s, the environmental movement gave new content to the government's concern.

8.1.2 The Diversion of Pollutants

In the 1960s, the Department of Interior, which had responsibility for surface water pollution control, wanted to extend its jurisdiction to include groundwater.[14] Other dramatic changes in the Federal Water Pollution Control Act (FWPCA) were being considered; the Administration bill would have extended its jurisdiction to groundwater,[15] and in the debates on what became the 1972 FWPCA Amendments, arguments were put forth for a federal groundwater pollution program patterned on already existing surface-water pollution legislation.[16]

The principal objections to the groundwater scheme were that it would set up a federal program of land-use planning, traditionally a local government preserve, and that groundwater was physically too complex to regulate in general terms. Some of the proponents of groundwater regu-

AN ENVIRONMENTAL PERSPECTIVE

lation indeed saw it as a federal land-use planning program; one of the then-current notions about pollution was that it stemmed from unplanned growth.[17]

The results of the fight over groundwater jurisdiction were inconclusive. Both sides claimed victory in the Senate Report on the 1972 Clean Water Act Amendments,[18] while the statute itself was ambiguous.[19] EPA continued to assert jurisdiction over groundwater in the Agency's surface-water pollution control program, but its view was rejected by the better-considered decisions of the courts.[20] When the Safe Drinking Water Act expressly gave EPA jurisdiction over some injection disposal wells in 1974,[21] the Agency — to the regret of many — gave up its efforts to extend all of the Clean Water Act to groundwater protection as such.

Yet some groundwater protection language was included in the 1972 Clean Water Act Amendments.[22] The states were required by section 208 of the Act to make plans for regulating water pollution from all sources, including those from sources outside the permit system administered by EPA.[23] The states' plans were to include provisions for controlling groundwater contamination from waste disposal:

> Any [state] plan . . . shall include, but not be limited to . . . (J) a process to control the disposition of all residual waste generated in [the planning area] which could affect water quality; and (K) a process to control the disposal of pollutants on land or in subsurface excavations within such [planning] area to protect ground and surface water quality.[24]

While the details of groundwater protection were left to the states, EPA was to provide technical information and criteria for groundwater quality, which the states were to employ in their section 208 plans.[25]

This was the first federal hazardous waste legislation and the first general groundwater protection statute. It remains in the Clean Water Act,[26] although EPA has never given the legislation life.[27] Section 208 created, for the first time, a distinction between ordinary trash and hazardous wastes. The statute used the term *pollutants* to distinguish the more hazardous waste, which was then subject to special regulation when disposed of on the land.[28] The term betrays its origins; the drafters of the statute believed, with some justification, that provisions controlling air and surface-water pollution might drive industrial polluters to put their wastes into wells and landfills. Section 208 plans were meant to keep pollutants from being shifted to groundwater.

Section 208 was defeated by its own ambition. A national land-use planning system was politically impracticable and, in any case, had no real constituency.[29] EPA, as always, was consumed by narrower issues

222 INSURANCE AND RISK MANAGEMENT FOR HAZARDOUS WASTE

and never gave much attention to the broad planning process that the Clean Water Act envisioned. Concern for groundwater focused on specific sources of contamination, while the broad prevention program languished. But the concern was not entirely forgotten; four years later, in 1976, similar bills were proposed as Solid Waste Disposal Act Amendments.[30] In what would eventually become the Resource Conservation and Recovery Act (RCRA)[31] the provisions for trash disposal plans and special regulation of more hazardous waste were repeated, but with far stronger federal enforcement authority than had been provided in the Clean Water Act.[32]

8.1.3 Waste Disposal Law

While the pollution control laws were flowing down one channel, waste disposal laws had moved down another. Partly because of the boom in disposable packaging, roadside litter was growing worse, and the cities were running out of space in their landfills. Bills were introduced in Congress in the 1970s to reduce the volume of wastes by making waste disposal more difficult and expensive, and less disruptive to the environment.[33] According to the theory behind these bills, the costs of product disposal were not being taken into the marketer's accounting; the environment was, therefore, being consumed as if it had no value.[34] The waste disposal bills were intended to internalize these costs, thereby making the costs of disposal part of the market price of the product. This would allow the marketplace to allocate environmental resources in the freest and most efficient manner. Consumers buying the least expensive products in a free market would ensure that resources were given adequate protection when compared to other values. As an afterthought, the bills also commanded the creation of state plans — copied from the air and water statutes — for better regulation of the disposal that continued; the main purpose, however, was to discourage the discarding of valuable products by making disposal costly and recycling more accessible.

One of the principal legislative proposals was for a required deposit on discardable beverage containers, then believed to make up a large part of litter and solid waste.[35] Most of the debate in Congress in 1975 and 1976 concerned mandatory deposit proposals, commonly known as *bottle bills,*[36] but the proposals became so contentious that their sponsors were forced to withdraw. The remaining provisions of the principal bills were put together in a compromise negotiated between staffs of the House Commerce and Senate Public Works committees. This was passed by both houses

AN ENVIRONMENTAL PERSPECTIVE 223

without a conference: no member of the House read the final RCRA bill before it was enacted.[37]

What remained were the regulatory provisions applicable to disposal facilities — the echoes of section 208 of the Clean Water Act requiring state plans for trash disposal and more stringent plans for the safe disposal of hazardous waste; the final waste disposal law was titled the Resource Conservation and Recovery Act of 1976.[38] The new statute gave some force to the law by requiring EPA to prepare criteria for many provisions of the state plans, including performance standards for disposal facilities.[39] The plans were to include a permit system which EPA was to administer until the states had enacted adequate legislation.[40] Manifest systems were to be used to ensure that hazardous wastes were sent only to permitted facilities.[41]

For the first time in waste disposal law, pollution control had become dominant; groundwater protection, never before mentioned in the Solid Waste Disposal Act, now became the single most important environmental purpose of the new amendments.[42] The old vehicles of the Progressive movement had been commandeered for groundwater protection.

8.1.4 Confusion and Delay

EPA had extraordinary difficulties implementing RCRA, not all of which stemmed from ineptitude. Problems persisted under two administrations and many changes of personnel; some of the difficulties, therefore, must be attributed to the statute and to the difficulty of the task it set.

In retrospect, it is evident that there were two serious and closely connected problems with the hazardous waste provisions, subtitle C of RCRA.[43] First, there was little in the statute or its history to indicate just what Congress wanted to have done in this field. Second, to the extent the statute gave explicit directions, it embodied a contradiction.[44]

The paucity of express direction is not surprising. If there was a problem stemming from hazardous waste disposal, that problem had not been crystallized in an adversarial setting; there was a vague belief that pollutants kept from the air and water by the command of earlier statutes were now being dumped on the land, from which the pollutants would eventually find their way back into the air and water by a more circuitous route. There was some testimony at hearings that this was indeed happening, and a few instances of groundwater pollution actually affecting drinking-water supplies were cited.[45] Industrial waste disposal practices were not systematically examined, nor was there much in the legislative record concerning the effects of regulation — or even what would be regulated and how. Yet

224   INSURANCE AND RISK MANAGEMENT FOR HAZARDOUS WASTE

one clear, practical judgment is discernible on the record; hazardous wastes in open dumps and municipal landfills were problems, and hazardous wastes should therefore be disposed of only in specially licensed facilities.[46]

EPA was directed to set standards for the disposal of hazardous wastes and to create a permit and manifest system to ensure that all waste went only to facilities that met the standards.[47] Waste disposal standards were the skeleton from which the rest of the regulations had to be suspended, but standards are just another way of stating legislative goals. What, then, does the statute intend for EPA to accomplish? The law says only, if grandly, to "protect human health and the environment."[48]

Individual members of Congress may have meant no more, if they were conscious of this provision at all, than to express some concern and leave it to EPA to determine the problem and form a remedy. Unfortunately, the words chosen to express this vague concern created a sharp contradiction that for years immobilized EPA.

The difficulty was that many wastes, including hazardous wastes, were disposed of on the land, in landfills, lagoons, dumps, and unconstrained heaps. There was no immediate alternative to land disposal for most wastes. Many were not flammable and could not be incinerated; the air and water pollution control laws limited releases into those media; and the activities that generated wastes, including the process of living, could not be carried on without them. By default, therefore, land disposal was a necessity. Yet the standard set for waste disposal on land was single-minded: EPA must protect health and environment — and do nothing more nor less. EPA had no authority to consult other values, or so the statute read on its face. There was little in the law's history to suggest alternate readings.

Toxic or hazardous pollutants rarely have any threshold of action in large populations: if they are released in a way that may result in many people being exposed, there is a risk of injury.[49] If EPA consults health and safety values alone, EPA can only set standards that either entirely prohibit releases of such pollutants or allow them only at some legally negligible level.[50] Hazardous waste disposal plainly was expected to continue under RCRA; EPA is directed to set up a complex permitting scheme for disposal facilities.[51] But hazardous waste put on the land will eventually leak away from the site of the disposal. Landfills eventually must leak, no matter how securely built they may be. Was EPA, therefore, to ban disposal of hazardous wastes on land? Was such a ban possible?

Faced with an environmental quality standard that ignored cost, and with the difficulty of banning land disposal altogether, EPA vacillated for years. It tried to require landfills that would not leak for some period of time.[52] The Agency also considered and then rejected the notion of banning

AN ENVIRONMENTAL PERSPECTIVE   225

certain forms of waste production. The difficulty was similar to the one EPA faced under section 112 of the Clean Air Act, which on its face seemed to require the shutdown of large industries if needed to end the release of toxic air pollutants.[53]

Perhaps worst of all, in the view of some of RCRA's sponsors, EPA deferred regulation of hazardous waste from *small generators* — waste generators that produced less than 1000 kilograms per month of hazardous waste. Wastes from small companies were most likely to go to municipal landfills; EPA thus seemed to ensure that the one clear mandate contained in RCRA, to keep hazardous waste out of ordinary landfills, would be greatly delayed. In fact, the exemption created by EPA persisted until the statute was amended in 1984.[54] EPA officials had worried about putting small businesses out of operation. The Agency had also feared the threat of massive noncompliance. These were reasonable concerns. It was not until the Reagan Administration inadvertently mobilized the public that political support for stringent regulation could be assured.[55]

At first, then, there was no constituency pressing for regulation of hazardous waste, and when the constituency finally arose, it had other items on its agenda.

8.1.5 Love Canal

In 1976, when RCRA was enacted, concern about hazardous wastes had been somewhat abstract and theoretical. But a few months later, the press began to report on an abandoned chemical dump in New York State.[56] A school and some houses had been built on the site of the filled dump, and the people who lived there were frightened for the health of their children.[57] The place had the memorable, incongruous, name of Love Canal; it gave life to the abstract concern over hazardous wastes.

The residents of the area were eager to protect their children and to recover for the damages they had suffered. Among other courses of action, they brought suit against the Hooker Chemical Company and other companies whose discarded chemicals were believed to be present at the dump.[58] But Hooker had filled in the land and donated it to the school district years before; the suits against Hooker faced other serious obstacles, and the school district had no significant assets.

EPA had little authority to assist the local residents in cleaning up the horror they had found under their feet; local governments had little money or expertise. EPA had some authority under the Clean Water Act to require cleanup of oil spills on or near surface-waters; costs of cleanup could be

226 INSURANCE AND RISK MANAGEMENT FOR HAZARDOUS WASTE

recovered from the originators of the spills, regardless of fault. The recovered money would then go into a revolving cleanup fund.[59] The oil-spill cleanup program had worked well; it provided a base of experience and some model procedures for a federal emergency response program.

Seizing on the precedent, in 1978 Congress quickly extended the reach of the Clean Water Act program to cover spills of hazardous chemicals, as well as petroleum, on or near navigable water. The National Contingency Plan was altered to allow EPA to respond when the spill occurred at an onshore facility threatening a navigable waterway; the Coast Guard continued to respond to oil and chemical spills from vessels.[60]

Building on this slender foundation, several bills were proposed to extend the combined response program to all onshore spills; anticipating them, EPA pushed its Clean Water Act jurisdiction to — and perhaps past — the limit, and began to respond to onshore spills.

There was some sentiment in Congress for keeping onshore oil- and chemical-spills response programs separate, but EPA favored combining the programs into a single Superfund, that would be replenished by recoveries from responsible parties.[61]

There was considerable opposition. David Stockman, then a Michigan Representative, and others contended that Love Canal was unique, or at least rare, and that states had both the capacity and the traditional authority for responding to such disasters.[62] EPA staff disagreed, and they set out to dramatize their view that abandoned waste dumps were a serious national problem; senior Agency officials helicoptered into suspect sites and held press conferences; sponsors of Superfund bills made lists of suspect sites in every Congressional district. The Justice Department set up a special unit to bring suits under the *imminent hazard* provisions of RCRA, the Safe Drinking Water Act, and the federal common law of nuisance to abate the hazard at abandoned sites. The suits named the generators of waste found at the sites, since the generators were often the only solvent parties visible. EPA, after some initial reluctance, agreed to prepare such suits; the Agency set informal quotas of suits for regional offices and established a new headquarters team to manage cases as they were prepared.

The result was two years of escalating publicity and pressure on members of Congress to adopt new legislation. In 1980, a Republican President and Senate majority were elected. In the closing hours of the session, a lame-duck Democratic Congress passed the Comprehensive Environmental Response, Compensation and Liability Act (CERCLA), still called Superfund.[63] Superfund was, however, only a chemical-spill program that contained an anomalous exclusion for onshore spills of petroleum products; these remained unregulated. The onshore oil-spill cleanup bill had re-

AN ENVIRONMENTAL PERSPECTIVE 227

mained separate and was never adopted.[64] Much later, Congress added to RCRA another cleanup program, for underground storage tanks — covering the worst source of petroleum spills omitted from Superfund.[65]

CERCLA ratified EPA's emergency response program, and the legal theories under which the Justice Department had brought approximately 60 imminent-hazard suits prior to the law's passage.[66] Superfund provided federal authority to respond to onshore chemical spills as emergencies, and made the generators and dumpers of waste proper defendants, despite their lack of present connection with the sites, and made them responsible for reimbursing the cleanup fund.[67]

8.1.6 RCRA Revisited

Whatever else may have been accomplished, the national controversy over orphan dumpsites prompted people to decide how they felt about hazardous waste disposal on land. The Reagan Administration at first misjudged the public's feelings about hazardous waste, which had been inflamed by the controversy over orphan sites. RCRA had always been slighted in EPA's planning; the new Republican Administration in 1980 was concerned with budget cutting and the reduction of government regulation, and so the neglect of RCRA enforcement was not remedied. The new EPA continued to vacillate over land-disposal regulations, and at first accepted the reality that landfills would leak. Then, when shocked into awareness of the depth of the public's concern, the Agency adopted standards for landfills that would keep them from leaking during their operating lives and for some time afterward.[68] EPA, however, preserved the small-generator exclusion that had so annoyed some of RCRA's original sponsors.[69]

By 1984, it should have been plain that landfills would inevitably leak, and that the contradiction contained in RCRA had to be resolved by banning land disposal of most hazardous wastes as quickly as that could be accomplished. California had already taken a step in that direction and had shown it did not lead over a cliff. Democratic Representatives introduced bills directing EPA to carry out a staged ban similar to California's. After eight years of EPA vacillation, many Congressmen had become impatient with the Agency; Reagan Administration challenges to Democratic-led House Committee authority inflamed the tempers of Congressmen and added to the public's perception that EPA could not be trusted to carry out the mandate of these laws.

The outcome was the 1984 amendments to RCRA[70] which set rigid schedules for EPA to carry out most parts of the land-disposal program,

228 INSURANCE AND RISK MANAGEMENT FOR HAZARDOUS WASTE

including the permitting of facilities and the gradual phasing-out of land disposal for most hazardous materials that might eventually escape from landfills.[71] Some compromises among goals and schedules were adopted, and some flexibility in the ultimate standards was allowed, but Congress took back to itself the authority to make these compromises; EPA's discretion was largely withdrawn. The small-generator exception was drastically cut back, so that only generators of 100 kilograms per month or less were entirely exempted.[72]

The prohibition of land disposal of hazardous waste was to be carried out for groups of wastes, on a schedule set by statute. If EPA missed any of the deadlines, the prohibition would take effect; EPA could act only to lift the ban. The only significant escape route from the prohibition was through treatment of the wastes. EPA was required to set national standards for waste treatment based on the best demonstrated, available technology (BDAT), and chose to establish performance standards similar to the BAT standards for toxic pollutant discharges under the Clean Water Act. Wastes treated to these standards could be land-disposed. In this indirect fashion, the hazardous waste program began to adopt national performance standards for treatment facilities that specified the discharge they could make into disposal facilities, similar to the limits on discharges into air and water. It is plain that BDAT is or should be a series of steps over the years toward complete elimination of land disposal as technology advances.

The 1984 amendments to RCRA and the debates that led to them resolved another question that had puzzled EPA. Both RCRA and CERCLA had some retrospective application; the hazardous waste laws were meant to clean up the pollution caused by improper disposal in the past, as well as to prevent new problems from developing.[73] RCRA, by now increasingly patterned after the earlier pollution control laws, directed EPA to protect health and environment and was silent as to cost or feasibility. The natural implication of the statutory language was that groundwater would be protected, regardless of cost. Cleanup, it seemed to follow, should restore the original quality of groundwater; EPA made this both the threshold and the goal of RCRA cleanup. Drinking water quality standards — of which only a few had been established — were to be used as goals when available. Acknowledging that either background or drinking-water purity might be impractical to attain, the Agency set up a procedure for permit holders to establish alternate concentration limits at particular sites. Congress ratified this approach to RCRA cleanups, and then turned to the bigger question of cleanup at abandoned waste dumps.

AN ENVIRONMENTAL PERSPECTIVE 229

8.1.7 Superfund Revised

By 1984, Superfund had become the focus of environmental protection — in part because of the Administration's stubborn resistance to the program — and all eyes turned to the needed reauthorization of the statute, whose funding authority would expire in 1985. The trust fund for site cleanup, initially established at $1.5 billion, was almost entirely committed for cleanup at a few dozen sites, while EPA estimated that 1500–2500 sites eventually would require a remedy. The original, hastily drafted statute was badly in need of clarification and amendment, and the interest groups affected by Superfund were well organized and intent on securing amendments when the statute was taken up again. Three years of effort were needed to reauthorize the Superfund program. The Superfund Amendment Reauthorization Act of 1986 (SARA)[74] expanded authorization for the cleanup program from $1.5 billion to $8.5 billion, and in numerous detailed ways held EPA to a strict accounting for rapid progress. There were no fundamental changes in the program's design or purpose; EPA's procedures were largely ratified.

8.1.8 Goals

The goals of the program were more clearly stated. Superfund had grown out of the emergency response campaign,[75] and the statute could be read to allow EPA to choose remedies on a case-by-case basis.[76] The Agency read the statute as authorizing and perhaps requiring that remedial responses — long-term soil and groundwater cleanup — be carried only to the point of cost-effectiveness at each site.[77] Dissatisfaction with this approach became acute when EPA faced the prospect that its own activities at Superfund sites were subject to RCRA requirements. Cleanup at the site of a storage or disposal facility would have to reach appropriate background levels (or drinking-water quality standards) to comply with RCRA, but would then exceed the site-specific requirements of EPA's Superfund cleanup policy.

The Environmental Defense Fund (EDF) sued to overturn the Agency's rules, and EPA decided to settle the suit by accepting a version of EDF's proposal. EPA tightened its Superfund policy to more closely approximate RCRA. Superfund cleanup would continue until it achieved standards for environmental quality established under other statutes — except where local conditions required some concession.[78] EPA backed away from this

230   INSURANCE AND RISK MANAGEMENT FOR HAZARDOUS WASTE

settlement for some time, but eventually, EPA accepted necessity, and quietly adopted a policy that set groundwater quality standards as the goals of Superfund cleanup. Applicable standards, borrowed from other programs where available, were incorporated into the Superfund feasibility studies for each remedial cleanup site. The Agency later attached a policy statement to its National Contingency Plan to reflect the new approach.[79] SARA ratified these goals and made them obligatory: Superfund cleanup in most cases must continue until drinking water quality standards, or other *applicable* or *relevant and appropriate* environmental quality standards are met at each site. At EPA's insistence, a procedure for setting alternate concentration limits, similar to the RCRA permit procedure, is authorized, and in narrow cases of impracticality the Agency may waive soil or groundwater standards. Once the goal is selected, of course, the cleanup remedy chosen must be cost-effective.

The contemporaneous 1986 amendments to the Safe Drinking Water Act required EPA to greatly increase the number of drinking water quality standards it sets, and as their coverage grows, these standards will come to serve as the most common goals of both RCRA and Superfund cleanup.

Despite complexity and confusion, the question "How clean is clean?" was on its way to being answered.

8.2 The Meaning of the Land-Disposal Ban

With this familiar history pinned up on the wall, we can see the main lines that lead to the present question.

The three original sources of national concern over hazardous waste — litter and open dumps; dumping of industrial wastes to circumvent pollution controls in other media; and the abandoned chemical dumping grounds symbolized by Love Canal — very naturally produced a national program with one central purpose: to do away with land disposal of hazardous waste, especially the off-site disposal that was perceived to have led to the Love Canals.

Why do away with land disposal entirely? Some eventual leaks seem inevitable, even at the best-managed sites, but surely they can be kept within limits. A thick layer of low-permeability clay, for instance, is self-sealing, and leaks only at a predictable rate and after a predictable period of time. There are highly insoluble compounds of inorganic wastes that seem little different in environmental risk from the ores that were originally mined. In air and surface-water pollution, we tolerate certain limited emissions; the standard of control tends to be the best achievable with available

AN ENVIRONMENTAL PERSPECTIVE

technology;[80] why not accept some pollution from land disposal, at least in places where it seems to do no harm?

The best-available technology standard has been imported into hazardous waste management, but please notice the role it plays. Land-disposal sites are not permitted to emit any statistically significant discharges, and must clean up any significant pollution found on their premises. The best-available technology comes in only at a different point, as a qualification to the land-disposal ban.

We therefore begin with a ban on land disposal, and an exception is made for the residue that remains after applying the best-available treatment technology. The ban takes effect just as quickly as treatment technology advances.

This is similar to the scheme of the Clean Water Act, which sets as a goal the elimination of all water pollution. This goal is to be met only as quickly as the advance of available control technology allows, with the added proviso that local standards of environmental quality must always be maintained. The Clean Air Act similarly, if in more complex fashion, requires that air quality standards be achieved as quickly as reasonable control measures allow. In both statutes, there are actual dates set for the elimination of pollution, but as we know these are steadily extended as we draw close to them, and by now we can consider the goal a distant one, to be approached but perhaps not to be achieved at any foreseeable date.[81]

In short, the national hazardous waste management program is now drawn on the same lines as air and surface water-pollution control, with this twist: land disposal of hazardous wastes is not treated as if it were an industry emitting pollutants. Land disposal is treated as if it were the pollution itself.

This is understandably difficult for the industry to accept. But it is vain to argue logic on this point, i.e., to argue that some forms of land disposal may be less hazardous than the pollution we allow from other sources, that active land-disposal facilities may be preferable to bankrupt Superfund sites, and so on, at least so far as general policy goes. The history of hazardous waste law has produced a massive political opposition to land disposal per se, that has its roots deep in thousands of communities. It seems to be idle to shake one's fist at the sky — to quote Justice Holmes once more.

In short, hazardous waste management has now taken pretty much the same form as other pollution control programs, assuming only that land disposal is itself a form of pollution to be eliminated as quickly as treatment technology allows.

The question with which we began — What role should formal risk

232 INSURANCE AND RISK MANAGEMENT FOR HAZARDOUS WASTE

assessment play in hazardous waste management? — therefore can be answered with another question: What role can risk assessment play in environmental protection generally?

8.3 Priorities and Consistency

One very frequent claim is that EPA does not spend its resources wisely, and that more would be accomplished if the Agency used risk assessment methods to set priorities. Of course, the statutes do not permit this now, but seem to require a sort of first-come, first-served system of control, with some pollutants slipped in at the head of the line by their friends in Congress. But if we could rewrite the statutes at will, would we use risk assessment to set priorities?

Two arguments are generally made on behalf of risk assessment in this context. The first of these seems to fall victim to history. This is the cost-effectiveness argument: EPA would get a bigger bang for its buck if it would just stop and think and set priorities before regulating.

The worst, first is an appealing motto. It is usually put forward as an argument to abandon some existing program, rather than to make a case for added controls, however, and that has given it a bad name. The only example that comes to mind of risk assessment being used to identify a target for added controls is the case of lead additives in gasoline. The Reagan Administration apparently proposed to complete the phase-out of lead additives after looking at the costs and benefits of regulation. It would be easier to be impressed by this if there had not been an earlier effort, also allegedly based on cost effectiveness, to relax these same regulations.

And that last point gets to the difficulty here. Risk assessment is not really a deterministic method that produces the same result no matter who applies it. Because of our ignorance in many areas, and because there are no universally accepted methods, risk assessment is at best just a way of analyzing and stating clearly the elements that go into a decision. There is rarely enough agreement on these elements for people to say, "These pollutants clearly should be the first to be addressed." And indeed this seems to be one of the reasons that pollution-control programs have evolved toward all-inclusiveness. If all toxic and hazardous pollution is to be eliminated as quickly as possible, it may not matter very much where one begins.

The second argument is more compelling, and has more pertinence to hazardous waste management.

National programs are set up pollutant by pollutant, one environmental

AN ENVIRONMENTAL PERSPECTIVE 233

medium at a time. EPA has one program for controlling the emission of volatile hydrocarbons into the air, and quite different programs, for differently named hydrocarbons in surface-water and land disposal. There is some evidence collected in EPA studies that because the hazardous waste program is more extensive in reach, some wastes are simply being diverted from land to air and surface water. Many of us have anecdotes of this happening. The Philadelphia sewage treatment plants were found to be the city's largest sources of toxic air pollution, for instance, presumably because industrial wastes were going down the drain, when arguably they might better have been sent to land disposal.

Perhaps the answer is to tighten up air and water controls. But even when EPA has extended the range of air and water toxic-pollutants designations, and has adopted more drinking-water standards, it will be a very long time before these pollutants have all been eliminated. Progress depends on the best-available technology, but at the moment this is being determined medium by medium, without much information or concern about the *relative* costs or benefits of controlling pollutants in different media.

If we can do a better job of controlling air and water pollution, by diverting some pollutants to waste treatment and eventual land disposal, perhaps that alternative is worth considering. Even here, however, there may be only a very limited role to play for risk assessment. People do not really consider risks to be fungible; that is the problem various *bubble* proposals run into. Few people are willing to accept a greater burden of hazardous waste in their neighborhood, solely because some calculation of net costs and benefits for society as a whole shows that those wastes are better disposed on land than into air or water.

Still, if there is a substantial problem here — if environmental protection would be better if all media were considered together in a formal risk assessment — then it might be a problem that industry and environmentalists and other people could work on together.

8.4 Approaching Zero

A final area in which risk assessment can and perhaps must be used has appeared more recently. The air and water pollution-control programs, like Superfund and RCRA, wrestled for a long time with the question of how clean was clean. Similar questions have come up in workplace safety, in protecting food from contaminants, and no doubt, in other contexts.

The ultimate goal expressed in statutes often has been to eliminate

234   INSURANCE AND RISK MANAGEMENT FOR HAZARDOUS WASTE

pollutants and contaminants entirely, either because of distaste or because there is no apparent threshold of injury when large populations are exposed.

An obvious, practical question arises in each setting, however. Absolute physical or mathematical zero often is not attainable. Some pollutants, like benzene and radiation, are ubiquitous; others, like vinyl chloride, are valuable products and cannot be totally eliminated without ceasing to manufacture them entirely, an option that might disrupt the national economy in drastic ways, quite out of any reasonable proportion to the gains to be achieved in the last increments of control. Of course, these are familiar questions. As some control programs have begun to move into these last increments, a consistent answer has been given. *Zero risk* is not a reasonable goal, except in unusual circumstances. As a general policy, we can only hope to reduce risks to *insignificant* levels.[82]

On this point, the courts seem to have been saying that under present legislation we are not to use cost–benefit analysis to determine what is significant.[83] Indeed, it may be that the costs of regulation are not to be weighed in any express fashion. The question seems to be: What level of risk is so low, that it is the psychological (or political) equivalent of zero?[84] Past that point, further controls would be gratuitous and supererogatory.

The goal of groundwater cleanup, therefore, is not zero risk, but only the achievement of drinking-water standards (or background levels) for those pollutants that have such standards; and this seems to be the kind of no-significant-risk standard that must be set in air and water pollution control. Drinking-water standards of course are not *logically* applicable to Superfund or RCRA cleanups, but they seem to serve the psychological role that is required. Many standards remain to be set, and since they must be set consistently for air, water, and groundwater, I would think the subject of general *de minimis* standards for toxic pollutants seems to have some practical importance for hazardous waste management. For environmentalists, there remains the task of articulating some express analysis on which standards may be based that will ensure that standards are consistent among media, so that we do not face diversion of pollutants into the area of greatest risk.

Risk assessment therefore might be helpful in setting the *de minimis* standards that now seem to be required. Indeed, it has been used in this way; all existing environmental quality standards, including MCLs for drinking water, are based at least in part on risk assessments. The surface-water quality criteria numbers that are sometimes used to set goals for groundwater cleanup are just 10^{-6} risk numbers, after all. Risk assessment might be used in more sophisticated ways, to set limits that are constrained

AN ENVIRONMENTAL PERSPECTIVE 235

by both individual and population risk, or to take still other factors into account.

But, these are plainly matters that do not affect fundamental policies; they are important only at the margin. In general, pollution-control programs aim to eliminate all significant pollution as quickly as the progress of technology allows, and that aim is not likely to change soon. Perhaps it is worth revisiting that fundamental purpose periodically, but each time we return to it, it seems to be affirmed more strongly.

Notes

1. See CERCLA § 121(d) (2) (B) (ii); 121(d) (4).

2. Holmes, O. (1881) The Common Law 1.

3. The United States Public Health Service provided assistance of about $5,000,000 per year to local governments for solid waste disposal research from the early 1950s under authority of the Public Health Service Act. 42 U.S.C. §§ 241 and 261(a) (public health research and vector control); see Kovacs and Klucsik (1976), *The New Federal Role in Solid Waste Management: The Resource Conservation and Recovery Act of 1976*, 3 Colum. J. Envtl. L. 3:205. Modern statutes begin with the Solid Waste Disposal Act of 1965, Pub. L. No. 89-272, tit. II, 79 Stat. 997 (assistance to states to develop solid waste disposal plans), *amended by* Resource Recovery Act of 1970, Pub. L. No. 91-512, 84 Stat. 1227 (guidelines and grants for demonstration facilities), *completely revised by* Resource Conservation and Recovery Act of 1976, Pub. L. No. 94-580, 90 Stat. 2795 (comprehensive regulatory scheme for waste disposal), amended by Quiet Communities Act of 1978, Pub. L. No. 95-609, § 7, 92 Stat. 3079; Solid Waste Disposal Act Amendments of 1980, Pub. L. No. 96-482, 94 Stat. 2334; Used Oil Recycling Act of 1980, Pub. L. No. 96-463, 94 Stat. 2055; Comprehensive Environmental Response, Compensation, and Liability Act of 1980, § 37, Pub. L. No. 96-510, tit. III, § 307, 94 Stat. 2767; Hazardous and Solid Waste Amendments of 1984, Pub. L. No. 98-616, 98 Stat. 3221.

4. 42 U.S.C. §§ 6921-6939a.

5. Bok, E. (1921). The Americanization of Edward Bok. pp 255-256.

6. Id. at 256-267. Bok wrote about himself in the third person.

7. Id. at 256.

8. See Gilliam, A. (1979). Voices for the Earth: A Treasury of the Sierra Club Bulletin xix-xxi, pp 499-500.

9. See, e.g., Novick, S. and D. Cottrel (Eds). (1971). Our World in Peril, an Environment Review.

10. See, e.g., Anderson, F., A. Kneese, R. Stevenson, and S. Taylor, (1977). Environmental Improvement Through Economic Incentives, pp. 4-6, 41-45; Thompson, D. (1973). The Economics of environmental protection. pp 8-11.

11. Commoner, B. (1971). The Closing Circle. pp 295-296.

12. See Kovacs and Klucsik, *supra* note 351, at 257-260.

13. RCRA § 1004(14), 14 U.S.C. § 6903(14) (definition of *open dump*).

14. See, e.g., Federal Water Quality Administration, Department of Interior (1970). *Clean Water for the 1970s: A Status Report*. pp. 16-17, 23.

15. Id. at 16.

236 INSURANCE AND RISK MANAGEMENT FOR HAZARDOUS WASTE

16. See, e.g., S. Conf. Rep. no.; 1236, 92d Cong., 2d Sess. 116 (1972) *reprinted in* 1 Committee on Public Works, a Legislative History of The Water Pollution Control Act Amendments of 1972 299 (1973) [hereinafter cited as Legislative History]; S. Rep. no. 414, 92d Cong., 1st Sess. 52-53 (1971), *reprinted in* 2 Legislative History, *supra*, at 1470-71. See also Legislative History 1, *supra*, at 275 (remarks of Representative Kemp) (noting that groundwater was being given the same emphasis as surface water "for the first time in history").

17. See S. Rep. no. 414 at 73, *reprinted* in 1 Legislative History, *supra* note 364, at 1491.

18. *Id.* at 98, *reprinted in* 1 Legislative History, *supra* note 364, at 1513 (supplemental views of Senator Dole).

19. The question is whether groundwater was included within the definition of waters of the United States. Eckert, (1976) *EPA Jurisdiction Over Well Injection Under the Federal Water Pollution Control Act*, Nat. Resources Law. 9:455, 456-458 (citing cases that support the proposition that FWPCA jurisdiction could include groundwater if underground waters would "flow into or otherwise affect surface waters").

20. Compare United States Steel Corp. v. Train, 556 F.2d 822, 851-53 (7th Cir. 1977) (EPA may regulate disposal wells under the Clean Water Act's § 402 permit provisions) with Exxon Corp. v. Train, 554 F.2d 1310 (5th Cir. 1977) (deep disposal well not required to obtain EPA permit). See also Eckert, *supra* note 367 (analysis of cases discussing extent of EPA jurisdiction over groundwater).

21. Pub. L. No. 93-523, §§ 1421-24, 88 Stat. 1660, 1674-1680 (1974) (current version codified at 42 U.S.C. §§ 300h-300h-4).

22. The 1972 amendments were, properly speaking, the Federal Water Pollution Control Act Amendments of 1972, Pub. L. No. 92-500, aa 2, 86 Stat. 816. The present form of the statute is commonly referred to as the "Clean Water Act" however, and to avoid confusion is referred to by this designation.

23. Clean Water Act § 208(b), 33 U.S.C. § 1288(b) ("planning process").

24. *Id.* 208(b)(2)(J) & (K), 33 U.S.C. § 1288(b)(1)(B)(2)(J) & (K).

25. See *id.* 304(a)(1)-(6), 33 U.S.C. § 1314(a)(1)-(6); Exxon Corp. v. Train, 554 F.2d 1310, 1325-26 (5th Cir. 1977).

26. See *supra* note 414 and accompanying text.

27. Id. See Wilkins (1980), *The Implementation of Water Pollution Control Measures — Section 208 of the Water Pollution Control Act Amendments*, Land and Water L. Rev. 15: 479, 480 (1979) (ineffectiveness of Section 208 attributable to Congressional naivete); Comment, *Enforcement of Section 208 of the Federal Water Pollution Control Act Amendments of 1972 to Control Nonpoint Source Pollution*, Land and Water L. Rev. 14: 419, 446 (section 208 not effective to control nonpoint source pollution due to EPA unwillingness to compel production of state programs); but see Mandelker (1976), *The Role of the Comprehensive Plan in Land Use Regulation*, Mich. l. Rev. 74: 899 (Section 208 is a useful planning tool).

28. See *supra* note 373 and accompanying text.

29. But see Train (1975), *The EPA Programs and Land Use Planning*, Colum. J. Envtl. L. 2: 255 (former administrator of EPA argues for rational land-use legislation to integrate all environmental laws).

30. Pub. L. No. 94-580, § 2, 90 Stat. 2795 (1976).

31. See *supra* note 350.

32. EPA could enforce the section 208 planning requirements only by withholding financial assistance; see Natural Resources Defense Council v. Costle, 564 F.2d 573, 580 (D.C. Cir. 1977); the Agency could enforce RCRA directly. See RCRA §§ 7002-7003, 42 U.S.C. §§ 6972-6973 (1982).

AN ENVIRONMENTAL PERSPECTIVE

237

33. See Kovacs and Klucsik, *supra* note 351 at 216-259.

34. See F. Anderson, A. Kneese, P. Reed, R. Stevenson and S. Taylor, *supra* note 358 at 3-4.

35. Kovacs and Klucsik, *supra* note 351, at 259.

36. See, e.g., CONG. REC. 122: 21393-21401 (1976) (floor debate on the Solid Waste Utilization Act of 1976, S. 2150, introduced by Senator Randolph of West Virginia). Although Senator Randolph's bill had many of the elements of the final statute regulating hazardous waste, the floor debate was almost solely concerned with amendments proposed by Senator Hatfield to ban disposable beverage containers. *Id.* at 21404-21728. There was no special discussion of hazardous wastes nor groundwater protection, although these were the focus of the final legislation. Aside from the interest in beverage containers, the Senate was absorbed by the Toxic Substances Control Act, Pub. L. No. 94-469, 90 Stat. 2003 (1976), *codified as amended* at 15 U.S.C. §§ 2601-2629, then under consideration by another subcommittee.

37. Kovacs and Klucsik, *supra* note 351, at 216-220.

38. Pub. L. No. 94-980, 90 Stat. 2795 (1976), *codified as amended at* 42 U.S.C. §§ 6901-6987 and 9007-9010.

39. RCRA §§ 3004-3005, 42 U.S.C. §§ 6924-6925.

40. *Id.* § 3005, 42 U.S.C. § 6925.

41. *Id.* § 3002(5), 42 U.S.C. § 6922(5).

42. See generally House Comm. on Interstate and Foreign Commerce, Subcomm. on Transportation and Commerce, Staff Materials Relating to the Resource Conservation and Recovery Act of 1976 94th Cong., 2d Sess. (1976) [hereinafter cited as Materials]; see also Cong. Rec. 122: 32597 (1976) (opening statement of Representative Rooney, floor manager of the House bill) ("At present EPA lacks the authority to protect health and the environment from those often lethal substances indiscriminately dumped in the nearest landfill without regard to the presence of underground water supplies or the danger of poisoning by contact"); J. Quarles, Federal Regulation of Hazardous Wastes xv, xix and 15 (1982).

43. RCRA §§ 3001-3013, 42 U.S.C. §§ 6921-6934.

44. The contradiction stemmed primarily from the directive that EPA was to protect health and environment, which, if carried to its logical conclusion, could require the prohibition of land disposal of hazardous wastes. Yet such disposal methods were plainly intended to continue under RCRA. See *infra* notes 397-402 and accompanying text.

45. See Materials, *supra* note 390, at 39-41.

46. See *supra* note 390; H.R. REP. NO. 1491, 94th Cong., 2d Sess. 3, 9-12, *reprinted* in 1976 U.S. Code Cong. and Ad. News 6240, 6246-6250 (House bill basis of final compromise with Senate).

47. RCRA §§ 3001-3005, 42 U.S.C. §§ 6921-6925.

48. *Id.* § 3004, 42 U.S.C. § 6924.

49. Ruckelshaus, W. (1985). *Risk, Science, and Democracy*, ISSUES IN SCI. AND TECH., Spring 1985, at 19.

50. In Industrial Union Dep't, AFL-CIO v. American Petroleum Inst., 448 U.S. 607 (1980), a plurality of the Supreme Court raised the interesting question of whether there was not some level of risk so low that it would be unreasonable to regulate absent express direction from Congress. A minority thought such regulation to be unconstitutional. *Id.* at 672-676.

51. RCRA § 3005, 42 U.S.C. § 6925.

52. *Fed. Reg.* 46: 28314 (1981).

53. Clean Air Act § 112, 42 U.S.C. § 7412; see Ruckelshaus, *supra* note 48, at 21-22. For a brief account of EPA's vacillations, see Smith, (1982) *EPA's Permitting Requirements for Land Disposal Facilities*, Nat. Resources L. Newsletter 15:1.

238INSURANCE AND RISK MANAGEMENT FOR HAZARDOUS WASTE

54. Hazardous and Solid Waste Amendments of 1984, Pub. L. No. 98-616, § 221, 98 Stat. 3221, 3248-51 (codified at 42 U.S.C. § 6921(d)).

55. The Reagan Administration effort in 1981 and 1982 to reorganize EPA and to conserve funds for hazardous waste cleanup produced a spectacular confrontation with the Democratic majority in the House of Representatives. EPA Administrator Anne Gorsuch was forced to resign and hazardous waste program chief Rita Lavelle served a prison sentence for perjuring herself before a Congressional committee. The scandals attracted public attention and gave irresistible force to Congressional proposals for strict regulation of hazardous waste. See, e.g., Lash, J., K. Gillman and D. Sheridan (1985), A Season of Spoils; Burford, A., and J. Greeya (1986), Are You Tough Enough?

56. Levine, A. (1982). Love Canal: Science, Politics, and People. pp. 2, 16-21.

57. *Id*. at 11-15.

58. *Id*. at 19. The U.S. Army was included among those who had allegedly disposed of materials in Love Canal, although this was denied by the Pentagon. *Id*. at 25.

59. Clean Water Act §§ 311(f)-(i), 33 U.S.C. §§ 1321(f)-(i).

60. *Id*. § 311(b), (c), 33 U.S.C. § 1321(b)(c).

61. See Stever, D. (1986). The Law of Chemical Regulation and Hazardous Waste § 6.02.

62. See, e.g., *cong. Rec.* 126: H9437 (daily ed. Sept. 23, 1980) reprinted 1 Superfund: a Legislative History 111 (H. Needham and M. Menefee (Eds.) 1984) [hereinafter cited as CERCLA Legislative History] (remarks of Mr. Stockman that only 65 of surveyed sites required response); Kovacs and Klucsik, *supra* note 351, at 212-13.

63. Pub. L. No. 96-510, 94 Stat. 2767 (codified as amended in scattered sections of 26, 33, 42 and 49 U.S.C.). CERCLA was passed a scant two days before the House was scheduled to adjourn. Senator Randolph remarked: "I am disappointed that such an important bill to help solve such a pressing problem must be addressed in the last days of Congress." *Cong. Rec.* 126: 30930 (1980).

64. See Cong. Rec. 126: H11795 (daily ed. Dec. 3, 1980) *reprinted* in 1 CERCLA Legislative History, *supra* note 452, at 8.

65. See Novick and Stever, *supra* note 21, at § 13.04.

66. About 30 were actually filed before the statute was enacted. See D. Stever, *supra* note 409, at § 6.02.

67. CERCLA § 107, 42 U.S.C. § 9607; see generally 1 CERCLA Legislative History, *supra* note 410, at 163-361 (liability); Novick and Stever, *supra* note 21, at § 13.06.

68. See *supra* note 400 and accompanying text.

69. See 40 C.F.R. § 261(a). EPA's small-generator exemption was superceded by the Hazardous and Solid Waste Amendments of 1984, Pub. L. No. 98-616, § 221, 98 Stat. 3221, 3248-51 (codified at 42 U.S.C. § 6921(d)). The exemption as it had existed was recognized as one of many gaps in the RCRA requirements. See H.R. Rep. No. 198, 98th Cong., 2d Sess. 19-20, reprinted in 1984 U.S. Code Cong. and ad.News 5576, 5578.

70. Hazardous and Solid Waste Amendments of 1984, *supra* note 417.

71. See *id*. 101(b), 98 Stat. 3224 (codified at 42 U.S.C. § 6902) (national policy is to reduce or eliminate hazardous waste as "expeditiously as possible"); See also id. § 201(a), 98 Stat. 3226-27 (codified at 42 U.S.C. § 6924(c)) (bulk or "noncontainerized liquid" not to be placed in any landfill, effective six months from the date of enactment; EPA was given 15 months to promulgate final regulations to "minimize" containerized as well as "free liquids"); §§ 201(d) & 213, 98 Stat. 3227, 3241-42 (codified at 42 U.S.C. §§ 6924(d) & 6925(c), (e) (schedules for terminating "interim status," issuing permits to existing facilities; banning land disposal of certain wastes); 201(a), 98 Stat. 3228-29 (to be codified at 42 U.S.C. §§ 6924(e)

AN ENVIRONMENTAL PERSPECTIVE 239

and (f)) (disposal of dioxins into deep injection wells prohibited unless EPA determines that such is not harmful to health).

72. Hazardous and Solid Waste Amendments of 1984, § 221(a), 98 Stat. 3248 (codified at 42 U.S.C. § 6921(d).

73. See, e.g. *id*. § 402, 98 Stat. 3271 (codified at 42 U.S.C. § 6973(a)) (imminent hazard); United States v. Northeastern Pharmaceutical and Chem. Co., 25 ERC 1385 (8th Cir. 1986).

74. Pub. L. No. 99-499, 100 Stat. 1613 (1986).

75. See *supra* text following note 65 (imminent hazard lawsuit campaign).

76. See CERCLA § 105(a)(7), 42 U.S.C. § 9605(a)(7) (remedial actions must be "cost-effective").

77. See 40 C.F.A. § 300.68 (1984), amended by *Fed. Reg.* 50: 5861, 5905-5907 (1986).

78. See Environmental Defense Fund v. EPA, No. 82-2234 (D.C. Cir. 1982); Fed. Reg. 50: 47912, 47917 (1985).

79. See, e.g., Sen. Moynihan's remark that EPA was "simply digging up toxic waste in one place and planting it in another" *Cong. Rec.* 131: § 11864 (daily ed. Sept. 20, 1985); *Cong. Rec.* 132: H9603 (daily ed. Oct. 8, 1986), remarks of Rep. Henry concerning lack of permanent remedies.

80. See generally, S. Novick, D. Stever and M. Mellon (eds.) (1987). *Law of Environmental Protection* 1: § 3.

81. *Id.*, §§ 2.05-2.06.

82. See Industrial Union Dep't, AFL-CIO v. American Petroleum Instit., 448 U.S. 607 (1980); Natural Resources Defense Council v. EPA, F.2d (1987) (*en banc*).

83. American Textile Mfgr. Instit. v. Donovan, 452 U.S. 490 (1981).

84. See Natural Resources Defense Council, Inc. v. EPA, *supra* note 82.

Summary: Sheldon Novick

I will focus on what seems to me an area of perhaps fundamental and certainly principled disagreement between environmentalists and the regulated community. Environmental statutes, in general, now set a goal of eliminating all significant pollution risks from the environment. *Significant risk* is defined differently in different settings, but what's consistent among the statutes — and we're talking about all of the environmental statutes administered by EPA, not just the hazardous waste laws — is that the significance of risk is determined without taking into account the cost of achieving that standard. By and large, that's okay with environmental groups and organized environmental organizations. That aspect of federal law is fine, and that goal is a reasonable one, so long as it's not required to be met with undue speed.

In my paper, I sketched briefly the history of how that goal came to be embedded in the hazardous waste statutes. It is the residue of a period of optimism in the 1960s and early 1970s when the federal government announced that one of its very ambitious goals was the goal of doing away

240   INSURANCE AND RISK MANAGEMENT FOR HAZARDOUS WASTE

with pollution. That was a bipartisan program. President Nixon was one of the more aggressive environmentalists of that age, and the bills that we have now were in fact drafted by a Republican administration in the early 1970s. It was a time of real optimism and technological ambition. It was thought then that the United States could direct itself toward simply eradicating many of the ills of American society. I don't know that we have outlived that kind of optimism. Yesterday the President announced that it is federal policy that this country has zero tolerance for drugs. I didn't notice any cost–benefit analysis there. That's how we do things in this country. We take aim at evils and we announce that they are to be eradicated. That's what we've done with pollution.

For various historical reasons — some of them good, some of them bad — pollution has been made one of the evils on our agenda for eradication, along with drugs, crime, and poverty. We have learned from the history of the other statutes that these goals are firmly held and cannot be explicitly compromised. It's as if in the war on drugs we'd announce that it's okay for one school child in ten to be addicted to cocaine because that's okay on a cost–benefit basis. That can't be announced as a goal. The only kind of goal one can have with a war on drugs — once there is a war — is to do away with drugs eventually, and if that can't be done soon, to go on moving in that direction and never expressly sacrifice the goal. Of all the compromises that are made, many are implicit and most of them have to do with timing and scheduling, the things that we will do first and the things that we will have to go on living with for a while, perhaps indefinitely. We have found that those compromises have indeed been made relatively tacitly in the pollution control statute.

In 1972, at a meeting just like this, people got together and said, "My God, if we have to comply with the Clean Water Act it will cost billions, hundreds of billions" — and we hadn't got to trillions yet. The express goal of the Clean Water Act was, remains, and was reaffirmed last year by Congress to be the elimination of all water pollution. The Clean Air Act's goal is the elimination of all significant pollution in the air. These are express goals. The two goals of hazardous waste regulation, however distant they may be, are to (1) eliminate all significant contamination of groundwater by wastes, and (2) eliminate land disposal of waste as soon as that can be done.

The *as soon as* is important. Many of the tacit compromises have already been made in the hazardous waste area, and it is probably counterproductive to go back and keep hammering on the goal, which, to a great extent, represents the national consensus. Seven years ago there was an effort to reopen that national consensus and to try to reexamine the goal

AN ENVIRONMENTAL PERSPECTIVE 241

of having, as a federal government program, the elimination of pollution. We all know how counterproductive that exercise has been. It is very distressing to see all over the country, now, an effort to reopen these topics of really fundamental national consensus that have been fought over for 20 years or more, and to polarize the discussion into a question of whether these are in fact national goals or national purposes.

Once we accept them as goals, however, there's a good deal to talk about and a good deal that can be done. Aside from the question of how ultimate cleanup standards are to be set, I didn't hear a lot of things that environmentalists would have disagreed with during this conference.

Having said that in my paper, I sketched out some areas in which risk assessments might be used from an environmental perspective to improve the functioning of hazardous waste programs. Those are consensus items, and Marcia Williams will talk some more about the areas in which it appears risk assessment can be used to improve the effectiveness and the functioning of hazardous waste programs in order to reduce the friction and make the system work more sensibly and more rationally, and improve the consensus among the people who indeed have to carry out these programs. They are optimistic programs, and they require general agreement and cooperation if they are going to work at all.

Chairperson's Remarks: Marcia Williams

Last night we spent the first 20 minutes trying to solve the problem of why it is that organized environmental groups don't come to conferences like this. We acknowledged that their participation would be helpful. We decided that one of the problems we were really trying to solve was how to get public participation in these kinds of issues. Then we got into the issue of whether there is a problem with the environmentalists' view of risk assessment. This, in fact, is the view that sees risk assessment often as undoing things rather than setting a proper framework for deciding things. Environmentalists tend to see pollution as morally wrong and to set a goal of zero risk. Then when we try to engage environmentalists in an acceptable risk argument, we've immediately lost them.

So again we came to a framing of our problem that I would express as follows: "How can we get the public involved more effectively in reaching acceptable solutions?" This does not necessarily involve discussing what acceptable risk is. We came up with nine action items, some of which are more specific than others.

(1) We should try to do a better job of coming up with a way to integrate

242   INSURANCE AND RISK MANAGEMENT FOR HAZARDOUS WASTE

some of the laws. By that we do not mean coming up with a single environmental statute, which everybody has talked about for 15 years and that we all can agree is never going to happen. But there is a way, as each law comes up for reauthorization, to put into those laws some means for achieving consistency and equivalence between them, so that we do not have different decision frameworks when we try to solve a problem under the Clean Water Act than when we try to solve the same problem under RCRA.

(2) We should try to convey to the public, more effectively, the kinds of real choices that are available. The one thing that the public can accept or begin to accept, if we do this in a constructive way, is that they have to do (a) or do (b) or do (c) in order to solve this problem, but they must do one of these. They can't just close their eyes and make the problem go away. We need to focus more on the communication of the choices.

(3) Along with the above, there is a strong education component, which means we have to do a better job of providing information to make the choices. We touched on a number of things here — for instance, the concept of perhaps labeling products for recycling that are recyclable, so that consumers have an ability to help make choices for themselves. I think there are other things, too. We touched on some other kinds of modes of getting information out, including things like working with the media and trying to provide credible sources of information for people, sources that are better than those available today.

(4) We identified the need for developing institutions whose purpose is to generate and then share information in a neutral and unbiased kind of way that will be better accepted.

(5) We had a very good discussion on the fact that it's probably much easier to get the public involved at a local level than at a national level. The idea is that we really need to get more involvement. We need to deal more extensively with local problems and then come up with local risk assessment and local solutions, and try to get those in place. The idea here is that if we get some successes at that level, we will begin to get a public that can then begin to participate more effectively at some of the national policy levels.

(6) Our belief as a group is that we need to come up with ways to provide assistance for doing more of this local decision making. We need to search for processes that empower local people. One of the big problems arises when people get angry and frustrated and withdraw from a process — it's a reaction because they do not have any power. I think Bill Ruckelshaus and some of the efforts he's making with siting are helpful. He's changing the discussion from "We're going to come to your town to site," to "We'll

AN ENVIRONMENTAL PERSPECTIVE

243

be glad to talk to you about how we build the plant and see if you want us in your community." This gives people a sense of choice as to whether they want to have a facility in their neighborhood or not. That's an example of the kinds of thing that we can do to begin to empower local people. There are many other examples.

(7) The next thing we came up with is the fact that we do need to try to come up with better ways to get data before laws are put in place. Once they're in place, it's very hard to change the basic philosophy attached to those laws. Right now the coalition on SARA is an example of such an effort, but obviously the ideal situation would be to have those kinds of information-gathering efforts long before we are in quite as deep a hole, or quite as reactive a position.

(8) We pointed out the need to come up with more technology development, in that we really ought to be looking for legislative solutions that better foster new technologies than some of the solutions that we have. Right now, if we look at many of the statutes and the way that they push to develop better technology, we can see that they're not really effective. In many ways, particularly if you look at the RCRA program, such statutes bet on today's technology, not on the ability to develop better technology in the future.We didn't come up with specific solutions, but some good policy thinking could be done in this area.

(9) The last point that emerged from our discussion is trying to work out ways to help develop more choices for the public. In other words, people may have to pick one of three or four things to do, but let's see what we can do to try to develop additional choices for them to have in the tool bag when they have to pick the one they want.

9 AN INDUSTRIAL PERSPECTIVE ON AN INTEGRATED WASTE MANAGEMENT STRATEGY

Isadore Rosenthal and Lynn Johnson

Introduction

This paper presents one company's approach to developing and implementing an integrated industrial waste management strategy spanning all wastes and all segments of the corporation in all geographical areas. The discussion will cover three major areas:

(1) Our view of the past, our future goals and the policy that we hope will guide us to these goals.
(2) Some specific programs which implement our goals in regard to both remediation of past practices and conducting current operations.
(3) Changes government could make to improve protection of the environment.

Background

Rohm and Haas has learned through the school of hard knocks that obeying current laws and regulations is not good enough in today's world. Em-

245

ploying good practice and complying with the law and regulations in past years does not protect the company today from either the wrath of the public or of the very governments who wrote the laws and regulations which they directed us to obey. The classic case is Superfund where in the 1960s and 1970s we used sites for waste disposal that were recommended to us by government agencies, licensed by government agencies, inspected regularly by government agencies and routinely given a clean bill of health. Today we are being held responsible for those sites because the owner and operator failed to provide the service we paid for and the government changed its standards and laws post facto. Perhaps even worse than the economic consequences of these post facto changes is the implication that our past actions were a result of moral failure. The public and the same governments who recommended the sites to us and regulated them now wish to treat us as criminals.

Similarly, in the tort field Rohm and Haas is not immune to litigation and extremely costly settlements, although we have obeyed the law as it existed when we acted. Had we been wiser 30 years ago, we would have anticipated the consequences and the present public concerns about landfills and toxic substances. We might have avoided some of the difficulties we and others face in remediation of landfills and in toxic tort actions alleging injury from past waste disposal practices. We were not wise enough, and we have not avoided the consequences.

Future Goals and Policy

Today the company recognizes that it must be ahead of society's rising risk perception/acceptance/standard-of-behavior curve. Our goal, therefore, is to conduct our operations in such a manner (prudently) that our principle stakeholders (employees, neighbors, customers, and shareholders) will feel now and in the future that the impacts on them and their environment are at least not unacceptable. Our aim is higher than just *tolerance;* we would like to be thought of as a valued corporate citizen in the community. We have tried to incorporate this thinking into a policy that will serve to guide us in making decisions that are consistent with this goal. The basic elements of our policy are

(1) The company is committed to effective environmental control of its operations worldwide as an intrinsic aspect of its business.
(2) The company will take prudent action to ensure that its operations are free from unacceptable risk.

AN INDUSTRIAL PERSPECTIVE

247

(3) Primary responsibility for executing our environmental control program rests with the management at each location.

There are two terms here, *prudent action* and *unacceptable risk,* that deserve more discussion. They are the guts of our policy and delineate what we believe our stakeholders expect of us. They represent the values that we believe will serve us better than reliance on law alone.

To us, prudent action means:

- Discovering safety/health/environmental risks
- Assessing safety/health/environmental risks
- Compliance with the law and good practice
- Rejecting unacceptable risks
- Reducing risks where viable
- Informing of risks
- Auditing

Prudent action means discovering safety, health, and environmental risks. It is active rather than passive. For example, it means taking those steps necessary to find and quantify our emissions by calculation or measurement. Prudent action means assessing risks. We must evaluate those emissions and the resultant concentrations in the environment against a risk standard. In most cases, communities around the world have not set numerical measures of risks to which we can compare our operations. Therefore Rohm and Haas has often found it necessary to create internal standards against which to compare its operations. Standards are set on the basis of the best information on risks currently available now. Future discoveries may unearth new information that will show that risks existed of which we were not aware. When and if this occurs and if injury results, we may have to remedy the injury, but we hope to avoid charges of negligence.

Prudent action calls for informing the people who may be at risk from our actions. This is relatively easy with our workers, who are paid to listen and learn. The public is much more difficult to reach. We have used advertisements in local papers, open-house days, and speakers at local functions. Some plants have public reading rooms available with information on the plant. Some have placed information in public libraries. Three of our plants now have citizen advisory committees. Permits and emissions documents are, of course, public information. We have also used press releases when, for instance, a groundwater study was completed.

We, of course, must comply with the law of the country in which we

248 INSURANCE AND RISK MANAGEMENT FOR HAZARDOUS WASTE

are operating. Laws and regulations around the world have very different bases: some are technology-based, some are health based, some are based on environmental considerations, and some require arbitrary percentage reduction. Our locations must know the laws under which they operate and must have an ongoing mechanism to learn of new laws and regulations. In the U.S. this is a massive undertaking; the process of compliance is sometimes more costly than the compliance requirement itself. RCRA permits for drum storage pads are a good example of this topsy-turvy result.

Prudent action means reducing risks further than required by law and good practice where this can be done without significantly affecting the economic viability of the operation. There is no simple formula to guide us in achieving this goal. The engineer must develop information of risk as a function of costs. The business manager must develop the impact of these incremental costs on the long-term viability of his business. The project manager must develop a consensus based on the impact of this information in the light of the company's values.

Prudent action means rejecting *unacceptable risks*. An unacceptable risk is a risk that is:

- Significantly greater than law or good practice
- Significantly greater than similar activity elsewhere in Rohm and Haas
- Significantly greater than similar activity outside Rohm and Haas
- Likely to damage company relationships with employees or community

Clearly our company is not in a position to decide what level of risk a citizen in a host community will find acceptable. However, we have decided that risks to our stakeholders greater than those described above are unacceptable to us based on business and value considerations, and we hope that lesser risks will at least be tolerable to our stakeholders in the light of the benefits we bring to the community. By making information available in advance of an incident, we hope to promote discussions with other stakeholders in regard to potential risks.

Finally, prudent action includes auditing ourselves to assure that what we intend to happen really does. We have an extensive set of audit programs covering safety, medical, industrial hygiene, and environmental programs. Audits are regularly scheduled, and they detail deviations from our corporate standards and areas where improvements can be made. Programs for correcting these deviations are negotiated with local management and are monitored quarterly until completion.

A copy of the policy statement is attached as Appendix 9-1.

AN INDUSTRIAL PERSPECTIVE

Management System

There are altruistic reasons why we are interested in good environmental control, and there are also very practical reasons; it is good business to be careful with the environment. Our corporate management believes that good environmental control is an intrinsic part of total quality, and we are committed to quality as the way to build a good business. Our environmental policy and the programs that flow from it have the backing of the Board of Directors and the Chairman of the Board through the President and the officers right down the chain of command. Management has been fairly successful in convincing everyone throughout the corporation that the company really means to have *quality* safety, health, and environmental programs, and is willing to bear the costs required to establish and maintain them. Ten years ago, some folks would shrug their shoulders and say, "Ah, when management sees the bills required to achieve performance better than that of our competitors or better than the law requires, they will quit this nonsense." Well, it is not true; when management sees the bills, they examine them critically, wince, and pay them. High as these bills may be, they pale when compared to the costs of remediation that are assessed against us under the new doctrines of law, especially joint and several liability. As anyone knows, pursuit of quality leads to lower ultimate costs because it reduces the amount of *rework* (e.g., remediation). Our line troops have come to believe in our general policy; in fact they are now sometimes ahead of the corporate gurus.

Rohm and Haas is organized into four geographical regions around the world. At the regional level, we have a coordinating staff consisting of one or a few professionals who provide expertise and coordination to guide plant programs. At the corporate level, we have a small staff responsible for setting policy, gathering information on corporate-wide programs for management, and auditing our facilities.

Line management at the location is directly responsible for implementing our programs and for seeing that we obey the law. The line manager has an environmental expert to assist him — this ranges from a part-time person at small, simple facilities to a considerable staff at large plants.

The real action is at our production facilities. Plant management, engineers, foremen, and employees are at least as important as proper facilities. Unless the production people are committed to a quality environmental operation, it won't work. All of the same elements that enter into reducing the risk of a Bhopal-type accident exist with regard to routine environmental releases, though of course with much less dramatic consequences. Engineering a facility that is capable of meeting the envi-

250   INSURANCE AND RISK MANAGEMENT FOR HAZARDOUS WASTE

ronmental risk standards we have set for ourselves is a necessary condition. Operating it well over a long period of time is by far the harder task. One needs a process hazards plan that deals with items such as:

- Training
- Adequate monitoring devices
- Maintenance
- Hazard and risk reviews
- Investigation of accidents and near misses
- Housekeeping

For us, environmental control must be an intrinsic part of a total process control strategy that also deals with achieving quality products, accident reduction, productivity, and developing people. The sections above have discussed the basic approaches, policies, and management system with which we operate. These are not much different from those used by many corporations around the world. The following sections will discuss some of the specific programs that we have undertaken.

Programs

Waste control can be logically divided into the traditional areas of air, water, solid waste, and groundwater, and our corporate compliance programs are divided along those lines also. We have divided them that way because these are the routes by which emissions and discharges can leave our plant site and reach the environment or our neighbors. In addition we want to discuss our overall waste minimization program, which will deal with all routes of waste emission.

Air Guidelines Program

The air guidelines program is one we began several years ago. The basic requirements are simple. A plant must have an inventory of its emissions to the air — an inventory by chemical for a specific list of chemicals. For each emission, the ambient concentration outside the plant is calculated using computer air-dispersion models.

For each chemical on that list, we have created an ambient air guideline. The ambient air guideline is a numerical value that is applicable in the community; it is a number derived from the toxicology data base available

AN INDUSTRIAL PERSPECTIVE 251

for that compound, using some extremely conservative assumptions and safety factors. The calculated community concentration is compared to the guideline and if the guideline is exceeded, additional controls are installed. As process changes are made, each plant must recompute emissions or remeasure the emissions and check them against the guidelines. The facility must also make an effort to establish whether there is also a significant emission of one of its chemicals from a nearby noncompany source and take this into account. We started out using the ambient air guidelines established by the city of Philadelphia for a list of 90 compounds. We were one of the principle participants in assisting the city in their program of establishing these guidelines and enacting them into law. About 30 of these were chemicals used around our company, and they provided the start for our program. A year ago we decided to expand the list to about 125 compounds and to review and revise the initial set. This is being done with the assistance of an outside consultant.

We find this system works very well; it gives our plants a numerical standard against which to compare their performance. It is an internal standard that we believe should not present an unacceptable risk to the community. It is not necessarily a standard accepted by the community, but it changes the boundary conditions of debate from a general feeling of *dread* that we are subjecting the community to an unknown risk to discussions about the numerical value of the standard and the details of the methodology used to derive it. Although we have had no community input on setting our newer standards, we do feel they have some credibility; there is a logical basis for each standard, and we think that if we are below that standard we will cause no harm to the community or the environment. In general, our new standards are stricter than those arrived at by the City of Philadelphia in a process in which the public and environmental activists had an opportunity to participate.

The air guidelines program applies worldwide, as do all of the company's health-based standards. We expect the guidelines to be evergreen. As new information becomes available, the guidelines will be changed.

Water Guidelines Program

We have begun a water guidelines program. This program is three years behind the air guidelines program and is in its infancy. We have divided this program into two categories: drinking water and aquatic life. We have set the aquatic life aside for the moment; we simply do not have the time and resources necessary to develop a set of internal aquatic-life protection

252   INSURANCE AND RISK MANAGEMENT FOR HAZARDOUS WASTE

guidelines that we feel would stand the light of day. For now, we are dependent on government standards in regard to protecting aquatic life, but we hope to get to this in another year or so.

On the drinking-water-guidelines front, we have established a screening value of one part per billion applicable to those plants whose discharges can reach drinking water supplies. There are about nine of our plants around the world that are upstream of drinking water intakes. Plants are measuring their discharges for over 100 compounds and comparing the results to the screening value of a part per billion at the intake of the nearest downstream drinking water supply. If the compound is measured or estimated using conservative techniques to be above one part per billion, then we will establish a limit for it. For some chemicals, EPA has established drinking-water criteria. They are generally in the range of a few parts per billion up to a few tens of parts per billion. Other governments have also established limits, and we may be able to use some of those. Where government has not established a limit that we believe is suitably protective, we will have to derive one of our own from the available toxicology data base.

Again, the philosophy is the same:

- Know what is being discharged
- Know where it goes
- Know what the concentration is going to be when it gets there
- Measure it against a standard or guideline
- Undertake corrective action when our internal standards are not met and
- Make the results of our program available to those potentially at risk.
  We expect first results of this program this year.

Groundwater Protection

Groundwater is another route by which chemicals might leave our plant site and reach our neighbors. Groundwater is out of sight, difficult to measure, and not well understood by most folks. Fortunately, in most cases groundwater moves very slowly, and there are mechanisms available that tend to attenuate spills that have seeped into the ground. Nearly five years ago, we set out to measure this route, quantify it, and monitor it. We established a program to define the geology and hydrogeology and to measure the water quality under each of our plant sites. From this information, we can predict where the groundwater is going and what risk, if

AN INDUSTRIAL PERSPECTIVE

253

any, is associated with that flow. The cost of this program ranges from a few tens of thousands to a few million dollars per plant site. To find out about underground strata, we hire a consultant who drills several wells on a site, takes core borings, has a geologist examine them, compares what he learns to what is known about the geology around the site, measures the water levels in the holes, and measures or calculates the flow rates — all standard hydrogeological techniques. We have now completed surveys at virtually all U.S. plant sites and most of our sites in Europe; we are starting on sites in Latin America this year, and about half the sites in the Pacific region have completed geology and hydrogeology studies.

The studies and the demographics of the area give us the data needed for risk analysis and risk management. We have done some of the risk analysis studies entirely in-house (generally the simple ones), but have also used some of the prestige consulting firms for others. We are not as expert in groundwater as we are in other pollution-control fields. Therefore, for each of our significant hydrogeological studies, we retain one of a number of independent academicians to assist us in the review of the contractor's results.

We are quite proud of this particular program. We have found only a few things that we need to worry about, even though all the sites are detectably contaminated by something. In our experience, if man has trod there, a chemical can be found in the ground underneath that arguably was not there before man came along. The contamination levels we have found are typically very low, and the rate of groundwater movement is very small. The water is typically moving to a local river or salt-water body at pollutant flow rates that are minuscule compared to our normal wastewater discharges. We think that, in almost every case, we have little or nothing to worry about from the standpoint of public or environmental health protection. In most cases, even before the report is completed, significant findings have been shared with local authorities and directly with the community where the local culture permits this. It is our intention to do this at all locations. Sharing data creates heartburn in some places, where such sharing of data is not the societal norm and is met with skepticism when it is done. Instead of saying "Thank you for sharing your data with us," some people say "What are you trying to hide?", "There must be something wrong if you are doing this sort of study," etc. We have had a few of these problems to deal with. Nevertheless, we feel it is the right thing to do. Groundwater is a resource to be protected, it is a route by which pollutants can reach the community, and it is a route that we must quantify and measure. The next step in our groundwater protection program is routine long-term monitoring to detect any significant changes that might occur.

254 INSURANCE AND RISK MANAGEMENT FOR HAZARDOUS WASTE

We are also extending this program to preventive measures. We have had spill-control plans in place at plants for many years to protect local waterways against spills. We are now refocusing these plans to ensure that they also protect against ground contamination. It does not take many pounds of a pollutant reaching the groundwater to make the water undrinkable.

Preventing future groundwater contamination has also led the company to institute a policy on underground tanks. The policy is simple: it says that we are not going to have them if we can avoid them. That would be a tall order for some companies that depend heavily on underground tanks. Two years ago, Rohm and Haas had about 100 underground tanks worldwide. By the end of a four-year period we expect to reduce that to about 20.

There are some situations where we are not going to be able to avoid having a tank underground. Those tanks left will be in vaults or they will be double-walled tanks. Along these same lines we are in the midst of developing a policy with regard to underground chemical lines and sewers. It is too early to tell where we will come out on this front, but our present thinking is going along the same lines we followed in regard to underground storage tanks.

Solid Waste Disposal

Solid waste disposal is another major area that has obviously received a great deal of attention. Earlier in this paper we noted that 20 years ago governments were telling us to put our wastes in landfills, telling us which landfill to use, and giving us permits. That has not proven to be satisfactory in the long run. Dealing with the problem of solid wastes has two dimensions: first, remediation of problems caused by past practices; second, what to do about current and future disposal needs. Regardless of the issues of equity or the rationality of public concerns about the size of the ecological problem associated with past landfill practices, it is a fact that deep-pocket companies associated with "poor" landfills are or will be held responsible for dealing with the problem. In this area our company philosophy is clear. Where a past landfill practice might cause ecological injury in the future or is judged unacceptable by present public standards, we will try to act proactively to address the problem. This we have done in some instances even when the best purely legal advice would have been to wait for the Superfund or some other government initiative to act. We have removed materials, organized study groups, built enclosures, supported Clean Sites,

AN INDUSTRIAL PERSPECTIVE

Inc., undertaken remediation measures, and communicated findings with the affected communities. By and large, we feel this proactive stance makes for good economic sense, besides building our credibility in the community. Unfortunately, we have specific instances in which our proactive actions have been blocked by the government and penalized by the courts, who, by refusing to acknowledge our voluntary contributions, have made us pay twice for the same cleanup. We face a long and continuing remediation effort, and we are concerned about the slow rate of progress being made by state and federal government Superfunds. Our policy will continue to be a proactive one to the extent the government allows.

Currently Rohm and Haas is doing all it can to minimize the use of landfills. We no longer landfill EPA hazardous wastes. Where we must use landfills, we treat them as toll contractors, inspect them before use, and recheck them every year.

Future Waste Disposal Programs

There is a need for a total waste control program; it must deal with the generation, emission, and disposal of wastes in all media in an integrated manner.

Our approach to future waste management is prioritized as follows:

- Elimination and reduction
- Recycle
- Destroy or treat to render it nonhazardous
- Landfilling and storage in facilities under our control or that are inspected by us to meet our standards.

Elimination and Reduction. Elimination is the first choice. Emphasis is put on low waste processes at the design stage, and existing processes are reviewed for changes that might eliminate wastes.

We are starting to handle wastes as if they are products. The company has a wide-ranging product quality-control program using statistical quality-control techniques, and these techniques are being applied to waste streams. We try to keep our waste streams in statistical process control, just as we do to the product that goes to the customer. Our managers are being taught to think of the waste stream as a product: we don't get any money for it, but it does have customers. The customers are our neighbors, and perhaps the government, and they care about the quality and quantity of our wastes. We must treat the waste stream as a product, keep it under

256   INSURANCE AND RISK MANAGEMENT FOR HAZARDOUS WASTE

control, and try to make a better product next year than we made this year because our customers have ever rising expectations. These ideas are beginning to flow around the company and will be part of our culture in a year or two.

Recycle.   If we cannot reduce waste (or once we have reduced it as much as we think we can reasonably do), the next step is some form of recycle or reuse. If we can find a use for it or recycle it to the process, that is the next most desirable thing to do with the waste.

Since one of the purposes of this conference is to suggest improvements in waste management, let us digress at this point and discuss one area in need of improvement. Recycle is quite easily done abroad, but here in the United States the government has chosen to regulate the recycle of wastes under RCRA. As any of you who have dealt with RCRA are aware, that program is designed to keep things from happening. The RCRA permit system does not work; we are eight years into the permit system, and not half of the permits applied for have been issued. It has been our experience with the very few permits that we have managed to get that there they are totally inflexible. Even the smallest of changes cannot be made without going through the whole permit process again. Equally frustrating, if a minor change, such as the name of the person handling RCRA waste, is not changed in the documents, we get fined. There is a great inhibition to doing anything that involves RCRA. Repeal of RCRA and starting over again on solid waste control is probably the best thing the government could do in encouraging recycling. Something short of that would be a drastic change in EPA's definition of *waste*. EPA has defined a waste to be whatever they believe it to mean today, and tomorrow it will mean whatever they believe it to mean tomorrow. This statement is a little facetious, but in fact, the definition is extraordinarily broad, includes lots of things that one would not normally consider to be waste, and continues to change with time. A simple definition, such as "a solid waste is a discarded material," would serve to greatly simplify the RCRA system and encourage recycle and reuse of materials by taking recycle and reuse out of the extremely cumbersome RCRA process.

Destroy or Treat.   If the material cannot be recycled and we cannot find a way not to make it, we proceed down the path to destruction of that waste. Incineration is the preferred alternative. Our typical solid and liquid process wastes are organic materials. We make few chlorinated or fluorinated materials, and our wastes tend to be carbon–hydrogen–oxygen mol-

AN INDUSTRIAL PERSPECTIVE

ecules. These materials are very amenable to destruction by incineration. We have a few incinerators internally, but at the moment we are dependent largely on contracting incineration outside. We are in the throes of trying to build an incinerator of our own, but we do not have a RCRA permit yet and probably won't for a year or two. If a waste cannot be incinerated, it possibly can be biotreated. Some of our major waste streams are biosludges from wastewater treatment plants. These are quite amenable to further biotreatment.

Landfilling. We generate a few waste streams that are inorganic and not suitable for any other use. They are going to be placed in landfills of some type now and in the long run. Landfill, or storage of any type, is our last choice for handling wastes. Where we must do so externally, we carefully inspect the facility and treat it as a toll contractor facility. We have focused our landfill reduction program on all process wastes worldwide with the exception of common trash. Common trash is not part of the program; each facility is permitted to dispose of common trash the way its local community does.

Waste Elimination

The company is planning a stronger emphasis on reducing wastes. In the short term, progress in reducing wastes can only come from doing a better job with the technology and processes in place. There is a limit to how much can be done in this fashion. The greatest payoff will come from new products and processes. The character and quantity of wastes produced by a new process will receive much greater scrutiny in the future, both because the costs of waste disposal are high and increasing and as a matter of principle. We shall emphasize waste minimization across the board, tying together air, water, and solid wastes in one waste minimization program. In jargon terms, this will be called a *cross-media waste-minimization program*. This particular concept is relatively embryonic at the moment, in terms of a corporate program.

Obviously, all waste-reduction programs depend on a good waste measuring system. We have had one in place tracking wastes to outside landfills and incinerators in the U.S. for several years. This has now been extended abroad, and we are trying to broaden it to cover all waste except trash. We are finding year-to-year comparison of results difficult, and clearly we

258 INSURANCE AND RISK MANAGEMENT FOR HAZARDOUS WASTE

must expend more effort in this area in the future if we are to have a successful program, and to know we have a successful program.

Other Programs

PCBs

There are some other programs that are worth mentioning. PCBs are an item of concern to all. We don't make them, but we use them, as everybody else does, in electrical equipment. We have been slowly phasing them out. This is one area where nature is taking a satisfactory course, and we have chosen to establish only one particular corporate target. We are eliminating PCB transformers in office buildings and production units where a fire could lead to huge cleanup expense and possibly loss of that production unit for a considerable time.

Major-Accident Prevention Program

Accidents of the Bhopal type (Hazardous Clouds) and Flixboro type (Explosive Cloud) are generally discussed under the rubric of safety, as opposed to accidental releases (spills) of the Sandoz/Swiss or Ashland Oil/Pittsburgh type, which are thought of as environmental problems and which we have discussed above. Nevertheless, major plant accidents of the Bhopal type do directly affect the community and are worth mentioning, since the same measures taken to prevent accidents that have acute effects also minimize the Ashland Oil type of spill event.

We have spent considerable effort on control of special hazard materials, particularly materials that can vaporize to form gas clouds or explosion hazards. We use a number of raw materials such as ethylene, propylene, and ethylene oxide that are hazardous materials. Immediately after Bhopal, we concentrated on reducing the risks associated with these operations using empirical methods.

While the empirical state-of-the-art approach served us well in the initial period of greater concern, we have since had the time to develop a more systematic/scientific program for dealing with this type of problem. We are now in the process of dealing with the hazards of handling and storing these types of materials at each of our plants. HAZOP studies, fault-tree analysis, and similar hazard and risk analysis techniques are in routine use.

AN INDUSTRIAL PERSPECTIVE

The path we are following in regard to the potential risk of injury from a major accident should be clear by now: quantify the potential risk involved in handling the material from an accidental release standpoint, and quantify the effect it may have on our neighbors. If there are significant potential consequences to members of the community, reduce that risk to levels that meet our guidelines and attempt to communicate that information to the community. Tied in with this Major-Accident Prevention Program (MAPP) is our joint effort with the community on emergency response systems. This combined communication/emergency response effort may be familiar to you under the Chemical Manufacturer's Association acronym of CAER. CAER (Community Awareness and Emergency Response) is in place in our company, and we have made significant progress in doing systematic risk and consequence analysis on all operations that have the potential for impact on the community. The advent of SARA Title III has lent a new dimension to this program, since the community may now ask the same questions we ask of ourselves.

Audits

No strategy would be complete without reviewing how we try to verify that we are doing what we think we are doing. We audit in a number of separate disciplines for compliance: medical programs, industrial hygiene programs, safety programs, and environmental programs, among others. We audit every facility, in each discipline, on average every two years, worldwide. The audit reviews compliance against regulations and laws applicable to that facility, and reviews compliance with corporate policies and programs. We do our environmental audits with environmental professionals from corporate and regional staff and staff from plants other than the plant being audited. The result is a simple list of action items that should be taken to improve the compliance program. We also examine management issues, soft issues, such as, "Is the plant prepared to comply with future laws, regulations and policies? Do they really understand what they are doing? Do they have the kind of systems in place that assure that they know what their discharges are? Are the systems sufficiently robust that we can have confidence in the data?" These are soft management issues and difficult to measure, but we try to assess them.

Neither our audit system nor our performance worldwide is up to the standards that we hold up for ourselves, but we are working on it and getting better each year. In our opinion, auditing is an essential element of managing.

Changes Government Could Make

One of the purposes of this conference is to suggest improvements government could make. Let us close by discussing some of the things that we think government might do to help us out.

In our opinion, government should be setting performance standards for use across the country. These standards should be numerical, applicable in the community, and set on a rational, scientific, health basis. This would tend to level out the playing field and stimulate business to come up with the most economical approach for achieving these standards. It would stimulate fundamental attempts to develop product/process ensembles that lend themselves to such goals. Congress and EPA have tended to avoid setting community standards. EPA has preferred to attempt to set equipment-based standards, so-called *best available technology* standards. BAT is a slippery concept in itself, and requires a tremendous amount of detailed engineering knowledge. Technology that is BAT for one process is not necessarily BAT for another process, or even for the same process operating under different circumstances at a different location. Nevertheless, that has been the thrust of our federal government. This policy has been a disservice to industry and the public. This policy has often required business to install technology where it would do no good, while at the same time ignoring problems on which business should have spent more time and effort. We would like to see the federal government set water-quality standards applicable out in the river and air-quality standards applicable out in the community. Given a firm standard that we can all accept and believe in, industry will figure out how to meet it, and if it really is necessary to protect public health, either we will meet it or we won't discharge it — we will go out of the business. The levels must be credible, of course. EPA has a tendency to pyramid safety factors, including hiding some of them in its development documents where they are hard to find. This can result in numbers that are not credible. Numbers that are not credible are no better than no limits at all, particularly when having set the number unrealistically, EPA backs off enforcement against politically powerful regions or business areas.

When there are several sources contributing to an air toxic or water-quality problem in a locality, then the community must decide how the allowed emissions will be distributed among the emitters. Whether this is best done by allotments on an economic basis, sold, done by grandfathering, or done via political power is not the subject of this paper. However, we believe that once given our allotment, we are much better at devising ways of achieving it economically than is the government. This

AN INDUSTRIAL PERSPECTIVE

is not to argue against the government per se; our plant people tell our central corporate groups the same thing when we offer solutions to their problems, rather than performance targets. In general, the person closest to the problem and responsible for it is in the best position to find the best solution.

We recognize that the situation becomes immensely more complicated when we deal with ubiquitous air pollutants such as SO_2, O_3, etc. However, here once again performance standards coupled with apportionments among industry segments appear to us to be a better approach than specified technology.

A second point is that EPA often regulates to keep an activity from occurring, rather than regulating to protect the environment. The RCRA permit system is a morass that almost doesn't work. The NPDES permit system for wastewater discharges is only a little bit better. We are 16 years into the NPDES permit system, and it still takes many months to negotiate a permit. Something that ought to be done in 10 days takes typically 10 months. RCRA permits take several years. The regulated community cannot respond and cannot do a good job of waste control if it cannot get permission to act. EPA should be a positive player, not a negative one. A permit application that meets the law and regulations should be approved in 30 days.

And that brings us to the third item: risk-reduction goals prioritized on a cost–benefit basis within available resources. All of us would like to aim for zero risk in regard to any particular activity. However, we do not have sufficient resources to do this for all risks, and we need to prioritize and set realistic goals across all media commensurate with the resources available. When we have reached the first set of goals — i.e., used \$200 million to achieve a risk reduction of $X\%$ for Y people — then we can set a stricter goal. EPA has a habit of biting off far more than it can chew, much less enforce. This behavior tends to muddy the waters, stymie EPA programs, and penalize firms that voluntarily try to comply with the letter of the law instead of waiting for the enforcement axe to fall. Under the RCRA program, for instance, EPA has defined solid waste to include recycled materials. The Agency had some good reasons for doing this, but the result has been to inhibit the recycle of hazardous wastes. We have on a couple of occasions found a nice, neat, unique way to recycle a hazardous waste at an outside company. The outside company's process fit very neatly with this bit of waste we had; our waste could substitute for a small part of their raw material at an attractive price. But the moment we talked about getting RCRA permits, the other company saw itself being branded in the community as a hazardous waste treatment unit. That ended all consideration

262 INSURANCE AND RISK MANAGEMENT FOR HAZARDOUS WASTE

of recycling our waste. The ease with which one acquires liabilities under Superfund has added another problem. Today, we think very carefully before sending any waste outside for recycling. If the firm doing the recycling goes bankrupt some day, we could be held jointly, severally, and strictly liable for any waste problem they ever caused from any operation, even if we conducted our usual toll contract review and accounted for the proper handling of our specific materials. EPA's expansive definitions have also brought in wastewater, and EPA is now regulating wastewater streams under both the Clean Water Act and under RCRA. This has been a duplication of effort from our viewpoint and a large drain on EPA's resources, which could have been better spent on regulating solid waste disposal.

We would like to see EPA do more basic research in such areas as atmospheric chemistry and less research on process control schemes. We have been campaigning against hydrocarbon emissions in this country now for over 20 years as a means of reducing ozone. We are little better off today, in terms of ozone in the troposphere, than we were 20 years ago, even though hydrocarbon emissions have presumably been reduced a great deal. Perhaps we don't know very much about how hydrocarbons, oxygen, and sunlight interact to form ozone, and we don't really know very much about how to control it. There is nothing to be seen on the horizon but more of the same old control schemes. Until we fundamentally understand what is going on, we will not see commensurate improvement in peak ozone concentrations.

We would also like to see EPA do more monitoring. The best controls in the world are useless unless we really know what is happening in the environment. An EPA monitoring network is an absolute necessity for gathering the basic data to prioritize the regulatory and standard setting process and to protect human health and the environment. We cannot afford to waste our resources in shotgun approaches requiring control equipment wherever it can be purchased. Both the community and industry need to know what we are accomplishing and what needs to be accomplished next. Long-term monitoring networks are the only way to obtain the basic truth data.

Government should establish a bias towards action by potentially responsible parties (PRPs) at Superfund sites where remediation is required. There should be both financial and moral encouragement to those PRPs who take good-faith actions to alleviate a problem pending an ultimate solution. Under the present system, government expects each PRP to take full responsibility for an entire site. A PRP who is responsible for a few percent of a site and is willing to lead a cleanup will be saddled with the entire cost and the problem of finding and suing other PRPs for reim-

AN INDUSTRIAL PERSPECTIVE   263

bursement of their share. Where PRPs cannot be found, the leader is stuck with the entire bill. This is a powerful disincentive to action on the part of companies, especially at large sites where there are hundreds of PRPs and where no private party generated a significant portion of the waste. Government could use shared funding more innovatively and could take an active role in pursuing all PRPs at a site. Government could also recognize voluntary efforts, instead of ignoring them as it does now.

We are in need of modifications in the legal framework to enable businesses that act in accordance with the law and good practice to obtain affordable insurance against claims of environmental damages that are not reasonably foreseeable or not under their control. When laws are changed retroactively or the courts and government redefine and expand definitions in unpredictable ways, our insurers tell us that there is no basis upon which to write insurance that makes economic sense.

Finally, we are in need of a better social infrastructure for the collection, treatment, and disposal of the residual waste product generated by individuals, municipalities, and small business. Saying "Not in my backyard" and expecting the waste to go elsewhere isn't working. One needs only look at the junk tires, old sofas, rusting stoves, and similar trash along out-of-the-way roadsides to see that we need a better waste-handling system.

The basic responsibilities for managing our wastes belongs with us. It would be nice to have government position itself to help firms trying to accomplish this goal while it goes about its other mission of enforcing the law against those who don't.

Appendix 9-1. Safety, Health and Environmental Policy

February 11, 1987

FROM: J. P. Mulroney

SUBJECT: SAFETY, HEALTH AND ENVIRONMENTAL POLICY

Consistent with the Rohm and Haas Company's established values, which include concern for its employees, its neighbors, and its customers, I wish to reaffirm our position regarding safety, health, and the environment.

The Rohm and Haas Company will take prudent action to ensure that its worldwide operations and products are free from unacceptable risks to the health and safety of its employees, customers, carriers, distributors, the general public and to the environment.

The Company will provide the people who handle or use its products with information concerning potential hazards associated with handling or using the products and information on how to handle and use our products safely.

Each Regional Director and Division Manager is responsible for having procedures in place to implement this policy. These procedures will be reviewed by the Corporate Director of Safety, Health and Environmental Affairs, who is responsible to advise on prudent practices and audit ongoing operations for risks in these areas.

Managers and supervisors in all Company locations, including manufacturing plants, offices, and all other places of Company business, are directly responsible for ensuring that their operations and procedures function in conformance with this policy. The introduction of new products and processes will be managed by procedures which include appropriate reviews and control of safety, health and environmental risks.

All employees are to be aware of this policy and incorporate safety, health and environmental practices in the conduct of their jobs.

J. P. Mulroney

JPM:das

GUIDELINES - Re: Environmental Control Policy

GUIDELINES

"Prudent actions" consist of at least:

1. Discovering and assessing health, safety and environmental risks.
2. Conforming with applicable laws, regulations, or generally recognized codes of good practice.
3. Rejecting activities whose risks are unacceptable, whether or not the risks meet legal standards.
4. Reducing even acceptable health, safety and environmental risks when further substantial risk reduction can be achieved with efforts that do not significantly affect the viability of the operation.
5. Providing the relevant information we have about health, safety and environmental risks and safe use and handling to those likely to be at risk as a result of a company activity.
6. Auditing for compliance with items 1 through 5.

AN INDUSTRIAL PERSPECTIVE   265

A risk is "unacceptable" if it meets any of these criteria:

1. The risk is significantly greater than that embodied in law, regulation, or generally accepted codes of good practice.
2. The risk is significantly greater than that for representative comparable activities within the Company.
3. The risk is significantly greater than that for representative comparable activities outside the Company.
4. The Company is not reasonably assured that those assuming the risk are informed concerning the risk and how to minimize it.
5. Company activities or the use of Company products are likely to significantly damage the Company's relationship with its employees or the community.

Safe implies "acceptable risk" rather than the total absence of risk or harm.

Summary: Lynn Johnson

I'm going to talk about how one industry approaches the question of integrated waste management. Though I know that our model would not necessarily apply to everybody, it is satisfactory for us. It's our way of handling this issue of broad integration across all waste lines. The policy I'm going to talk about this morning, and some of the programs that we have, came to us via a rather a honest route: we got stung a number of times. A decade or two ago, we were under the impression that if we obeyed the law and the regulations, that was good enough; but we turned out to be wrong. Obeying the law and regulations doesn't protect us from society's wrath when something goes wrong — it doesn't even protect us from the regulators who told us what to do and how to do it!

One of the essential elements of the policy we have is a company committed to effective control of environmental affairs worldwide. We're a worldwide company. We have probably 30 facilities abroad, and we apply these policies worldwide. We will take prudent action to ensure that our operations are free from unacceptable risk. These are the two buzzwords: *prudent action* and *unacceptable risk*.

Primary responsibility for executing the environmental control policy rests with the local management. It has to rest with the local management to get the job done. We cannot have some central corporate group accomplish a worldwide objective across 30 or 40 facilities.

The term *prudent action* means a number of things to us. It means

266   INSURANCE AND RISK MANAGEMENT FOR HAZARDOUS WASTE

discovering safety, health, and environmental risks. It means assessing those risks, once we've discovered them. It means compliance with the law and good practice, of course. It means rejecting any risks that we decide are unacceptable risks. Finally, it means informing others — the public, our employees, our stockholders — of what those risks are, and then auditing to ensure that what we think is happening really is happening.

The term *unacceptable risk* means a risk that is significantly greater than allowed by law or good practice, or significantly greater than that of a similar activity at a competitive or peer plant, or significantly greater than a similar activity within the company. A plant in France is not going to operate significantly differently from one in Colombia or in the United States.

Finally, a risk is unacceptable if it is going to damage the relationship with either our employees, or our customers, or those who may be around us, even if the risk is legal.

We have implemented a number of programs under this philosophy, and I am going to discuss three or four of these and then close with two suggestions on what the federal government can do to make things easier.

Several years ago we instituted a new ambient air guidelines program. Under the air guidelines program, each facility is expected to know what it emits on a chemical-specific basis; to assess that concentration by calculating the downwind concentration in the community around it; and to compare that concentration to a standard in the guideline. We've developed our guidelines in-house. We started off using a set of guidelines developed in Philadelphia by Air Management Services. Those are about seven or eight years old now, and a program has been undertaken to revise and update them. Finally, we install controls if we can't meet the guidelines, and we then make that information available to the community.

With regard to water effluents, we have undertaken a program to assess the effects of discharge on downstream drinking-water supplies. We have about nine plants around the world whose effluents are upstream of somebody's drinking water. Again, the philosophy is the same — an awareness of what we're discharging, measurement and calculation of that discharge, and a calculation of the concentration downstream.

In the case of water, we have found we don't have the resources to develop guidelines for 100-odd chemicals, so we have chosen to cut off above a part per billion. If the effluent is above a part per billion, we're going to develop a guideline, and then decide if something has to be done. If it is below a part per billion, that falls below our list of things to do today; we'll come back to that tomorrow. Finally, we're going to install controls if we can't meet our own guidelines, and make that information available to the public.

AN INDUSTRIAL PERSPECTIVE

Groundwater is another way pollutants can get out of our plants to the public. In the area of groundwater protection, several years ago we set out on a program to determine the geology, the hydrology, and the water quality under our plant sites. We calculate the potential impact of that site on neighboring drinking-water supplies, and in most cases we found that our water is very slowly moving to a river — most of our plants are located on rivers. If there is a potential impact present, and if it is of concern, we'll undertake remediation, and in a couple of places we have done that. Finally, long-term monitoring is required to monitor ground water.

In the area of solid waste disposal, we undertook a program to reduce the use of landfills to the extent practical. Again, the basic premise is the same: know what you're sending out, keep track of it, know what the chemicals are, know where they're going, destroy that waste if possible. We use a hierarchy for handling waste: after identifying the waste, we ask these questions: "Can the process be changed? Can the way that we are doing it be changed? If it can't, can we recycle the waste internally? If not, how do we destroy it — biologically, or by incineration?" We're fortunate that most of the materials we make are polymers and organic materials — carbon–hydrogen–oxygen compounds — and they're very easily destroyed by incineration. Finally, if there's absolutely nothing else to do, the waste is stored in the ground. We prefer to avoid that course, because long-term storage is just a long-term liability for us.

There are two things that government might do to make life easier for us. We have found that we need numerical performance standards reflecting how to protect the health of the community around us. We would like to see government establish performance standards based on ambient concentrations in air or water. We would like EPA to abandon the technology controls system. Technology controls are not connected with public health. They avoid the public health issue, and we think that's wrong.

Secondly, EPA should streamline its permitting processes; this point was made by the prior speakers. If we can't get the permits, we can't get things done. When the government has taken over the responsibility for permitting us to do something, the system stops when the permitting stops, and many of the permitting programs are just not working well.

Chairperson's Remarks: Paul Arbesman

I really want to commend the paper that's been presented as part of this session, because I think that for those of us who want to get a job done in terms of risk management — i.e., in a way that is broader than integrated waste management — the types of programs outlined in that paper need

268 INSURANCE AND RISK MANAGEMENT FOR HAZARDOUS WASTE

to be implemented. We discussed those issues in our panel session and came up with a number of points.

(1) There needs to be a corporate commitment to health, safety, and environmental programs that brings about a change in basic decision making on how problems are approached. What does that mean? What would be behind such corporate commitment?

Well, it would mean a strong policy that everyone understands and that is articulated throughout the company, with resources to back up the policy and program we are talking about. It involves things like an audit function, so that we see whether things are happening the way you tell everyone it's happening. It is fine to say we have a program worldwide. But has anyone checked to see if it's working?

Corporate commitment is concern about things like health, safety, and an environmental review of new projects; if we are really serious about having projects designed in the future that cut down on waste problems, then we have to get in on the front end and work as a team to try to make sure that those issues are addressed. Also, we need to look at existing plants from a risk reduction standpoint. We have got a large base out there. What can we do to improve it? That's a tough task for those of us who have worked on it.

We need to reemphasize the hierarchy of waste management. I'm talking about air, water, and solid and hazardous waste — all forms of waste discharges. The hierarchy has to be articulated in the company's program: Do not produce waste in the first instance. If you have to produce it, try to accommodate recycling in your program so that you are reusing it. If you cannot recycle it, try to treat and destroy it so that it does not bother anyone else. And finally, for any of the residues that cannot be handled in that fashion, you've got to look at permanent storage, because society is not going to tolerate, over the long term, the types of problems we're facing now under the cleanup programs. Everyone understands that. We're also smart enough in industry to know we do not have all the answers.

(2) There is a need to develop a way to keep track of waste reduction performance. This is a big issue today: EPA is working on it, Congress is working on it, and there are proposed bills addressing it. We outlined it as an issue for this group because someone needs to know how to track the numbers to gauge progress. It is not easy to carry out a mass balance approach across processes. It is somewhat more straightforward to look at it on a project basis. However, the numbers don't always add up for us to be able to conduct a year-to-year analysis that people would like to have to show the general trend of waste reduction programs.

(3) Then there is the small-company issue. I think it's interesting that the two prior panels hit on this issue. Someone needs to start a better

AN INDUSTRIAL PERSPECTIVE

bridge-building program between the big companies and the small companies. We could not quite figure out what that bridge looked like, but we did identify it as an area that needed to be pursued. Trade associations have a key role here, along with the responsibilities of the individual companies themselves.

(4) In the area of permitting, we emphasized the need for streamlining and simplification. We talked about time limits. A number of states have time limits for permits to be decided, and if they are not decided within the time limit, they get approval by default, and then the companies live under that permit. They are accountable for their waste and for meeting standards. I've never found that industry is bothered by being accountable; it is more the rigors of getting through a process that bothers them. So we recommended consideration of that type of approach — time limits, where no action means approval. Maybe that would allow regulatory agencies, as Irv Rosenthal pointed out, to focus their efforts on the permits that really need a detailed review in order to improve their performance.

(5) There should be a general increase in the permit-by-rule program. Marcia Williams hit on that yesterday, when she commented that there are lot of facilities that are clogging up the permit process so that things can't get done, whether it is storage tanks or small modifications that don't really need to go through that procedure. We're still accountable in the final analysis anyway for meeting the standards, so if this process is separated out it allows more important projects to move through.

The result of all these efforts would allow a transfer of resources out of the paperwork arena and into enforcement. I really think that this result would be a plus for the environmental area, because if there are more people that are out there looking at what is being done, the effort is more productive than trying to dot all the i's and cross all the t's on pieces of paper, not knowing whether the program is working in the field.

(6) On the issue of standards, we talked about setting more ambient standards, and also about the need to consider regional application. We do have regions in this country that deserve a sort of consolidated approach, where an individual plant, in order to meet a standard, needs to account for other plants in its vicinity. We also have to accept the fact that these standards are going to change. We do our best with today's knowledge, but 10 years from now we may move on to another level. We hold industry accountable, but we minimize governmental involvement in the how-to considerations. Government's true role in the long term is establishing safe limits, and industry's role and strength is figuring out how to do it on a cost-effective basis. We have shown we can do so if we are given the standards, and I think that's a positive development.

(7) In the area of waste site cleanup, we focused on risk-based, site-

specific standards for cleanup that minimize the Agency's involvement in the how-to questions. If the government spends all its time figuring out the required cleanup level, let the industry people who have already said they will pay the bill and are accountable for it for 30 or 40 or 50 years figure out the most cost-effective way to do it. I think a real road exists here to improve the program and to cut costs.

(8) Discussing the issue of standards, we identified as a weakness an area that I will call *combination effects* — for example, a number of pollutants in the same airshed or in the same watershed, or synergistic or antagonistic combinations. We do not know enough about these interactions to be as smart as we would like to be in that area.

(9) Finally, on the issue of cleanup there is a sense, generally, that those of us who sign the checks, and agree to do the cleanups are treated in a very cavalier and sometimes in an almost criminal manner. We feel that we take the brunt of the abuse from the Agency even though we're there, I won't say voluntarily, but knowing our responsibility. It will be four to five years down the road before any type of government effort is made to go after the people who never show up. That is just not fair; it is just not right. It sets in motion a whole mix of wrong attitudes to get the cleanup program moving. This area needs to be addressed.

10 AN INSURANCE PERSPECTIVE ON AN INTEGRATED WASTE MANAGEMENT STRATEGY

Neil Doherty, Paul Kleindorfer,

and Howard Kunreuther

10.1 The Nature of the Insurability Problem

The generation, transport, storage, and disposal of hazardous wastes all entail significant risks. Policymakers in government and industry typically foresee important roles for insurance as a prudent instrument for mitigating these risks and for assuring appropriate compensation in the event of damage. Policymakers and others differ in their views as to what role insurance might properly play in an integrated waste management strategy. This paper is concerned with analyzing this issue from both a theoretical and an institutional perspective.

Insurance may be defined as a mechanism for pooling the risky exposures of a number of individuals or firms. The risky exposures define the prospects for financial loss facing each individual or firm, and the pooling process permits the risk facing each to be reduced or removed through a process of diversification. Insurance markets have arisen on a voluntary basis to

---

Special thanks to Rajeev Gowda for his assistance in preparing this paper. We benefited greatly from the comments of Joan Berkowitz, Leslie Cheek, David Havanich, and John Morrison.

272   INSURANCE AND RISK MANAGEMENT FOR HAZARDOUS WASTE

cover many types of exposure. Common examples are fire; weather-related damage; operation of automobiles and other means of transport; common law liabilities for malpractice, defective products, ownership of property, etc.; and loss of life or health.

The essential feature of such voluntary markets is that they offer both parties some advantage from trading. From the individual's viewpoint, insurance reduces risk, enabling her to face a more predictable financial future. Similarly, the firm will benefit from risk sharing, thus permitting longer-range planning and a less costly operation. Owners of the insurance company are offered an expected return on their investment that compensates them for putting their capital at risk. A healthy insurance market requires such mutual gain.

Some insurance markets in the United States, such as homeowners, automobile, and life, do appear to function effectively with minimal intervention from state regulators. Other markets are more troublesome and sometimes disappear altogether. A recent example is the limited interest by the insurance industry in marketing earthquake insurance in areas of vigorous seismic activity. Recent proposals have the federal government serve as the principal bearer of the risk of catastrophic losses. Another example is the medical malpractice insurance market that has suffered a sequence of crises in the 1970s and 1980s. Additionally, the U.S. commercial-liability insurance market as a whole appeared to be in a state of crisis in the mid-1980s.

No market has been more troubled recently than the market for environmental impairment liability (EIL) insurance. Coinciding with the general liability insurance crisis, a sequence of mammoth and controversial court awards, and recent Congressional legislation related to potential liability from operation or cleanup of facilities, the pollution insurance market has all but disappeared, despite the fact that major legislation had defined a clear financial-responsibility role for the insurance sector as part of a social policy for managing environmental impairment risk.

The pathology of insurance markets may be addressed by identifying the conditions under which they do function efficiently. Such conditions have been described as the *preconditions for insurability*. The expression *preconditions* is strong, since insurance markets do exist and flourish even if all the stated conditions are not met. However, serious breach of several conditions can lead to stress in the insurance market, and in some cases, such as EIL, may lead to only a trickle of available coverage. The conditions for insurability are widely discussed in most insurance textbooks, and were recently formalized (from an actuary's viewpoint) in a book by Baruch Berliner (1982). These conditions provide a basis for discussing feasible

AN INSURANCE PERSPECTIVE

and appropriate roles for insurance in hazardous waste management and serve as the foundation for our discussion of insurance in an integrated waste management strategy.

Our paper will proceed as follows. First, we discuss recent legislative, judicial, and regulatory developments in the environmental area. We then outline the general *preconditions for insurability*. We then generalize this discussion by developing an insurability grid to classify current insurability problems in hazardous waste management. Against this backdrop of insurability, we discuss the central role of insurance in an integrated waste management strategy. A set of alternatives for correcting the current problems in EIL insurance markets in this area are offered in conclusion.

10.2 Insurance Implications of Environmental Legislation

Social concern for the consequences of environmental pollution is reflected in recent legislation and in the decisions of courts. If judicial decisions provide a barometer of social values, then the recent relaxations of conditions for a successful court suit and the increase in the size of awards is compatible with an increased social emphasis on compensating those affected by toxic waste and, arguably, in deterring or curtailing activities that generate these wastes. But a more direct expression of social concerns lies in legislation. The main issues can be illustrated by two important recent Congressional enactments: the Resource Conservation and Recovery Act (RCRA) of 1976 and the Comprehensive Environmental Response, Compensation and Liability Act (CERCLA) of 1980.

RCRA has been described as a *cradle-to-grave* approach to the regulation of hazardous wastes. It focuses on the treatment, storage, and disposal of hazardous wastes, requiring (among other things) that owners and operators keep adequate records and show financial responsibility for the operation and closure of sites and for the postclosure care of sites extending to a period of 30 years after closure.[1] A principal means of satisfying the financial responsibility requirements is the demonstration of adequate insurance protection.

In contrast, the emphasis in CERCLA has been on the cleanup of existing sites that have been deemed to pose serious public health problems. While this Act provides some funding for cleanup activities, it nevertheless imposes liability for such cleanup costs on "responsible parties." Courts have consistently held that CERCLA imposes strict liability, meaning that the government need not prove negligence, or failure to exercise due care, in order for defendants to be liable for cleanup costs (see GAO, 1987).

274 INSURANCE AND RISK MANAGEMENT FOR HAZARDOUS WASTE

Judicial decisions have also imposed *joint and several* liability on responsible parties. The recent amendments to CERCLA, the Superfund Amendments and Reauthorization Act of 1986 (SARA), continues to maintain the existing structure of liabilities. However, while indemnification of cleanup contractors if they are unable to obtain adequate insurance in the market is authorized under SARA, this has not yet been implemented. Insurance companies like the American International Group and Environmental Compliance Services are now offering liability insurance coverage for Superfund cleanup contractors on a case-by-case basis.

These legislative enactments reflect a joint concern that hazardous wastes be controlled and that severely hazardous sites be abated. In the debate surrounding these enactments, a set of implicit roles for achieving these ends has evolved. Some of these roles have generated considerable controversy:

(1) *Insurers will act as policemen*, monitoring the activities of those who generate, transport, and dispose of hazardous wastes. The policeman role encompasses the monitoring of activities, the measuring of the degree of risk, and the provision of incentives to abate the risk. The monitoring and measurement activities may potentially be achieved through a combination of risk assessment and actuarial assessment, the former concentrating on engineering tools to measure risk and the latter concentrating on statistical methods. Of course, monitoring and measuring risk is insufficient to influence behavior. Insurers also have incentives at their disposal. Poor risks can be denied insurance, may have restrictive conditions imposed on their coverage, or may be subject to higher premiums. Such incentives can potentially induce economically efficient decision making with respect to the generation and disposal of toxic wastes.

(2) *Insurers will act as regulators*, setting standards for the generation, storage, and disposal of hazardous wastes. This role (see Abraham, 1982) is much more extensive than the somewhat passive role of policeman described above. Under the regulatory role, the insurance industry could use its facilities for risk assessment, pricing, and underwriting to determine and enforce acceptable standards of generation, storage, and disposal. This role for insurance seems to have been assumed by default. One theory of financial responsibility envisages a partnership between government and the private sector (see Cheek, 1982): government sets appropriate standards of safety and protection, while insurers, protected by the imposition of those standards, underwrite the third-party liability assumed by the regulated operators. At issue is whether the EPA has fully assumed the regulatory powers granted under RCRA and CERCLA, or whether it has passed this mantle to the insurance industry.

AN INSURANCE PERSPECTIVE

(3) *Insurance will act as a risk-spreading mechanism.* The costs of toxic accidents may be large, perhaps catastrophic to those who bear them. For example, individuals and their families affected by toxic accidents may be financially devastated if they are not compensated for their injuries or losses. But compensation may cause financial ruin to the firm(s) deemed liable to pay for the costs of an accident. Through insurance, the financial impact of accidents is spread over all parties involved in the generation and disposal of toxic wastes. Insofar as insurance premiums are passed on to consumers through price increases or to the Internal Revenue Service through tax deductions, the costs of accidents are spread even further over the consumer base and over the tax base.

(4) *Insurers will act as guarantors of financial responsibility* should those who are deemed to be liable for damages have insufficient resources to pay for compensation or cleanup. It is in the nature of toxic accidents that the potential costs may exceed the net worth of those held liable or may only become apparent after the responsible firms cease to exist. Small firms controlling highly toxic substances have the potential for creating damages amounting to countless billions of dollars. Limited liability and bankruptcy laws are being overridden in environmental pollution cleanup cases, with courts holding that cleanup costs take a first claim on the assets of a company in bankruptcy. (See Midlantic National Bank vs. N.J. DEP, U.S., 106 S. CT. 755, 88L. Ed.2d.859, 1986). The states of Massachusetts, New Hampshire, New Jersey and Connecticut have enacted superlien laws that put government claims for cleanup ahead of all preexisting creditors, both secured and unsecured.[2] Yet insurance companies may be held responsible for cleanup costs over and above the value of the assets of bankrupt companies.

(5) *Insurers may be viewed as a deep pocket.* This potential role is highly controversial and is unrelated to the risk- spreading role. The risk-spreading role envisages a passive role for insurance. By appropriate pricing, the accidents of the few are spread over the premiums of the many, with the expected losses covered by premiums. Thus the role of the equityholders of the insurance firms is to use their capital to cover any random mismatch between aggregate premiums and losses. With risk spreading there is no (ex ante) subsidy between equityholders and other stakeholders.

But the deep-pocket role is confiscatory insofar as it envisages (ex ante) subsidies from insurance stockholders to outsiders. Such subsidies appear to be implicit in the retroactive, and potentially unlimited, nature of the liabilities imposed under CERCLA, since the policies on which these liabilities might fall could never have been priced to anticipate this risk.

(6) *Insurance will play an economic role as a free-standing capitalist*

276   INSURANCE AND RISK MANAGEMENT FOR HAZARDOUS WASTE

institution creating value. This role envisages insurance as an interlinking set of contracts voluntarily entered into by capital providers and those interested in risk-spreading through insurance. The various parties to this arrangement expect fair economic returns (investors) or service (policy-holders). Their expectations are based on a presumption of reasonable stability in the laws and regulations governing commercial activity in general and insurance in particular.

While some combination of these roles appears to be implied by social policy, it does not follow that a voluntary insurance industry will or can participate. Indeed, we will argue in the next section that toxic torts and the liabilities imposed through legislation perform very poorly against the criteria for insurability. These problems of insurability are partly technological and partly judicial in nature. But the problems have been severely aggravated by the federal legislation on operating and cleaning up facilities. Indeed, the legislative initiative to coopt the insurance industry into a social policy for hazardous waste management has probably been self-defeating. It was predictable that the insurance industry would opt out of its policy role in managing hazardous wastes as soon as this role was thrust upon it.

10.3 Insurability Criteria and Environmental Pollution

We shall now examine the general criteria for insurability of risks and analyze their performance in the context of environmental pollution. Appendix 10-1 applies these criteria to a hypothetical scenario involving a chemical firm and an insurance company. A more detailed discussion appears in Kunreuther (1987).

10.3.1 Uncertainty

Insurance policies attempt to limit coverage to events that are unintended and uncertain. Traditionally, insurance has been viewed by the buyer as the demand for certainty. Thus highly uncertain events tend to generate a demand for insurance, whereas events that are quite likely do not. An insurer will not provide coverage to firms who face a certain loss over the next year except at a premium that approaches the magnitude of the loss itself.

In the case of insuring against environmental pollution liabilities, judicial decisions and environmental legislation have caused certain risks to be uninsurable according to this criterion. Specifically, the insurance industry

AN INSURANCE PERSPECTIVE 277

feels that clean-up costs for unsafe waste disposal sites are uninsurable because the damages have already occurred.

Insurance policies attempt to limit cover to events that are unintended and uncertain. Thus they exclude coverage for all pollution incidents that are not *sudden and accidental*. However, some courts have interpreted this clause to cover gradual pollution incidents, stating that the damages which occurred were *unintended and unexpected* from the standpoint of the insured.[3]

10.3.2 Low Correlation

The possibility of a catastrophic loss may make a risk uninsurable. The successful pooling of risks, which is the basis of an insurer's activity, requires a low correlation between the individual exposures. The higher the average correlation between exposures, the greater the chances that claims will exceed reserves from the insurance pool due to a catastrophic loss. Such catastrophes can be avoided when the risks covered by the insurer are independent of each other.

Environmental risks are not highly correlated at one point in time. However, the effects of legal developments and judicial interpretations are felt across the system and affect all insurance contracts. Insurance contracts are written at a given time under a particular regime of liability rules and contract interpretations. Changes induced by legislation or adjudication simultaneously change the probability of a claim for all contracts. If legislative or judicial changes could be correctly anticipated at the time contracts are written, the impending changes could be included in the premium calculations.

For example, the introduction of joint and several liability by courts in their interpretation of CERCLA introduced a new liability retroactively for all companies involved in the various stages of the waste lifecycle.[4] This problem could conceivably arise even in the case of an existing facility if there are leaks from these facilities that cause damage. Then these facilities might be closed and treated under the auspices of CERCLA, thus providing scope for the imposition of joint and several liability.

10.3.3 Identification of Losses

Losses must be well defined as to time and place. In order for contracts to be effective, it must be possible to determine when and where such

278   INSURANCE AND RISK MANAGEMENT FOR HAZARDOUS WASTE

losses occurred. Problems of identifiability for environmental risks are well known. First, the long latency periods for the development and recognition of damages are clear. A landfill may leach for many years before identifiable effects are noticed. If the firm liable for the landfill had several insurers during this period, severe identification problems arise. The imposition of joint and several liability by courts in their interpretation of CERCLA introduces a new liability for all companies involved in the various stages of the waste lifecycle.

Second, court decisions have clouded the issue of identifiability of losses. For example, courts have redefined the meaning of the term *occurrence*. In the now famous Jackson Township case,[5] the alleged negligence of the landfill operators resulted in the pollutants seeping into an underground aquifer with resulting contamination to 97 privately owned wells. In this case, the court decided that the contamination of 97 wells resulted in 97 different pollution occurrences and proceeded to apply the coverage accordingly.

Third, in the Jackson Township case the court went on to award damages for nonpecuniary losses. A total of $15.9 million (reduced on appeal to $5.4 million) was awarded for emotional distress, impact on quality of life, and cost of medical surveillance, to the plaintiffs who were victims of pollution.

10.3.4 Estimating the Probability of Loss

Ideally, for the sound financial operation of insurance, the probability distribution of future losses should be capable of accurate measurement. If insurers are ambiguous as to the chances of certain events occurring, they will want to charge higher premiums for coverage (Hogarth and Kunreuther, 1989).

In the case of environmental pollution insurance, the recent work of Henri Smets and the Organization for Economic Cooperation and Development (OECD) deserve mention. In a very complete analysis of environmental pollution damages in several areas (including air, noise, oil, transport, and hazardous waste), Smets concludes from the available data for OECD countries that the frequency and severity of losses in these areas to date are well within the range of insurability in the sense that these empirical distributions are similar to those encountered for accidents such as airline crashes, where insurability has not been a problem. (Smets, 1986).

The key problem in the United States regarding environmental pollution lies in the hazardous waste area. Judicial decisions and legislative enact-

AN INSURANCE PERSPECTIVE

279

ments have a major impact on escalating the uncertainties as to whether a claim will be filed and awarded. For example, the introduction of strict liability for cleanup of waste sites for all responsible parties, and the subsequent extension of joint and several liability to these cases, increases the likelihood that any party involved with the damage-causing waste site will be held partially liable.

10.3.5 Moral Hazard and Adverse Selection

Both these conditions have an adverse impact on the provision of insurance. *Moral hazard* refers to intentionally careless behavior on the part of an insured who knows that if an accident occurs it is covered by insurance. *Adverse selection* refers to the inability of the insurer to differentiate between good and poor risks because of imperfect information on the quality of the firm. If a premium is based on the average risk across all firms, then the poor ones will purchase a disproportionate amount of coverage and the insurer will suffer unexpected losses.

One attempt by insurers to combat the problems of moral hazard is the institution of the *owned-property exclusion* clause in their contracts. In theory, this ensures that the costs of maintaining a policyholder's property would not be borne by the insurer. However, in one recent case, the owned-property exclusion was narrowed.[6] The court held that the defendant insurer not only had a duty to defend and indemnify, it had a duty to abate further injury from improper waste disposal on the plaintiff's land. Since the insurance company was contractually obliged to pay for all past, present, and future damages for which the plaintiff might be found liable, and because the plaintiff in this case was unable to afford abatement measures, the court held that the insurance company must undertake to abate the hazard.

In another case,[7] the court overruled the defendant's contractual provisions regarding the pollution exclusion and owned-property exclusion for removal of toxic waste from the plaintiff's property. It ruled that the insurer must cover these damages for reasons of public policy and held that "the health, safety and welfare of the people of this State must outweigh the express provisions of the insurance policy in issue."

Adverse selection can arise as a result of the retroactive imposition of liability, as has occurred under CERCLA. All parties connected with the wastes disposed of at a particular site have been held jointly and severally liable for the costs of cleanup. Thus those parties that have maintained accurate records of the wastes that they handled and have been covered

280 INSURANCE AND RISK MANAGEMENT FOR HAZARDOUS WASTE

by insurance would be listed as potentially responsible parties and held liable, while those who have handled wastes but show no records of having done so might not be found liable.

10.4 An Insurability Grid for Hazardous Waste Problems

The previous section outlined general conditions for insurability of environmental-pollution liability coverage without distinguishing between the timing of events and the nature of the liability incurred. If one decomposes the problem, then there are a set of risks that may be insurable. Table 10-1 presents a simple grid designed to determine what are insurable and uninsurable pollution risks.

The table distinguishes between past events, such as disposal of toxic wastes in a landfill or accidents which have caused injury or environmental damage, and future events, such as planned disposal or potential accidental leakages of toxic materials during production, transportation, disposal, or storage. It further distinguishes between the costs of cleanup of existing sites and the costs of settling suits for injury and/or damage. A further distinction is made according to policyholder characteristics. *Multi-user* refers to firms that share transportation and disposal facilities with other producers. *Vertically integrated firms* are those that handle waste disposal in house.

The grid is meant to be illustrative rather than exhaustive. Certainly there are other dimensions that are of importance, and there are environmental issues that do not fit comfortably in this grid. However, the grid is useful for highlighting insurability problems.

Liability for cleanup cost for existing multi-user sites comes within the top left cell of the grid. Insurability problems here are the most severe for

Table 10-1.   Insurable and Uninsurable Pollution Risks.

	Past	*Future*
Cleanup		
Multi-user	Least insurable	
Vertically integrated		More insurable
Compensation		
Multi-user		
Vertically integrated	More insurable	Most insurable

AN INSURANCE PERSPECTIVE

281

reasons outlined in the previous section. Insurer involvement was envisaged under the CERCLA legislation. However, certainty associated with cleanup, combined with the retroactive liability provision, imply that the role for insurers was defined to be that of taxpayer rather than risk bearer.

The confiscatory nature of the tax arises since the insurers had no opportunity to price this into their premiums when the policies were originally issued. Such a tax may be enforced against insurers in two ways. The tax may be obtained by expropriation of shareholders' surplus. Alternatively, the tax may be raised by generating subsidies from other lines of their business, such as fire or automobile insurance, written by pollution liability insurers.

The scope for cross-subsidies from other lines is limited by the knowledge and willingness of other policyholders to provide this subsidy. In a competitive insurance market, those insurers writing fire or auto business who were not encumbered by an existing pollution insurance portfolio would be able to underprice others who had to extract a subsidy from their fire or auto insurance lines to support the CERCLA-imposed cleanup costs. In such circumstances, the EPA would effectively have no economic power to tax those other lines. This implies that the economic power to tax is limited by the extent of the policyholder surplus; beyond that point shareholders are protected by limited liability.

Current estimates of expected cleanup costs from designated Superfund sites is likely to far outstrip the aggregate surplus of all insurers who write general liability (Katzman, 1988). Given the severity of insurability problem in this cell, it seems somewhat superfluous to distinguish between vertically integrated and multi-user cases. Nevertheless, insurers of firms using multi-user facilities have the aggravating problem that they may end up as the deep pockets should other users be untraceable, insolvent, or uninsured. Thus for the top left cell of the grid, the problems of insurability are excessively severe. Should the full potential for cleanup liability fall on existing insurers, the effect on future supply could be calamitous. Not only would such liabilities deplete existing capital (thereby affecting all lines), but investors would be extremely wary of future investments in the insurance sector lest future legislation impose some new confiscatory tax.

We will not go through all cells in the grid shown in table 10-1. However, it is instructive to compare the top left cell, (cleanup costs for past losses) with the bottom right cell, (compensation for future events). In the bottom right cell, exposures create the least stress on the insurability conditions; problems, however, can still be serious. The uncertainty test for insurability is likely to be satisfied, but many other violations may arise. For example, judicial instability can still give rise to high correlations over time as future

282 INSURANCE AND RISK MANAGEMENT FOR HAZARDOUS WASTE

precedent-making cases redefine coverage on existing policies before they expire.

Insurers already have taken some steps to reduce this problem by developing contract forms, such as *claims-made* forms, that effectively reduce the time period over which policies are vulnerable to such changes. (Moreover, while some states have revoked joint and several liability (see GAO, 1987), others still have this ruling.) Vertically integrated facilities that would like to insure their future environmental risks are the firms most eligible for insurance. Trends indicate that many large, financially stable, and well-managed firms are moving in this direction with respect to waste disposal activities in order to avoid the problems associated with joint and several liability. Some underwriters have suggested to us that the accidental discharge from new, state-of-the-art, underground storage or incineration facilities, which are subject to intensive risk assessment and which are used only by the insured, could well be considered to be insurable. A limited amount of coverage is now written on future risks by several companies in the United States and in Europe.

10.5 Role of Insurance in an Integrated Waste Management Strategy

Insurance is one of several policy instruments available for influencing economic, environmental, and health consequences of risk management for hazardous wastes. In considering the role of insurance in an integrated risk management strategy, it is important to consider a few examples of how insurance influences decisions related to hazardous waste.

In the transport, treatment, and disposal of hazardous wastes, insurance and financial responsibility requirements imposed by RCRA affect who will do business in these areas. Given the insurability problems arising from multiple-party arrangements and the risks involved for small-asset firms, it is difficult for anyone other than high-tech firms with considerable assets to obtain insurance in the hazardous waste area.

The financial consequences of liabilities arising from hazardous wastes could be spread by insurance, enabling risky but economically beneficial production of goods with hazardous by-products. Insurance provides further social benefits by promoting the timely remediation of the consequences of environmental incidents involving hazardous wastes and in providing compensation to those suffering losses. Anticipation of the net liabilities remaining after insurance coverage and the desire to maintain continuing coverage with an insurer provide substantial incentives for firms

AN INSURANCE PERSPECTIVE

to engage in proactive risk management, including the reduction of waste volume generated and transported and the development of safer methods of disposal.

The dictates of insurability and controllability of hazardous substances are complementary in promoting increased precision in classifying wastes, monitoring their movements, and generating standards for hazardous waste equipment and facilities. Thus both insurers and responsible industry officials agree that a continuing strong commitment to health and safety regulations and standards is absolutely essential, both for insurability reasons and for the monitoring and enforcement of regulatory policy.

The above points suggest a central role for insurance in managing hazardous wastes. Nonetheless, in discussing an integrated perspective, it is important to see insurers as only one of several interacting parties in the hazardous waste area. In figure 10-1, we show the interaction among in-

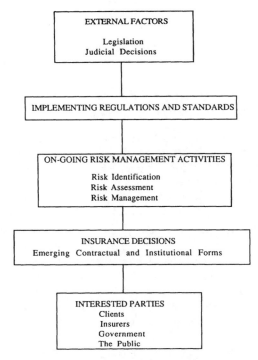

Figure 10-1. Conceptual framework for analyzing insurability problem.

284 INSURANCE AND RISK MANAGEMENT FOR HAZARDOUS WASTE

terested parties, highlighting insurance decisions. Insurance conveys both information and incentives to waste generators, transporters, and disposal facility operators. Depending on the nature of the insurance contract and the liabilities actually covered, ex ante decisions (e.g., how much waste to generate or what transport technology to employ) and ex post consequences will be affected. As noted in previous sections, the features of insurance contracts and their interpretation by the courts also strongly affect the feasibility of having private insurers and industry agree to an insurance contract in the first place.

The availability and price of insurance will interact with firm characteristics and policy constraints in the choice by firms of the risk management alternatives to be undertaken by them. And these choices will determine the environmental, health, and economic consequences of risk management activities. Several specific areas of hazardous activity have been analyzed by determining the effects of insurance requirements and various liability rules on risks to exposed individuals and society, under the assumption that an industrial firm minimizes its expected costs; for example,

- The reduction of hazardous waste flows (Kleindorfer, 1986);
- Transport decisions related to hazardous waste (Kleindorfer and Vetschera, 1987); and
- Site remediation decisions for treatment and disposal facilities (Kleindorfer and Kunreuther, 1987).

In each case, several general features arise from the analysis.

(1) Private and social costs often diverge in the sense that what is best for firms may not be best for society. This leads to various second-best tradeoffs in regulating hazardous risks and in setting insurance requirements. The nature of these tradeoffs is significantly affected by the level of assets at risk by the firm (the higher the level of assets, the stronger the incentives for risk reduction) and by the nature of the liability rules in force (see Shavell, 1987).

(2) Insurance coverage, residual liability, and the level of protective activity are joint decisions by firms and insurers and must by analyzed jointly to understand their financial and other (e.g., environmental) effects.

(3) Social equity in risk bearing and the efficiency and scope of victims compensation also are strongly related to insurance and liability requirements.

Insurance has a central, but mediating, role in the risk management process. On the one hand, insurance assures compensation for damages should they arise. On the other hand, insurance provides incentives to risk

AN INSURANCE PERSPECTIVE 285

managers of firms through premium differentiation and coverage availability for potential liabilities. When confronted with these incentives, risk managers can and will make corresponding decisions.

Competitively available insurance may be viewed simply as a part of the normal cost of producing goods whose by-products are the hazardous wastes in question. The operations themselves must, of course, be compatible with established standards and regulations embodying considered social judgments on the appropriate tradeoff between cost (including insurance cost) and environmental and health consequences. This view of insurance implies that its integration into risk management and operations is no different than that for other factors of productive activity. In this sense, the most important problem in resolving an appropriate role for insurance in an integrated risk management strategy for hazardous wastes is to establish conditions under which a competitive market for EIL insurance can be assured. Therefore, we turn our attention now to appropriate measures that might move us from the present, very thin market for EIL insurance to a more stable and viable setting.

10.6 Expanding the Environmental Insurance Market

The principal benefit of industry's failure to obtain pollution insurance is that firms are now cleaning up their act. Chemical companies have been modifying their production processes so as to reduce the amount of toxic wastes to be disposed. In addition, they are relying on cleaner means for dealing with their wastes, e.g. incineration. Yet, with the exception of the very largest industrial firms, most companies have a need for pollution insurance. In this section we will outline a strategy for helping to restore the environmental pollution market.

As we discussed in the previous section, strict standards and careful monitoring procedures are essential if insurance is to be used as a policy tool in developing a meaningful hazardous waste strategy. But changes are needed if insurance is to emerge as a meaningful policy tool in hazardous waste management.

10.6.1 Changes in Contract Design

Given the rapid and retroactive evolution of liability rules, it is inevitable that insurers have sought to redesign their contracts in order to insulate themselves from new undiversifiable and unpriced risks. The obvious ex-

286 INSURANCE AND RISK MANAGEMENT FOR HAZARDOUS WASTE

ample of such a change is the substitution of the *claims-made* contract for the *occurrence* policy. The old policies gave the policyholder protection not only for his personal or idiosyncratic risk as assessed against a given set of liability rules, but also, in effect, against the common risk of changes in the rules themselves. The idiosyncratic risk is diversifiable and is the type of risk for which insurance is intended. But the risk associated with changing liability rules is not amenable to the insurance process.

The claims-made policy gives some protection to insurers from assuming the second type of risk. The essential difference between the two types of contracts is that the insurer can defer its underwriting and pricing decisions for latent claims, i.e., those that are not discovered and filed until after the policy period.

Attempts to introduce the claims-made policy form for general liability cover have had a checkered history among state insurance commissioners. While much commercial liability insurance and, to our knowledge, all EIL cover currently sold is on a claims-made basis, some states have resisted this concept as a degradation of cover.

It is true that the claims-made policy passes some of the risk of changing liability rules back to the policyholder, thus forcing him to coinsure. On the other hand, the prospect of insuring undiversifiable risk under the occurrence policy forces the insurer, if it were to offer coverage, to load its premiums severely to protect its financial stability (and thereby to offer its policyholders a reasonable expectation that their claims will be honored). This loading of the premium will itself deter insurance purchase and force the policyholder to coinsure or self-insure.

At least by decomposing risk into that which is diversifiable and that which is not, the claims-made policy achieves a more rational sharing of risk between policyholder and insurer (see Doherty, 1987). This reasoning leads to the conclusion that the revival of the insurance market will proceed more smoothly if policyholders and insurers are left to negotiate contracts that offer mutual gains from trade.

10.6.2 The Enforceability of Insurance Contracts

The willingness of insurers and policyholders to negotiate contracts that offer mutual gains from trade depends also on their expectation that the contracts will be enforced as written. The history of the difficulties in the pollution insurance market is one in which insurers have complained that the courts have found cover in circumstances in which the intention to provide cover was not present at the time the contract was written, nor

AN INSURANCE PERSPECTIVE

could any reasonable interpretation of the contract conditions support such settlements. Of course, it is the prerogative of courts to interpret contracts and resolve disputes. Nor do three academics have any comparative advantage over courts in imposing reasonable interpretations of contract wordings. Thus we are not in a position to evaluate decisions such as that of Jackson Township described earlier.

But the issue here is not whether particular court interpretations are reasonable or correct, but whether they convey to parties engaged in the writings of future contracts an expectation that those contracts will be enforced as written. Both the words and the actions of the insurance industry (their protests and their widespread withdrawal from the EIL market) testify that the courts have been collectively unable to fulfil this expectation. In their efforts to extract compensation from insurers, the courts may well have killed the goose that laid the golden egg!

Interpretation of policy wordings is one problem; the apparent disregard of those wordings to pursue social objectives is another. The Summit Associates case mentioned earlier may well turn out to be an anomaly. But if the precedent set by this case turns out to be robust, the consequences for the revival of a voluntary insurance market are bleak (see Cheek, 1987). The issue of concern in this case rests on a policy wording that excluded liability under the policy for damage to property owned by the policyholder. In overriding this exclusion, the court stated

> The underlying Public Policy in this case is quite clear when the potential for damage to the environment is this great. Consequently, the health, safety and welfare of the people of this State must outweigh the express provisions of the insurance policy in issue. As a result, the exclusion clause in the policy which pertains to excluding coverage where the damage is to the policyholders land, must be held inapplicable where the danger to the environment is extreme.

The issue seems to be that, by the very fact of issuing a policy, the insurer becomes liable to finance public policies, as determined by future courts. Such a liability is in the nature of a tax, since the insurer can only reasonably price its contracts on the basis of the cover that it offers and that is affirmed in the contract provisions.

10.6.3 Changes in Organizational Form

As indicated above, traditional designs of insurance contracts may not be optimal for the environmental pollution market. It may also be the case that traditional insurance organizational forms may be ill suited to this

288 INSURANCE AND RISK MANAGEMENT FOR HAZARDOUS WASTE

market. The superiority of the claims-made policy over the occurrence policy lay in its facility for decomposing risk into that which is diversifiable and that which is not.

Another method for achieving the same end is the mutualization of insurance. In a mutual or any other pool that is collectively owned by its policyholders, the functions of policyholder and shareholder are combined. This implies that risk that is not diversifiable within the pool is passed back to the owners of the pool (i.e., the policyholders) in the form of a risky dividend. Thus, in the absence of reinsurance, diversifiable risk is spread amongst members of the pool, but undiversifiable risk is retained by the policyholders. A reinsurance contract for covering this undiversifiable risk (which may lead to unusually large losses) should effectively decompose the risk. The fact that undiversifiable risk will command a risk premium affects only the amount of reinsurance purchased and not the integrity of the pool (see Doherty and Dionne, 1987).

Given these advantages to pooling, it is not surprising that several pools have recently been started to cover gradual pollution losses. To date, none have thrived. There is a good reason for this, namely the problem of adverse selection. Each company contemplating joining the pool is concerned that the other members will be more likely to have an accident than they will. If you feel you are the safest company on the block, then there is no interest on your part in having others join you. An obvious solution in theory would be to have on-site inspections at the outset to determine relative risk, and careful monitoring procedures after the pool is established. Rates would then be based on how safe the plant is considered to be, relative to others. In practice it is not easy to rank companies on the safety and risk dimension since each firm is unique.

10.6.4 Integrating a Scheduled Compensation System with the Tort System

Nicholas Ashford and his co-authors (1987) suggest the setting up of an administrative scheduled compensation system that will function in conjunction with the tort system. Once the exposure to toxic substances is discovered, a potential victim may apply to the administrative system for interim measures of support for medical surveillance and rehabilitation. The funds to support these payouts could come from a combination of feed-stock taxes, waste-end taxes, and insurance premiums for victims' compensation.

In the event that no responsible party capable of paying the losses can

AN INSURANCE PERSPECTIVE 289

be discovered, the victim will be paid out of the administrative system according to a multitiered schedule of awards. In the event that an insured or defendant is identifiable, the victim can, by his own choice, proceed through the victims' compensation system or through the tort system. If the victim chooses to take the administrative route, payments would proceed according to the predetermined schedule. Immediate payment for the victim, with subrogation of the fund against the insured, could be a part of this scheme. The principal advantage of the administrative system is that it sets up a schedule of damage awards with immediate payment as against the open-ended awards that would arise under the tort system.

This proposed system may pave the way for the reintroduction of pollution liability coverage since insurance companies can now offer coverage for claims filed under the administrative system, even if they determine that coverage under the tort system settlements is still uninsurable.

10.7 A Proposal for Insuring Hazardous Waste

To conclude this paper, we will develop the features of a program where insurance is an integral part of a hazardous waste management strategy. We recognize that there is a tension between the role that both the legislative and judicial system would like insurance to play in encouraging risk reduction by industry and the insurability criteria that influence what type of coverage the insurance industry is actually willing to offer.

The objective of an insurance program for hazardous waste would be to encourage producers, transporters, and storers of this waste to reduce the potential damage to the environment while at the same time providing a source of protection to such parties and compensation to the potential victims. In developing such a system, we need to recognize explicitly three features of the pollution problem today that require some type of private–governmental combination in providing insurance protection.

Feature 1. There is considerable uncertainty and ambiguity regarding the probability that specific events (e.g., leakage from a landfill) will cause specific diseases to individuals. Hence standard actuarial techniques cannot be used by the insurance industry in setting rates.

Feature 2. Current tort law does not define explicitly the precise nature of the liabilities to be borne by different interested parties involved in environmental pollution cases. Concepts such as retroactive liability and joint and several liability have created considerable uncertainty in this regard.

Feature 3. Liability rules have been unstable. Major changes in liability rules have arisen both through legislative changes and through the setting of precedents in civil cases. Moreover, such changes often arise between the time a policy is written and the ultimate discharge of all claims under that policy. Insurers cannot effectively price this risk.

With these three features in mind, what type of viable insurance program will encourage manufacturing industry to reduce the risks associated with hazardous waste and to provide adequate compensation to potential victims? Figure 10-2 depicts the different layers of coverage that can be provided against losses of different magnitudes by a well-specified risk-sharing arrangement between industry, insurers, and the government. As a whole, this three-layer program appears to be a useful starting point for combining insurance with other features of an integrated hazardous waste management program.

Layer 1: Self-Insurance. In order to encourage industry to adopt safer production and disposal practices, they must be required to bear the first level of pollution losses. This is equivalent to having a deductible on an insurance policy. There is an incentive for an industrial firm to consider safer production techniques if it knows that by so doing it can reduce the potential losses it will have to bear itself. The actual amounts of self insurance must be tailored to the differing needs of large and small companies.

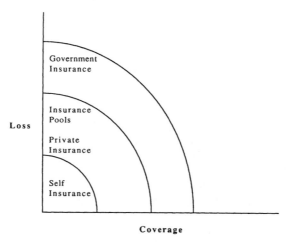

Figure 10-2. Insurance markets and layers of coverage.

AN INSURANCE PERSPECTIVE

Layer 2: Private Insurance or Insurance Pools. Private insurers should be willing to provide standard coverage against the next layer of losses as long as there is a well-specified upper limit that *will not be challenged in court.* By using risk-assessment techniques for appraising an industrial firm's operation, premiums for this layer of coverage can reflect the relative magnitude of risks. Today only two or three insurance firms have been utilizing such an approach in marketing coverage. We see these formal risk assessments as becoming standard practice in the future. As long as the upper limit of liability is well-defined, then insurers should be willing to offer environmental pollution coverage for this layer.

As a substitute for insurance provided by a conventional insurance carrier, layer 2 could be provided by an industrial pool, or group captive-insurance arrangement. Under such an arrangement, a group of firms simply pools their pollution exposures. Each firm is simultaneously a policyholder and a shareholder of the pool. The coverage afforded by the pool will lie within a well-specified band (e.g., between $20 million and $50 million), and the pool will often be set up and managed by an insurance broker. The advantages of such a pool are obvious: a number of firms are contributing, so that if one of them is unlucky it will be protected. It is not necessary to estimate explicitly the probability of losses of different magnitudes except to provide guidelines as to the size of the contributions by each firm. The pool is not designed to make a profit but rather to have sufficient resources to cover losses in the future. The principal difficulty in forming such a pool is the concern by every firm that enters that they will be the safest one on the block. Hence the importance of risk assessment procedures to delineate differences between safe and not-so-safe firms and the importance of using deductibles and other risk-sharing arrangements (e.g., coinsurance) for encouraging each firm to be concerned with future risks without relying on extensive monitoring and control procedures.

To date, insurance pools have not been successful in the environmental pollution area although they have been attempted by several large insurance brokers. We are more optimistic regarding the formation of such pools in a system such as the one we are proposing, since the upper limits of coverage are well defined. There may thus be little need for any type of reinsurance arrangement for the pool — a source of funds that has dried up for pollution coverage during the recent insurance crisis.

An additional problem that has been encountered in attempts to set up insurance pools or risk-retention groups for environmental liability is institutional. In many companies, risk managers have the authority to pay premium dollars but not to invest capital funds. This hurdle needs to be overcome.

292   INSURANCE AND RISK MANAGEMENT FOR HAZARDOUS WASTE

A sign of encouragement is the recent success of the ACE program, an insurance pool developed in 1986 by Marsh and McLennan for covering losses from general liability. It has a large deductible $100 million and a well-specified layer of coverage (losses between $100 and $200 million) so that a very select group of firms will want to join. The lessons from ACE might be applied to the more difficult problem of forming an industrial pool with a lower layer of coverage for environmental pollution.

Layer 3: Government Involvement. Given the current status of legislative rulings on hazardous waste as well as current tort law, it appears to us both essential and appropriate to turn to the government for the last layer of coverage. The uncertainty in the magnitude of the potential losses makes it highly unattractive for insurers to provide pollution coverage except on a highly selective basis. Private reinsurers are reluctant to provide excess loss coverage given recent toxic tort settlements in this area. Hence some form of government reinsurance for losses above a certain magnitude may be an essential element of any large-scale insurance program for managing hazardous waste.

We propose that a governmental agency similar to the Federal Insurance Administration be responsible for levying a fee on industrial firms to cover the potentially large losses which exceed the limits of layer 2. This fee would be partially based on the degree of risk faced by the particular firm (as determined by risk assessment procedures) as well as on the desired and/or required levels of coverage by the firm. Such a system has several advantages over private-sector reinsurance arrangements. There is less need for the government to be as concerned with developing actuarially sound premiums, since the principal objective of having them assume this layer of coverage is to ensure cleanup and perhaps provide compensation to victims, rather than to make a profit. The incentives for encouraging risk reduction measures by industry lie in the first three layers of coverage. It also seems appropriate for government to handle large liability losses, since it is the party principally responsible for the current state of affairs through CERCLA and RCRA legislation. Government also can have an impact on judicial reform (if that is deemed appropriate) in a much more direct manner than either industrial firms or the insurance industry.

By having the government directly involved in an insurance program, it has the opportunity to wield a big stick in facilitating an integrated waste management strategy. It can impose regulations and standards on industrial firms in return for providing layer-3 protection to them. It can also coordinate and help centralize a data base on losses and actual claims from pollution policies and thus help provide data for rate-setting across layers 2 and 3. Finally, the government will be directly involved in providing

AN INSURANCE PERSPECTIVE 293

compensation to victims rather than having such payments come indirectly through general tax funds.

On the other side of the ledger is the potential danger that government involvement in this final layer of coverage against losses will open up a Pandora's box in a fashion similar to workers' compensation. The principal hedge against this moral-hazard problem is the presence of the other layers of coverage, all of which explicitly encourage industry to operate safer facilities and reduce their side production of hazardous waste.

In summary, we see insurance as a vital part of a program for managing hazardous waste, but feel that industry, insurers, and the government have to join forces to make it work. Future legislation, such as the renewal of Superfund in 1991 and reform of the current tort system, will shape the relevant roles of each of these parties. However, we feel that even if there are radical changes in the legislative and judicial rulings, there will be a need for all three parties to each play a key role in the development of an insurance program that meets both societal objectives and conditions of insurability.

Appendix 10-1. Insurability Criteria and Environmental Pollution: A Scenario

The following hypothetical scenario motivates a discussion of insurability issues as they relate to environmental pollution:

> Two industrial concerns, the Alpha Chemical Company and the Beta Paint Company, dispose of their toxic wastes at the Wellbent Waste Management Inc.'s landfill. This is a state-of-the-art landfill that is lined with polyethylene, covered with a thin layer of clay, and sited in a large trench. There is still a low probability of groundwater leachate that could contaminate the municipal water supply of the neighboring town of Voxpopuli.
>
> Alpha would like to purchase insurance coverage against liability for environmental pollution from the Cappa Insurance Company. Cappa knows that Beta is uninsured and has been studying whether it should provide coverage to Alpha. In fact, Cappa has been seriously investigating whether or not to provide this type of pollution coverage to other industrial firms as well. What action should it take?

The decision by Cappa as to whether it wishes to provide insurance coverage to Alpha and other firms will be determined to a large extent by the nature of the environmental risk. The criteria for insurability that we have laid out will now be examined in the environmental pollution liability context.

294   INSURANCE AND RISK MANAGEMENT FOR HAZARDOUS WASTE

1. Uncertainty

Cappa is unlikely to provide insurance to Alpha because the uncertainty criterion is not satisfied. Its insurance policy limits cover events that are unintended and uncertain. Thus they exclude coverage for all pollution incidents that are *not* sudden and accidental. The courts have interpreted the phrase *sudden and accidental* to cover gradual pollution incidents, stating that the damages that occurred were *unintended and unexpected* from the standpoint of the insured.

2. Low correlation

Environmental risks are not highly correlated at one point in time. However, the effects of legal developments and judicial interpretations are felt across the system and affect all insurance contracts. For example, the introduction of joint and several liability by courts in their interpretation of CERCLA created a new liability retroactively for all companies involved in the various stages of the waste life cycle.

The Cappa Insurance Company, in attempting to draw up a contract to cover Alpha Chemical, may be reluctant to do so because it fears that judicial and legislative changes would make all the companies in its pool suddenly liable for unanticipated damage.

3. Identification of losses

Problems of identifiability for environmental risks are well known. First, the long latency periods for the development and recognition of damages are clear. A landfill may leach for many years before identifiable effects are noticed. If the firm liable for the landfill had several insurers during this period, severe identification problems arise.

Second, court decisions have clouded the issue of identifiability of losses. For example, courts have redefined the meaning of the term *occurrence*. In the now famous Jackson Township case, the court decided that the contamination of 97 wells resulted in 97 different pollution occurrences and proceeded to apply the coverage accordingly.

Third, in the Jackson Township case the court went on to award damages for nonpecuniary losses. A total of $15.9 million (reduced on appeal to $5.4 million) was awarded for emotional distress, impact on quality of life,

AN INSURANCE PERSPECTIVE   295

and cost of medical surveillance to the plaintiffs who were victims of pollution.

Such drastic interpretations could make the coverage offered by Cappa Insurance to Alpha Chemical somewhat different from that previously envisaged. In particular, nonmonetary damages cannot be covered by insurers because they are not amenable to clear identification.

4. Estimating the probability of loss

Regarding the estimability of loss probabilities in the case of environmental pollution insurance, judicial decisions and legislative enactments have had a major impact on escalating the uncertainties as to whether a claim will be filed and awarded. For example, the introduction of strict liability for cleanup of waste sites for all responsible parties, and the subsequent extension of joint and several liability to these cases, increases the likelihood that any party involved with the damage-causing waste site will be held partially liable.

For this reason, Cappa Insurance is wary of providing coverage readily to Alpha Chemical, even if the basic probability distribution can be estimated.

5. Moral hazard and adverse selection

These problems dog the environmental pollution liability area, and insurers have come up with various schemes to tackle them. *Moral hazard* refers to intentionally careless behavior on the part of an insured who knows that if an accident occurs it is covered by insurance. *Adverse selection* refers to the inability of the insurer to differentiate between good and poor risks because of imperfect information on the quality of the firm. If a premium is based on the average risk across all firms, then the poor ones will purchase a disproportionate amount of coverage and the insurer will suffer unexpected losses.

In our hypothetical example, the difficulty in monitoring Wellbent Waste Management's landfill over time presents Cappa Insurance with a moral-hazard problem once it issues insurance to Alpha Chemical. If Alpha were utilizing the landfill alone, then it would be possible to reduce such problems by instituting coinsurance and deductibles as part of the insurance policy. With Beta, an uninsured party sharing the landfill, however, there is little that can be done to prevent this other firm from being careless in

296 INSURANCE AND RISK MANAGEMENT FOR HAZARDOUS WASTE

disposing of its waste, since it knows that any damages caused would be handled by the insurance coverage obtained by Alpha.

The problem of adverse selection also arises in this case. Trends indicate that many large, financially stable, and well-managed firms are moving towards the vertical integration of waste disposal activities to avoid the problems associated with joint and several liability. This might mean that the only clients for the insurance coverage offered by Cappa Insurance are those that would be high risks.

Notes

1 RCRA third-party liability requirements for Treatment, Storage, and Disposal Facilities (TSDFs) are coverage against *sudden and accidental* occurrences of $1 million per occurrence, with an annual aggregate of at least $2 million, exclusive of legal defense costs. Similarly, land-disposal facilities must have coverage against *gradual accidental occurrences* of at least $3 million per occurrence, with an annual aggregate of at least $6 million. See United States General Accounting Office, *Hazardous Waste: Issues Surrounding Insurance Availability,* October 1987. Similar requirements for facilities within CERCLA's purview are being developed by the Environmental Protection Agency.

2 Joan Berkowitz provided us with the data on these recent court cases.

3 See Waste Management of the Carolinas Inc. v. Peerless Insurance Company (323 S.E.2d 726 (1984)), and Jackson Township Municipal Authority v. Hartford Accident and Indemnity Company (451 A.2d 990, 1986 (1982)).

4 Joint and several liability was established as the standard that *may* be applied in hazardous waste cleanups under CERCLA by the court decision in U.S. v. Chem-Dyne (572 F.Supp. 802, 808).

Joint and several liability refers to a situation in which more than one party is found to have caused or contributed to certain, often indivisible, damages. *Joint* means that all such parties are together liable for all of the damages. *Several* means that each of such parties is liable for all the damages. Joint and several liability, then, means that a party entitled to a recovery can recover the entire amount of his damages from all, some, or any one of the parties found to have caused or contributed, in some way, to the damages in question.

5 Robert Ayers, et al. v. Township of Jackson (493 A.2d 1314 (1985)).

6 See Riehl v. Travelers Insurance Company (Civil Action No. 83-0085 W.D. Pa., (1984)).

7 See Summit Associates, Inc. v. Liberty Mutual Fire Insurance Company (No. L-47287-84, Superior Court of New Jersey, February 25, 1987).

8 This point was brought to our notice by Joan Berkowitz.

REFERENCES

Abraham, K. (1982). Cost internalization, insurance, and toxic tort compensation funds. *Virginia Journal of Natural Resources Law* 2:123-148.

All Industry Research Advisory Council (1985). *Pollution Liability: The Evolution of a Difficult Insurance Market.*

AN INSURANCE PERSPECTIVE

Ashford, N.A., S. Moran, and R.F. Stone (1987). *The Role of Changes in Statutory/ Tort Law and Liability Insurance in Preventing and Compensating Damages from Future Releases of Hazardous Waste*. Center for Technology, Policy, and Industrial Development, Massachusetts Institute of Technology, Cambridge, MA.

Berliner, B. (1982). *Limits of Insurability of Risks*. Prentice-Hall, Englewood Cliffs, NJ.

Cheek, L. III (1982). Risk-spreaders or risk-eliminators? An insurer's perspective on the liability and financial responsibility requirements of RCRA and CER-CLA. *Virginia Journal of Natural Resources Law* 2:149.

Doherty, N.A. (1987). Insurance with random premiums. Working Paper, University of Pennsylvania.

Doherty, N.A., and G. Dionne (1987). Insurance with undiversifiable risk. Working Paper, Center for Research in Risk and Insurance, University of Pennsylvania.

Hogarth, R., and H. Kunreuther (1989). Risk, ambiguity, and insurance. *Journal of Risk and Uncertainty*. 2:5-35.

Katzman, M.T. (1988). Pollution liability insurance and catastrophic environmental risk. *Journal of Risk and Insurance* 55:75-100.

Kleindorfer, P.R. (1986). Environmental liability insurance: Perspectives on the U.S. insurance crisis. Working Paper, Risk and Decision Processes Center, University of Pennsylvania.

Kleindorfer, P.R., and H. Kunreuther (Eds.) (1987). *Insuring and Managing Hazardous Risks: From Seveso to Bhopal and Beyond*. Springer Verlag, Berlin and New York.

Kleindorfer, P.R., and R. Vetschera (1987). Risk–cost analysis model for the transportation of hazardous substances. Working Paper, Risk and Decision Processes Center, University of Pennsylvania.

Kunreuther, H. (1987). Problems and issues of environmental liability insurance. *The Geneva Papers on Risk and Insurance*, Vol. 12:180-197.

Shavell, S. (1987). *Economic Analysis of Accident Law*. Harvard University Press, Cambridge, MA.

Smets, H. (1987). Compensation for exceptional environmental damage caused by industrial activities. In Kleindorfer and Kunreuther (Eds.) *Insuring and Managing Hazardous Risks: From Seveso to Bhopal and Beyond*. Springer Verlag, Berlin and New York.

United States General Accounting Office (1987). *Hazardous Waste: Issues Surrounding Insurance Availability*, Washington D.C.

Summary: Neil Doherty

Our paper is a summary paper. We didn't try to come up with any very definite solutions to the insurability problem, but we did try to put a framework on it.

298   INSURANCE AND RISK MANAGEMENT FOR HAZARDOUS WASTE

The first thing we did was to look at the roles that seemed to be ascribed to insurance in the policy debate. A number of these roles were discussed yesterday either with some amount of affection or otherwise. I think most people agreed that insurance was a risk-spreading mechanism. I think some people may feel, but a lot of people would disagree, that insurance companies are properly considered to be a deep pocket. But when we're going through this list, we've got to remember at the end of the day that insurance companies are free-standing capitalist institutions that do have to attract capital. In order to do that, they have to promise their shareholders an expectation of a reasonable rate of return on their investment, and if they can't do that they're just not going to be there in the future. Our paper was largely conditioned by that particular thought.

In our approach to the problem of insurability, we borrowed from Baruch Berliner's book on the criteria for insurability. It seems that most insurance markets require certain preconditions to be in place before they're going to function, and in that sense the market for EIL is no different. A lot of these points have been discussed in the previous papers and in previous discussions, but there has to be some measure of predictability before an insurance policy can actually be priced. We need to be able to estimate and identify the types of events we want to insure. There shouldn't be any moral hazard or adverse behavior on the part of the policyholders as a result of getting the insurance.

One issue that is central to insurability problems in the EIL area is that of low correlation. A quick analogy will make things clear. An insurance market where you have problems of correlation is the earthquake insurance market. It's the same slip of the seismic fault that gives a claim on everybody's policy, and that's not the sort of risk an insurance company can diversify away. EIL insurance has similar sorts of problems in one respect, i.e., there's typically a long time gap between the writing of a policy and the setting of a premium and the final resolution of the claim. During that time delay, the legal rules — the liability rules that determine whether a settlement is going to be made — may have changed substantially. If it were just a coin-tossing operation — if juries or judges were going to flip a coin and decide these things randomly — it wouldn't be a problem. The problem is that once we have a change in those liability rules, the same change affects every policy in the same direction, and that's not the sort of risk that can be diversified. I think this point is central to the insurability problem. Liability rules are inherently unstable.

Having talked about the various criteria for insurability, we can see that all EIL problems are not created equal. The insurability problems are greater for some types of hazardous waste insurance than for others.

AN INSURANCE PERSPECTIVE

There's a past and a future dimension for the insurability problem. It's the difference between paying for the cleanup of existing sites and the payment of liabilities for future damage to people. That's a big dimension. The insurability problems differ immensely over that dimension. Cleanup of past environmental damage from multi-user sites is where the really serious problems arise. I think there are serious problems everywhere, but they tend to lessen as we go towards future compensation arrangements for pollution damages arising from vertically integrated facilities. The problems never disappear, but insurability is not uniform over the whole EIL area. I think that's an important problem, which actually came up several times yesterday.

To have any sort of revival of the insurance market, we thought there were two basic types of things that could happen: (1) Either you change or avoid the liability rules in place at the present time, or (2) you change the way in which insurance is actually organized and sold. Swiss Reinsurance Company has introduced an innovative change in contract design, discussed yesterday, which is very interesting and is a promising direction. I have some problems with it, but nevertheless I think those are the sort of innovations that might go places. This is an example of a change in contract design.

We can also change the type of organization that sells insurance. It's interesting that there were some experiments but not too many successes with mutual pooling types of arrangements so far. In other liability insurance areas, these mutual pools of various flavors have in fact been successful. The medical malpractice area is a good example of a market where this solution has been adopted with considerable success and the problems have been to some extent lessened. Most of our attention focused on the legal system and the instability in the liability rules. Whatever the specific solutions here, I think that for a revival of the insurance markets the judicial system and/or legislatures have to convey somehow to the investors who are going to put their capital up front in insurance operations that there's a reasonable expectation that the contracts they write will be enforced as written.

I don't want to get into any debate about whether *sudden* is *gradual* or whether an *occurrence* really is an occurrence or not an occurrence. The investors that put the money up front are the ones who have to be convinced, and it doesn't seem as if the courts have been able to do that with any great success in the past. As a significant precondition for a revival of the insurance market, reduction in the instability of the legal system is necessary.

Finally, any solution needn't involve all-or-nothing choices. We can have

300   INSURANCE AND RISK MANAGEMENT FOR HAZARDOUS WASTE

self-insurance, we can have regular insurance, we can have pools, and we can have the government involved either by spreading the cost we talked about over the tax base or possibly as an insurer of last resort. These approaches might be complementary rather than mutually exclusive. They might form various layers of coverage. In other words, we do not require all-or-nothing solutions to these types of problems.

Chairperson's Remarks: Jack Morrison

We had a pretty stimulating discussion last night with a very diverse and also a highly participative group. Before I start our specific discussion of recommendations this morning, I thought I'd share with you some of the parameters that we all agreed on in our group. They were as follows:

(1) We would try to deal with issues of the future as much as possible. In dealing with insurance, it is not very meaningful to deal with either the insurance problems that already exist, or for that matter to deal with the storage sites or the pollution that's already out there and can't be dealt with in any new way.

(2) Insurance is not and should not be used to cover the actual cost of cleaning up waste sites. What we were trying to deal with were the liabilities for personal injury and/or sickness and/or property damage to third parties that arise from the storage or cleanup of sites or various locations. Also we inevitably dealt with some other related environmental issues.

(3) In spite of the diverse group we had, we unanimously agreed that the evaporation of the insurance market for pollution liability may not have been too bad a thing for our country and for our desire to clean up the environment. Probably solutions such as self-insurance, movement of major organizations toward vertical integration of generation, storage, and disposal, and some of the EPA-mandated cleanups that have moved us closer to a clean environment might not have happened quite as effectively if we had the insurance mechanism out there fairly widely available, helping to absorb those costs and those problems. Particularly in the area of self-insurance and vertical integration, if there had been a totally viable insurance market, a lot of entities would not have had the incentives that they had in the absence of insurance.

So, while I personally feel a certain amount of embarrassment or a certain concern that we, the insurance industry, haven't addressed this particular liability problem, there is good reason to think that it might have been worse if we had, both for us and for the U.S. environment.

Now let me turn to three or four specifics that we would like to rec-

AN INSURANCE PERSPECTIVE   301

ommend or suggest. I think we were unanimous that these were ideas that could possibly help the insurance side of the pollution problem.

(1) The first suggestion has to do with the problem that cleanup contractors have in the liability world, where they are held to the same strict liability standards that the polluters are. One of our members suggested — and one of our cleanup contractors wished for something similar as well — that instead of the cleanup contractors being subject to strict liability, we get the government to rule that a pure negligence standard would apply to this specific group of people. Thereby they could probably buy liability insurance, if they are pretty sure that they would be held to a standard that's based on the current state of the art, that they would not be blamed for the pollution that was already there, and that the standard would be really a pure negligence standard. If this were the case, we would think that there would be more cleanup contractors who would be available, and they would probably do a better job. I think the suggestion is fairly reasonable.

(2) The second suggestion has to do with the theory of joint and several liability. We thought that, as a precondition to insurability, joint and several liability for the future would not be the optimal situation. As a compromise to the absolute elimination of that theory, we might get the government to modify it so that legislation would provide a rebuttable presumption by which a company might prove the proportion of liability for which it would be responsible. This would need to preempt the state laws in this respect and would really permit a generator or a transporter, if he kept adequate and approved records and if he could prove the extent of the proportion of his involvement in a storage site or whatever, to then be judged *not* on joint and several but on his proportion of the pollutants that were there and therefore bear only his proportion of the responsibility. There is something like that in federal law today, but not in many of the state laws. We think that this position might move us toward a situation where generators who satisfied the insurance market that they were pursuing acceptable practices and keeping adequate records might find the insurance industry move toward insuring them more readily. The generators would have to certify the extent of their pollutants, and the government would have to have accepted their certification as a precondition for insurance.

(3) We might investigate the usefulness of arbitration or alternative dispute settlements and get away from taking everything to the tort liability system, which is time-consuming and a waste of money. What I keep seeing and hearing, both here and in the newspapers, is that we are spending all our money on arguing about pollution rather than spending our money on cleaning it up. That's not just the insurance industry and it's not just the

generators. It's our whole country and our whole system. We're spending all the money and all the resources on the system, when we should spend it in clearing up the damn pollution! Anything we all could do to improve that would be a great help. Maybe arbitration and other settlement provisions as opposed to the tort system could help the situation quite a bit.

(4) Our final suggestion is that we investigate the possibility of some form of scheduled, administered compensation system as a replacement for the tort system or as something somebody could turn to as an alternative, rather than always using the tort system for pollution-related bodily injury claims. Such a system would be like a workers' compensation system for the people who think they are damaged by pollution. This system would require a lot of study. As we got into this last night, we could think of more problems than solutions, but again it's something that possibly ought to be looked at in order to simplify this whole problem and to see that the people who are really damaged are recompensed fairly quickly.

11 A LEGAL PERSPECTIVE ON AN INTEGRATED WASTE MANAGEMENT STRATEGY

Michael Baram

11.1 Legal and Economic Consequences of Hazardous Wastes

Industrial firms that produce, process, use, or store hazardous substances also create hazardous wastes. The wastes are either stored and dealt with on-site, or routinely discharged off-site, or transported off-site for final disposal. In any case, these wastes create health risks for workers and community residents as well as threaten property and resources.

The firms that produce these wastes fall into three broad categories — primary producers, intermediate processors, and end users. Thus, as a hazardous substance moves through its commercial life cycle from a primary producer to several downstream processors and numerous end users, the waste generated may endanger workers, residents, property, and resources at many industrial facilities that employ diverse methods of waste management.

Because of the risks, these very different types of firms face increasingly stringent regulations set by federal, state, and local authorities to govern the discharge and the disposal of their wastes; they also face fewer treatment and disposal sites, and escalating costs of regulatory compliance. In

303

addition, the firms face the prospect of cleanup litigation and liability under CERCLA[1], as well as civil penalties, costs for necessary studies and resource damage, and the transaction costs of fees paid for attorneys and experts, and similar burdens under state laws.

Each firm also faces the prospect of controversies and costly tort litigation for actual and potential harms to private parties arising from the discharge and from other waste disposal practices. Since tort theories of liability have been extensively modified over the past decade by the courts to do justice for persons injured by industrial wrongdoing, previously reliable defenses such as statutes of limitations have been eroded.[2] Further, the joint and several liability doctrine is now applied by many state courts to reduce the difficulties traditionally faced by injured persons who are unable to identify which of several firms caused their injuries, enabling these victims to fully recover from any one of the several firms jointly engaged in the tortious activity.[3] In addition, new rules of evidence liberally allow the introduction of epidemiological studies and other types of circumstantial evidence to prove causation, further enhancing the prospects of recovery by the victims.[4]

Tort claims may be settled or fully litigated. In either case, transaction costs are high. When compensatory damages are awarded, they are often accompanied by high punitive damage awards as well.[5] Even if no actual disease results are established, some courts now permit plaintiffs to recover for medical surveillance costs and for diminution of their environmental quality.[6] These new remedies are of particular concern to industry, since more tort claims are now being filed by persons without actual disease, but on the basis of their emotional distress and increased risk of disease following exposure to hazardous wastes.[7]

New federal and state *right-to-know* laws, which require firms to disclose information on hazardous materials to workers, state and local officials, and community residents, promise to fuel additional controversies.[8] For example, the Federal Emergency Planning and Community Right to Know Act of 1986 requires firms to make immediate disclosure of an accidental release of any of some 300 substances when the amounts exceed one pound or other designated amounts.[9] Manufacturers must also report annually to EPA on all routine releases of several hundred hazardous substances and the waste treatment methods employed;[10] many anticipate that local environmental leaders will use this public information to stimulate local opposition to further industrial releases.[11]

Finally, the recent insurance crisis, alleged to be due in large part to unforeseen environmental liability and unanticipated insurer responsibilities to defend such claims and provide coverage, has led to the reduced

A LEGAL PERSPECTIVE   305

availability and higher costs of environmental impairment insurance. Thus, many firms handling hazardous materials are unable to purchase sufficient insurance and must either self-insure to a greater extent than desired, join in the creation of a captive insurer, or go naked and become highly vulnerable to tort litigation and liability.[12]

Given these legal developments and their adverse economic consequences for industry, many corporate managers are taking prudent actions to reduce risks in order to thereby prevent liability and other losses. Thus, corporate risk management, which has traditionally been limited to achieving regulatory compliance and securing adequate insurance coverage, is now being transformed to achieve a more difficult goal, namely loss prevention by means of risk reduction.[13]

Many of the recent company initiatives have sought to reduce hazardous waste risks on an ad hoc basis and include, for example, the redesign of liability-prone products or facilities, the installation of a new *end-of-pipe* control to reduce toxic air emissions, and use of substitute materials.[14] But as regulatory requirements tighten further and losses continue to grow, more and more ad hoc measure are needed. And many of the ad hoc methods prove to be disruptive or inefficient. Some merely transfer risk and loss from one sector to another (e.g., from off-site to on-site, or from air to water).

The need for a more systematic and integrated approach to risk reduction and loss prevention is now seen by large firms. As a result, some of these firms are now defining and implementing megaconcepts of risk management such as *product stewardship, waste minimization*, and *source reduction*. When these more coherent approaches are taken, greater efficiency or cost-effectiveness can be achieved, as in the case of a source-reduction program that requires that new products be reviewed well before production to determine their waste, risk and loss consequences, and options for reducing these detriments.[15]

Integrated waste management has now emerged as the megaconcept that offers the best opportunity for corporate officials to use a systematic approach for addressing regulatory compliance and loss prevention.

11.2 The Integrated Waste Management Concept

The integrated waste management concept encompasses a diverse set of techniques. Some of these are focused on reducing off-site risk and liability consequences, such as

306   INSURANCE AND RISK MANAGEMENT FOR HAZARDOUS WASTE

- Retention or storage of waste on-site
- Recycling and reuse of waste on-site (e.g., solvent, recovery)
- Treatment of waste on-site (e.g., fixation, detoxification)
- Destruction of waste on-site (e.g., incineration, oxidation)

Other techniques are focused on reducing the amount, types, and toxicity of wastes generated at the source, such as

- Material substitution
- Product reformulation
- Process redesign and modernization
- Improved operations and maintenance
- Off-site recycling for reuse[16]

But the concept also encompasses the restructuring of management and adoption of new policies and decision-making processes to assure that a systematic approach to hazardous wastes is taken by the firm; the installation of new computerized information systems to provide data on waste streams needed for decisions on a continuing basis; the use of waste auditing and other accounting systems; the use of risk and loss analysis techniques and expert systems as aides for management decision making; and finally, the training of employees and the provision of incentives to workers and management.[17]

Obviously, integrated waste management will vary from firm to firm. In each firm, it will involve both a static set of formal techniques and technologies, and a dynamic management process. Management must have the analytic capability and authority to resolve internal disputes and to use those methods of waste and risk reduction that are most consistent with the firm's commercial activities and objectives and with the firm's risk- and loss-reduction goals. To succeed at integrated risk management, the firm must therefore set clear policy, provide sufficient authority and resources for managers to exercise business judgments in innovative ways, and hold management accountable for its performance over a sufficiently informative time frame.

Following the adoption of an integrated waste management policy, the firm will face the threshold issue of determining the sequence of methods to consider. From a corporate standpoint, it is usually more desirable to start with those methods that reduce off-site risk and loss, such as on-site storage and treatment or destruction of wastes, since these involve minimal disruption of manufacturing processes and business as usual. From this starting point, major investment in the on-site treatment technologies must follow. But the result of these first two steps is that interest in true waste-

A LEGAL PERSPECTIVE 307

minimization methods, such as deeper changes in processes and materials, are likely to be stifled. Why? Because these would now disrupt the processes that produce the wastes and make obsolete the costly technologies newly installed to destroy or treat the wastes on-site. Conversely, starting with or giving primacy to waste-minimization methods will lead to less waste generated, and could diminish the need for on-site storage and treatment options.

Therefore, a firm embarking on integrated waste management should simultaneously consider both sequences, and choose one on the basis of cost-effectiveness. But this may conflict with societal preferences for waste minimization, particularly where the corporate decision does not resolve significant off-site risk controversies.

Numerous other issues must be considered when a firm adopts an integrated waste management policy. Obviously, a small downstream end user that generates waste faces a different set of variables than a large upstream producer of hazardous substances. But all firms will face several common problems. One is the difficulty of modifying the existing corporate management structure so that safety and waste management personnel can jointly engage in integrated waste management together with product design and marketing personnel, process engineers and industrial hygienists, purchasing agents, legal counsel, and others, many of whom may be geographically dispersed across the country in the case of a large firm.

Since defensive attitudes and internal conflicts are likely to arise, and short-term economic benefits from less comprehensive measures are likely to dominate, integrated waste management will fail unless the corporate initiative is led by a senior official with broad authority to effectuate new measures despite internal opposition.

Several reforms linking corporate strategic planning to risk management will also be needed — for example, adoption of a company-specific long-range plan that can be used to demonstrate the potential net gains to be achieved from integrated waste management over five- to 10-year periods due to reduced costs of regulatory compliance and tort litigation. In addition, internal budget review processes and criteria must be modified to provide necessary funds to facility managers.

Finally, the roles of key personnel must be modified to serve integrated waste management. For example, the defensive use of corporate counsel for litigation should be modified so that they are also used to develop a *preventive counseling* function, one which provides estimates of potential civil penalties, litigation costs and awards, and other losses from the proposed integrated waste management approach in comparison with the losses likely to arise from doing business as usual.

From a societal perspective, however, the major issue at this time is

308 INSURANCE AND RISK MANAGEMENT FOR HAZARDOUS WASTE

what policy instruments will be most effective and acceptable for promoting integrated waste management in diverse industrial firms.

11.3 Policy Instruments to Promote Integrated Waste Management

Congress has expressed support for many of the elements of integrated waste management. For example, in 1984 Congress amended RCRA[18] to provide that it is "national policy . . . that, wherever feasible, the generation of hazardous wastes is to be reduced or eliminated as expeditiously as possible. . . ."[19] Reports by the Office of Technology Assessment, EPA, and state agencies have also recommended many of the elements of an integrated approach.[20]

Nevertheless, clear federal authority to require firms to adopt an overall strategy for integrated waste management is lacking at this time; and many even view existing waste management regulations as obstacles to the voluntary adoption by firms of integrated approaches.[21]

At this time, at least four legislative proposals are before Congress that would authorize waste reduction programs for industry, but the prognosis for enactment is uncertain.[22] Each raises difficult policy questions. For example:

- What qualifies as waste reduction (would waste treatment qualify)?
- Should waste reduction require both a reduction in waste volume *and* toxicity?
- How would compliance and progress be measured?
- Should integrated waste management deal with hazardous wastes only, or all environmental wastes generated by a firm (e.g., emissions, discharges)?[23]

But the most critical issues are which policy options to use, and who should be responsible for their implementation.

A universe of policy options is available to choose from. These range from the enactment of regulatory controls (e.g., performance standards) enforceable against industry by EPA or the states, to mere authorization of training and technical assistance measures and tax credits to stimulate voluntary adoption of integrated waste management by industry.[24]

Similarly, responsibility can be vested in federal, state, or local authorities, or left to the private sector. Proponents of the *new federalism*, now

A LEGAL PERSPECTIVE 309

in the ascendancy, would obviously prefer state, local, or private responsibilities rather than further enlargement of the federal bureaucracy.

Given the multitude of manufacturing and service facilities that generate wastes, as well as their incredible diversity, it is doubtful if a single federal agency could develop a cookbook of generic standards suitable for all firms, and then have sufficient resources and expertise to monitor and enforce compliance. Thus, from a pragmatic viewpoint, which happens to be consistent with the new federalism, primary responsibility for implementation should be vested in state or local authorities and private firms, with the federal role reduced to an oversight and technical assistance function. This model is already provided by the Emergency Planning and Community Right to Know Act of 1986, which deals with industrial accidents and the emergency response preparedness of state and local officials.[25]

Thus, it is instructive to look to recent findings by state agencies to determine their preferences. For example, a recent Massachusetts report on source reduction, which in many respects is synonymous with integrated waste management, provides the following:

> Source reduction is not likely to become widespread without the adoption of new policies which give greater emphasis to the need for reducing hazardous waste before it is generated. Implementation of a stringent source reduction policy might include fees and taxes on waste or virgin products likely to become waste, tax incentives such as credits and accelerated depreciation for source reduction equipment, technical assistance on reduction technologies, educational activities, awards for effective waste reduction, research and development grants to appropriate institutions, below-market loans to generators, and grants for feasibility studies or process changes. Bans on the use of substances or processes which generate unacceptable wastes and for which adequate substitutes exist could be considered. Experience with programs in other states suggest an integrated approach. In Massachusetts, a potential mechanism for evaluation and implementation of these approaches could be the creation of a "Center for Excellence" on waste management or the use of the University of Massachusetts based Cooperative Extension. Without a strong initiative from the State, an integrated source reduction program is not likely to go forward.

Assessment of the likely effect source reduction policies will have on waste generation rates is difficult to determine. Based on an analysis of source reduction activities in Massachusetts, DEM estimates that it is technologically feasible to reduce waste generation up to 48%. However, the rate of reduction may vary from a low of 13% to the high of 48% by 1997, depending on state policies and incentives for generators.[26]

To facilitate positive corporate responses, particularly among small firms, federal and state agencies will have to provide training and tech-

310   INSURANCE AND RISK MANAGEMENT FOR HAZARDOUS WASTE

nology transfer to clusters of similar firms in particular industrial sectors. In addition, these agencies could also educate insurers as to how insurance availability and pricing could be adjusted to favor firms that establish integrated waste management programs, and thereby create a strong economic incentive within the private sector.

This brief review of a complex subject indicates that state and local officials will have major roles to play in promoting integrated waste management. In addition to possibly enacting some performance standards and reporting requirements, these agencies will also have to provide small firms with training and technology transfer programs, help manager and owners determine potential net gains from the investments that will be needed, and educate insurers to bring about the incentive of favorable insurance rates for firms that adopt integrated waste management.

11.4 Conclusion

By carefully considering the variables that shape corporate decision making, some of which have been dealt with in this paper, legislators, agencies, and the firms themselves can take the steps needed to move integrated waste management from its current status as a concept to actual practice.

Notes

1 Comprehensive Environmental Response, Compensation and Liability Act, or "Superfund Law" 42 USC 9601.

2 Baram, M. "Chemical Industrial Hazards: Liability, Insurance and the Role of Risk Analysis." Paper presented at Geneva Association, Twelfth General Assembly, Oslo, Norway(June 23-25, 1985), and published in part in H. Kunreuther et al. (1987). *Insuring and Managing Hazardous Risks*. Springer-Verlag, Berlin; New York.

3 *Id*.

4 Federal Rules of Evidence, P.L. 93-595 (1975), codified at 28 USC app. 678(1982), and adopted by many states. For a critical view, see Weinstein, J. (1985). Improving expert testimony. *U. Richmond Rev.* 20:473.

5 Walker, V. (1986). Punitive damages and management of environmental information. *Environmental Analyst* (Sept 86):8.

6 Baram, M. (1987) Fear of disease fuels flurry of toxic tort cases. *National Underwriter* (July 20, 1987):31, citing recent cases.

7 *Id*.

8 Federal enactments include EPCRA, the Emergency Planning and Community Right to Know Act, 42 USC 11001, and the Occupational Safety and Health Administrations Hazard Communication Rule, 29 CFR 1910.1200 (Nov. 25, 1983) Over 30 states and hundreds of

A LEGAL PERSPECTIVE 311

municipalities have enacted similar laws. See discussion in M. Baram, *Corporate Risk Management*. Commission of the European Communities, EUR 11555 EN (April 1988).

9 EPCRA, Section 304

10 EPCRA. Section 313.

11 Chadd, C. C. and J. O'Malley (1986). Superfund amendments offer hope for plaintiffs in toxic tort actions. *Nat. Law. J.* (March 21, 1986):16.

12 Baram, M. (1986). Insurability of hazardous materials activities. Paper presented at Ninth Symposium on Statistics and the Environment, National Academy of Sciences (Oct. 28, To be published in *Statistical Science: Conference Proceedings* (1988).

13 See M. Baram, Note 8 supra.

14 See generally, *Minimization of Hazardous Waste*, Office of Technology Assessment, U.S. Congress (Oct. 1986); *Cutting Chemical Wastes*, Inform (1986); *Proceedings of the Fourth Annual Massachusetts Hazardous Waste Conference*, Mass. Dept. Environmental Management (Oct.1987). Some have reviewed various foreign approaches to support proposals for changing practices in the U.S. See Geiser, K., et al., *Foreign Practices in Hazardous Waste Minimization*, Tufts University Ctr. for Environmental Management (Aug. 1986); and Williams, A. (1987). "A Study of Hazardous Waste Minimization in Europe," *Environmental Affairs* 14:165. Also see *Hazardous Waste Management in Massachusetts*, Mass. Office of Safe Waste Management (August 1987) for information on waste reduction and management initiatives by large and small firms in the state; and *Waste Reduction: The Ongoing Saga*, U.S. EPA and Tufts University Center for Environmental Management (1986), which provides information on waste minimization initiatives at Rohm and Haas, Olin, Kodak, 3M, IBM, Hewlett-Packard, and other large firms.

15 *Id.*

16 *Id.*

17 *Id.*

18 The Resource Conservation and Recovery Act, as amended, 42 USC 6901. 42 USC 6902(B).

19 42 USC 6902(b).

20 Note 14 supra. Also see Lacroix, C. (1988). Waste minimization: Debating the meaning of a laudable goal. *Toxics Law Rptr.* (April 6, 1988):1214.

21 C. LaCroix, note 20 supra.

22 *Id.*

23 *Id.*

24 For a theoretical discussion of major policy options for dealing with risks, see Baram, M. (1982). *Alternatives to Regulation*. Lexington Books. For policy options to deal with hazardous waste minimization and source reduction, see note 14 supra.

25 Note 8 supra.

26 *Hazardous Waste Management in Massachusetts*, Mass. Office of Safe Waste Management (August 1987), p. 23.

Summary: Michael Baram

My paper was written on integrated waste management from legal and corporate management perspectives. Integrated waste management is a concept that means many different things to many different people.

312 INSURANCE AND RISK MANAGEMENT FOR HAZARDOUS WASTE

Obviously, it's a megaconcept. It incorporates efforts to reduce hazardous waste streams that leave plant sites; it incorporates other efforts at waste minimization at the corporate source; it incorporates the reduction of routine discharges. As a megaconcept, it includes many people, concepts, and practices.

There are many lesser concepts of hazardous waste management, such as on-site treatment and destruction of waste which are feasible. But recommendations to promote integrated waste management throughout American industry are rational, since such a program is more cost-effective in the long run than ad hoc tinkering with routine emissions or reducing on-site storage. So the megaconcept of integrated waste management has great appeal.

Designing some sort of policy instrument that would promote integrated waste management in American industries, large and small, is a complex problem, especially for the small firms. Can we design such policy instruments or should the matter be left entirely up to technology transfer within industry?

From the little research that I was able to do for this meeting, it appears that there is a lot of technology transfer within industry, but most of it is horizontal. The producers share a lot of information with each other through the Chemical Manufacturers Association and other producer trade associations about ways of minimizing waste or successful methods that are being used. On the other hand, there is little vertical transfer of technology down to intermediate processing firms that are small or medium-sized, and even less to the end-user firms, many of which may not be chemical firms at all, but may be electroplating firms or semiconductor firms. Thus, many of the firms that buy chemicals are really not chemical companies, and receive little information since they operate in different industrial sectors.

The task of integrated waste management is essentially contextual: In a Dow Chemical or in a Du Pont, it will be very different than it will be at Digital or at some downstream users of chemicals.

The problem, therefore, is how to develop some sort of policy instrument that will at least promote this good idea downstream through vertical technology transfer. Obviously, regulatory requirements, the liability system, and the insurance crisis are promoting waste minimization and methods of integrated waste management in large firms, but small firms and intermediate firms purchasing and using chemicals downstream are generally believed to be incapable of carrying out this kind of initiative. It seems clear that EPA or another federal agency could not possibly develop a cookbook approach to this problem for all industrial sectors. There is

A LEGAL PERSPECTIVE

313

simply no way, given the diversity of the facilities and activities involving toxic waste generation, that EPA or any federal agency could possibly have the knowledge or resources to set generic standards or licensing for integrated waste management. Therefore, government must carefully take the lead if the goal of integrated waste management is to be achieved.

In Massachusetts, California, and other states, there have been a number of thoughtful reports and proposals for a variety of policy instruments ranging from tax incentives to technical assistance programs run by state government, industries, and trade associations. These state initiatives may be feasible for specific facilities within the state, but I can think of another model if one really wants to pursue this further. The new model, suitable for federal legislation, is the one that we now have in the federal Community Right to Know law, a law that exemplifies the new federalism; in dealing with industrial accidents, it puts the authority to do emergency planning at the state and local level, and requires that information from industry be provided to the state and local officials.

A similar model could be used in integrated waste management. Each firm already reports its waste stream to state officials as required by Section 313 of the federal Right to Know law, and state officials could immediately develop programs in which they would then consult with the firms to jointly develop a set of integrated waste management approaches. This seems to be a modest policy initiative that could be useful.

There are other points that I raised in my paper, particularly those that involve the restructuring of management within companies and the difficulties involved. This is especially the case for a firm having many facilities across the U.S., particularly a firm that is rapidly buying up other firms. The question is: How does one make sense out of management structure and vest authority and power in senior officials to do integrated waste management throughout the firm? That is probably the greatest challenge in the area.

Chairperson's Remarks: Dennis Connolly

We took our assignment, which was to develop a set of points on Michael Baram's paper, very seriously, and developed nine.

(1) For the companies represented in our small group that have to deal with this kind of problem, the solution really seems to be driven by economics. In fact, economic pressure is really a tremendous incentive to integrate waste management.

(2) We drew a distinction between what happened in the biggest chem-

314   INSURANCE AND RISK MANAGEMENT FOR HAZARDOUS WASTE

ical companies, which have all sorts of resources and knowledge to deal with these problems, and smaller companies or those companies with limited chemical involvement. We felt that the smaller firms might be hurt both by their lack of knowledge and by the withdrawal from commercial operations of the largest and most knowledgeable hazardous substance users.

(3)We found it wasn't possible to generalize even about the big companies. Thus, even within the largest companies there might be instances of uncoordinated toxic substance practices. An example we discussed was the purchase by a giant of a relatively small company that was subsequently discovered to be using the yellow pages to find its local waste hauler.

(4) There are some problems with definitions as people look for guidance to what to do. We had discussed an example of a firm that used a large quantity of wood to fuel its incinerator. Since wood is classified as a flammable solid and thus as a hazardous material, this company was number one in the state in waste production. When it eliminated the wood, it fell below 200. So definitions can lead to erroneous perceptions.

(5) Our group felt that there was a problem with risk perception in the United States; there may be a significant difference between actual and perceived risks. There was some feeling that it might be better to take some of this cleanup money and devote it to something like AIDS research or hospitals or something with more direct health benefits.

(6) What is the most effective way to distribute knowledge to the lesser or smaller user? We came up with some ideas, but I can't say that we came up with a solution. We thought trade associations would be a very good way to distribute knowledge. We considered and found a lot of plusses and minuses in the use of incentives and disincentives arising from various tax, regulation, and legislative sources. We thought that insurance could be used to perform this role, in particular through its risk assessment and risk management roles, but that has become unworkable because of the liability system.

(7) We then got to what should have been our first point: What does integrated waste management mean? It was a little odd for us to come to it at the end of our session, but we were not really sure. Does integrated waste management mean reduction of waste? Is it elimination or source reduction of waste? Does it mean incineration? We thought that what it really meant was a coordinated approach and an educated decision process as to how firms are going to handle hazardous substances through the entire commercial stream.

(8) We talked about the problem and how improvements were going to be accomplished. We had some reservations about the use of regulation.

A LEGAL PERSPECTIVE 315

We thought that regulation causes a number of problems, for example, having to be *permitted* to do the right thing — that is, the situation where a firm wanted to do the right thing but had to go through a permitting process that made it so difficult that it was really easier not to do it. We also went back to our generalization point, namely that it is hard to generalize because the substances had different uses, substitutes, values, etc.

(9) Finally, we agreed that it was possible in some limited circumstances to use what might be called a *cookie-cutter* approach to disseminate technical support. Examples would be specialized operations like gas stations or electroplating.

12 A LEGISLATIVE PERSPECTIVE ON AN INTEGRATED WASTE MANAGEMENT STRATEGY

Jack Clough

I have balked at my assignment.

I was asked to contribute a paper for this conference on *A Legislative Perspective for Developing An Integrated Waste Management Strategy*. In the paper, I was asked to focus on the general objective "to develop recommendations for future legislation with respect to an integrated waste management strategy that recognizes the role of risk assessment and risk management as part of the policy process." The letter of invitation states "A key objective of the meeting is to develop an integrated strategy for hazardous waste management with the intent of influencing future Congressional legislation."

Congress is awash with policy papers that purport to solve problems like Integrated Waste Management. Most are never read. And if the truth be known, I have no solution.

I do understand, however, the process this nation will use to develop a solution over a period of years, i.e., the legislative process. I felt the most valuable contribution I could make to this conference was not to expound at length on a solution I do not have, but rather to describe the legislative process so that the various ideas generated by this gathering can be placed effectively into the solution process.

318 INSURANCE AND RISK MANAGEMENT FOR HAZARDOUS WASTE

I will make four assertions about the legislative process. They are personal observations which, I hope, will give the reader a different insight into a process that is fundamental to our collective lives yet poorly understood. Then, based on the assertions, I would like to draw four conclusions about how a group with specific concerns, such as this one, can have an effect on the legislative process.

12.1 The Four Assertions

Assertion 1: *We have a remarkable system of government that was designed first and foremost to protect the rights of the individual.*

When many people enter the public policy arena, they forget the purpose of the Constitution. They want statesmanship; they want efficiency; they want decisiveness. They frequently fail to understand that the Constitution purposely thwarts these qualities to protect the rights of the individual.

The framers of the Constitution realized that the power to govern is a zero-sum game; when one person exercises power in the form of decisions that affect the lives of the governed, other persons are precluded from making those decisions. To protect against the possibility that any one individual could amass too much power at the federal level, the Constitution establishes Constitutional equals who serve interlocking constituencies. By creating equals, the Constitution notifies an individual that he is in competition with others for his slice of the sum total of federal power. By creating interlocking constituencies, the Constitution guarantees conflict between the equals. Currently, these equals include the 535 members of Congress, the 9 justices of the Supreme Court, and the President. All these equals serve the same constituency, the citizens of the United States, but because of the different grouping each serves (a local population of roughly 500,000 for each House member, a state population for each Senator, and the entire population for the President and the Justices of the Supreme Court), they frequently have different views of the decisions they believe you want them to make on your behalf. Yet to make a binding decision, a majority of the Constitutional equals must overcome their differences and agree.

Because of the number of people involved, the net effect is purposeful inefficiency in the policy decision process. The benefit is that no one person or small group of persons can acquire sufficient power to make quick, unilateral decisions that permanently strip away individual rights.

Assertion 2: *The primary purpose of Congress is to resolve social conflict.*

A LEGISLATIVE PERSPECTIVE   319

We live in a diverse society — economically, geographically, ethnically, and racially. Different groups frequently have different needs, different ideals, and different values. Congress is the grand safety valve. It is the institution where differences are brought together, where everybody is offered a hearing, and where compromises that allow us to proceed as a nation are negotiated.

In seeking compromises to social conflict, Congress is forced to make value judgments. Although many would like to believe that Congress operates in an atmosphere where the best available information is sought and where logic and reason form the basis of each decision, the reality is frequently quite different. The simple fact is that Congress ends up with the social conflicts that lack clear-cut solutions, and Congress is forced to make its major decisions resolving social conflicts based on nonquantitative factors — what is best for my constituents, what is best for the nation, and significantly, what is best for my career as a member of Congress. In other words, when a member makes decisions resolving social conflict, he makes a series of value judgments.

Where information clearly resolves a problem, Congress generally exercises the wisdom to stay out. There are no bills under consideration requiring water to run uphill as a way of saving money on public works projects, or bills requiring perpetual motion to be incorporated in machinery as a way of conserving energy, or bills requiring men to conceive and bear children as a way of overcoming sexual biases. Those issues are clearly resolved by the information we have available.

Congress does get involved, however, when our best information is incomplete or irrelevant, yet immediate action is demanded.

Take, for example, our current social conflict over acid rain. All acknowledge that what goes up must come down. The exact steps between sending pollutants up and having acid rain come down and the relationship between where the pollutants go up and where the acid rain comes down are still under dispute. Our factual information is incomplete.

But the factual information is not the issue. The issue is how precautionary this nation should be while we continue to develop the information that will give definitive answers. Those living in the Northeast are concerned that by the time we develop definitive information, irreversible damage will have been done to their resources. Those living in the Midwest are concerned, however, that the potential solutions to acid rain will cost the Midwest large sums of money. Midwesterners want more information than is currently available to ensure that they will not invest in expensive equipment that may not contribute to a reduction in acid rain.

Congress will not make a factual judgment similar to judgments about

320 INSURANCE AND RISK MANAGEMENT FOR HAZARDOUS WASTE

water running uphill; Congress will make a series of value judgments based on economic impacts.

As another example, take time. Here, our factual information is near perfect but irrelevant. We can predict with great accuracy over millions of years when the sun will rise and set. Yet every two years the Congress addresses a major social conflict over daylight savings time — do we want that extra hour of daylight in the morning or the evening? Do we want to favor the farmer, who works in the morning, or the urbanite, who gardens after work? The scientific information is so irrelevant that during the last debate one Senator informed his colleagues that Eastern Standard Time is God's time and any other standard of time is immoral. (I guess the Senator was answering the question of what happened to southern California.)

Again, Congress will not make a factual judgment; Congress will make a series of value judgments based on the expressed desires of the various constituencies.

The point to remember here is that social conflicts arise because of differences in opinion. It is the function of Congress to resolve those differences by arriving at a collective value judgment.

Assertion 3: *If the function of Congress is to resolve disputes, then members of Congress are the dispute negotiators.*

By its very nature, negotiation and resolution of social conflict imply value judgments that distribute benefits to the parties involved. The benefits can be economic, psychological, aesthetic, or whatever, but in the end, each party to the conflict usually wants a distribution of the benefits that favors him. Resolution of conflict, therefore, necessitates both compromise and adversarial behavior by the various negotiating parties.

Under our system of government, the members of Congress are the federal negotiators for the various constituencies which exist in the United States. Constitutionally, a member's function is to represent his constituents aggressively in the Congressional process. The Constitution does not mandate statesmanship or even leadership. (In Washington, the definition of a statesman is a defeated politician.) Members are representatives and, as such, they follow their constituents.

If one thinks of members as negotiators, it becomes obvious that career advancement depends on the ability to win. Obviously, a member needs to win at the polls to retain his job. But just as important, he needs to win the issues he backs within the Congress if he is to succeed in his role as negotiator. There is little reason to negotiate if you do not want your position to prevail — and if a member does not win legislative contests,

A LEGISLATIVE PERSPECTIVE

321

he is viewed as a poor negotiator, and other members, who also want to be effective negotiators, will not follow his lead. If, however, the member is perceived as a consistent winner, he gains respect among his colleagues, the ability to influence others, and, eventually, a position of leadership within the Congress.

Members of Congress are negotiators who must build a winning record to survive.

Assertion 4: *Legislation is an ongoing social process.*

Legislation is like scientific experimentation. There is a perceived problem. A legislator proposes an hypothesis to solve the problem in the form of a bill. His hypothesis is introduced and makes its way through the legislative process, where others criticize it and suggest changes. As in science, most legislative hypotheses fail because of criticisms that are raised. The few that survive the criticism become laws, or the equivalent of an experiment that is tried out on the nation. Consensus as to the potential effectiveness of the experiment is measured by the size of the number of members who support enactment of the hypothesis. As information from the experiment becomes available, the conflicting interests reevaluate their positions, the compromise eventually disintegrates, and new hypotheses about the problem are considered. The legislative process, therefore, is an ongoing learning process of problem, hypothesis, experiment, and correction.

The process is a social process because, as discussed above, it involves opinions and value judgments. Formulating opinions and making value judgments is not a strictly rational process. It is a social process. Members sell opinions when they need support in their negotiations; members develop their value judgments by comparing their judgments against those of others. Hence, legislation is a social process and, hence, Congress is built around a very elaborate set of rules for social conduct. The rules for social conduct provide the basic framework and the flexibility that can accommodate the ongoing problem–hypothesis–experiment–correction process.

12.2 Practical Consequences of the Four Assertions

If you accept my assertions, then there are several corollaries, or practical consequences, that should be followed whenever one wants to affect the legislative process.

322 INSURANCE AND RISK MANAGEMENT FOR HAZARDOUS WASTE

Corollary 1: *The first assertion is that we have a remarkable system of government that was designed first and foremost to protect the rights of individuals. The corollary is that one needs to build a broad base of public support before major legislative changes can occur.*

Think for a moment what a conference like this one tells a Congressional type like me. It says that there is a group of people who believe the environmental experiment on the nation needs to enter the corrective phase.

I have talked to a lot of you over the past several years, and the argument for change goes something like this: Most of our major environmental laws were formulated during the early 1970s and are based on a concept of zero risk to public health. Since then, major changes have occurred that prove conclusively that zero risk can never be reached short of shutting down industry.

One of the changes has been the development of scientific instrumentation that can detect extremely low levels of contamination. In the early 1970s, it was extremely difficult to detect a few parts per million and the concept of zero contamination; hence zero risk to public health was a workable goal because *zero* equated with *undetectable*. This is not the case any more. Now technicians routinely analyze samples for pollutants in parts per million and parts per billion, and in the near future will be able to analyze them for parts per trillion. The improved instrumentation increases dramatically the costs of the environmental laws that are written to force us to zero contamination.

Another dramatic change has been computerization. In the early 1970s, computers that were less powerful than the ones we now hold in our laps took up entire rooms; it took hours to keypunch data onto cards; and the investigator was left to write his own program — in a computer language — to manipulate the data. Today, for a few thousand dollars, anyone can buy a desk-top computer with more capability than money could buy in the early 1970s; the data is entered directly from the equipment used for measurements; and for a few hundred dollars, the investigator can buy sophisticated statistical programs he does not even understand to manipulate his data. The result has been a dramatic increase in the number of studies establishing correlations between pollutants and potential health problems. While correlation does not prove cause-and-effect, it raises questions about precaution.

And, of course, a major, ongoing change is population growth. Factories that used to be on the outskirts of cities now have housing developments with $250,000 houses and up-and-coming attorneys pushing right up against the fences. Areas that were once abandoned and were ideal locations for

A LEGISLATIVE PERSPECTIVE

dumping of hazardous wastes now have schools or recreational parks built right over them. Virtually uninhabited areas in the middle of nowhere out west where aggressive mining was allowed are now important parts of the water catchment basins for Los Angeles or Denver. People are everywhere, and opportunities for exposure to pollutants continue to increase daily.

From a legislative perspective, this group is taking information that has become available during our legislative experiment with environmental laws and is asking very fundamental questions: Are the value judgments that were used to forge a majority coalition that supported zero risk environmental laws during the 1970s still valid? Is there an opportunity to form a new majority coalition around the idea that zero risk to public health is economically incompatible with manufacturing activities?

In my own view, those seeking a change in environmental laws to a risk management system still have a way to go to build a broad-based consensus among a majority of our Constitutional equals. Again, in my view, if this nation is to implement risk-based laws and preserve its manufacturing industry, we will need to accept some form of land-use planning. A majority coalition supporting land-use planning simply does not exist. It is totally unacceptable to the Westerners, and it is important to remember that approximately one third of the membership of the U.S. Senate, where an individual member can stop legislation from moving forward, is made up of Westerners.

Their concern, legitimately, is that because water is so scarce, allocation of the water dictates development within a given state. The idea that a GS-13 in the Washington office of EPA would exercise final approval over development in their states drives them wild.

I firmly believe we will edge toward risk-based environmental laws and land-use planning over the next decade. In fact, it is already starting to happen as part of the regulatory decisions EPA is making. But because the coalition for these changes will not exist until there is greater public education about acceptable risk, we cannot expect to see dramatic changes in environmental legislation over the near term.

Corollary 2: *The second assertion is that the primary purpose of Congress is to resolve social conflict; resolution of social conflict involves value judgments. The corollary is that whenever you enter the legislative arena you are engaged in an argument, and you must distinguish between information that is accurate and information that is persuasive. While the two subsets of information overlap, they are not identical.*

Business makes money by using accurate information. It is the idea of the better mousetrap. The firm that navigates its oil supertanker with

324   INSURANCE AND RISK MANAGEMENT FOR HAZARDOUS WASTE

satellite fixes stands less chance of running it aground and losing money than the firm that uses ancient star maps. The financial banker that uses computers to analyze stock-market movements has a significant economic advantage over the banker using slide rules. The farmer that uses crops bioengineered for pest resistance has a major advantage over the farmer who depends on farm labor to hand pick pests off his plants.

Unfortunately, the businessman tends to think that because accurate information is so important to him, it must be equally important to the politician. It is not. Social conflicts are resolved on the basis of argumentation, cajoling, salesmanship, and persuasive information. If we have a policy dispute about the effect of an environmental law, and you have a table of accurate numbers and I have a picture of a child who is dead or seriously crippled as a result of exposure to pollutants, I don't care what the table of numbers demonstrates, I will always win the argument, if for no other reason than I raise doubts about the degree of precaution we must exercise as a nation to protect innocent people.

If one is to enter the policy arena, one needs to remember that he is entering an argument, and he needs to frame his information to persuade: Have I treated others fairly? Have I been treated fairly? Think, for example, of the last time you had an argument with your spouse. Did you agonize over the accuracy of the information you were using, or did you shape the information for rhetorical impact? The same must be done in the policy arena.

Corollary 3: *The third assertion was that members of Congress are national dispute negotiators and they need to win to advance their careers. The corollary is that some decisions are not made directly, but indirectly, so that members do not set up losing situations for themselves.*

In a perfectly rational world, Congress would evaluate the benefit to the United States of various manufacturing activities that pollute the environment and pose a threat to public health, and then set levels of risk based on the benefit derived. Setting a specific level of acceptable risk — one cancer in a million, one birth defect in 10 million — would focus the debate on the intent of environmental laws, streamline the regulatory process, and reduce the level of uncertainty for the regulated community.

Do not hold your breath waiting to hear Congress debate acceptable risk on your televised coverage of the proceedings. For one thing, there are all the value-judgment problems associated with establishing benefits and risks. But more importantly, no member could survive the TV ad: "Your Congressman (or Senator) voted to expose his constituents to cancer. Was your mother's death by cancer preventable? Did the neighbor's

A LEGISLATIVE PERSPECTIVE

child come down with leukemia because of Mr. X's vote? Vote for Joe Blow; he won't let you down."

To avoid creating a losing situation, Congress has shared the responsibility to set levels of acceptable risk with the other two branches of government. It is an iterative process. Congress passes a bill with the general statement of intent. On becoming law, an executive agency creates regulations that set de facto levels of risk. If the levels of risk are too high or too low (e.g., too many people still are affected adversely by a pollutant, or industry makes a persuasive argument that the costs are too high in relation to the number of people protected), pressure is exerted on Congress and the agency to set new regulations and, indirectly, to set new levels of risk. Superimposed on this Congress–agency–public dialogue is judicial review of the procedures used to set the regulations.

Zero public health risk is a fiction we will continue to pursue in legislation until the public is much better educated to accept risk and until the members believe the public want them to address the questions of acceptable risk explicitly.

Corollary 4: *The fourth assertion is that legislation is an ongoing social process. The corollary is that the legislative process never renders a final decision.*

A piece of legislation, no matter how carefully crafted, will never be the final resolution of a significant social conflict. Once the legislative experiment is initiated, the correction phase will eventually become necessary. If the legislative experiment is initiated without broad public support, the correction phase will come sooner than if the experiment is initiated with the benefit of broad public support.

What I am trying to say is that a legislative trick to force risk management into environmental laws without public backing does not stand a good chance of surviving. Congress is made up of 535 CEOs, any one of whom is able to initiate a change in a decision that has been made by a previous Congress. All a member needs to do is get a majority of other members to agree to his change. The environmental risk management and integrated waste management systems that are the topics of this meeting will take several years of work to gain proper public support, starting with the iterative process, where some changes are taking place now, and culminating in legislative changes.

The ongoing nature of the legislative process assures you the opportunity to change laws you disagree with, but it also assures those with an opposing view the opportunity to change your legislative successes if you fail to generate or maintain broad public support.

326 INSURANCE AND RISK MANAGEMENT FOR HAZARDOUS WASTE

12.3 Conclusion

The most remarkable quality of the legislative process is that it wears people down. People approach the process frequently with some wild notions. They quickly learn they will need to compromise if they are to accomplish any of their goals. The process of compromise drives their thinking toward the political center. Knowing the ground rules for the process helps to gain certain advantages, but the advantages can be squandered quickly by rigidity and unwillingness to allow solutions to evolve.

In this context, one of the central issues this group should be addressing is not only the question of whether you can engineer legislative changes to the zero public health risk basis of environmental laws, but also whether or not you need to change your underlying assumptions about the business climate and the way you do your business.

If you include self-evaluation as part of your decision process, it will help you avoid producing policy papers that are never read and will help focus your efforts on the evolution of new solutions. Frequently, it is the groups that can accomplish breakthroughs in their own underlying assumptions that can find themselves at the center of coalitions for changes that truly benefit society.

Summary: Lee Fuller (representing Jack Clough)

The author of the paper on the legislative issue is Jack Clough, who is on the Energy and Commerce Committee of the House of Representatives. The premise of his paper was somewhat different, in the sense that it did not try to suggest how to argue about legislating an answer to these problems, but addressed the problems that have to be dealt with in legislation. He developed four assertions and four corollaries, which formed the basis for our discussions. I will describe those briefly.

The first assertion is that we have a remarkable system of government that is designed first and foremost to protect the rights of the individual. The corollary to that is that one needs to build a broad base of public support before major legislative changes can occur.

The second assertion is that the primary purpose of Congress is to resolve social conflict. Resolution of social conflict involves value judgments. The corollary to that is that whenever anyone enters the legislative arena, he is engaged in an argument, and he must distinguish between information that is accurate and information that is persuasive. While the two subsets of information overlap, they are not identical.

A LEGISLATIVE PERSPECTIVE

The third assertion is that members of Congress are national dispute negotiators and they need to win to advance their careers. The corollary here is that some decisions are not made directly but indirectly, so that members do not set up losing situations for themselves.

The fourth assertion is that legislation is an ongoing social process, and the corollary is that the legislative process never renders a final decision.

As a conclusion to this paper, Jack Clough suggested that what in fact happens when you go into the legislative process is that you become involved in a process that ultimately leads to a compromise. People start at different positions. In order to get a broad-based solution they, at some point, come together.

Chairperson's Remarks: Lee Fuller

I want to add a personal observation with regard to Jack Clough's last point. It is true that when you are dealing with legislation you will work towards a compromise; but that compromise will be among the parties involved in the compromise. If one is not a participant in the legislative process, then he will be outside the compromise. So the issue that, in large measure, becomes a question for a group like this, or for anyone, is how to get to the table — how to be part of the legislative process. That is where we focused a lot of our attention.

In some of the earlier presentations, there were some examples of issues like this that surfaced. For example, the industry group talked about the need to set standards — ambient standards, a regional approach in some instances — as a way of trying to define the process by which industry would then comply; and industry would be given a lot of flexibility to comply. Someone observed that the approach is a fine philosophy unless the standard happens to be zero. The difficulty is that when we are operating in a legislative context, the consensus may much more easily fall towards zero because the credibility of those of us who come in and try to present an alternative is not high enough to convince Congress, to convince the key members who are involved in the process, that in fact zero is not where the goal ought to be. Sometimes this failure is because the past history of industry in general, on these types of issues, has not been one that Congress is willing to take to the table and endorse as the reason why they think something other than zero is the right solution.

The insurance group observed that perhaps we need to be looking at different mixes of insurance options to deal with some of these complicated compensation dilemmas. One of those was the perception that perhaps,

328 INSURANCE AND RISK MANAGEMENT FOR HAZARDOUS WASTE

at some point in the process, we needed to have some type of government insurance come in. This was an issue that we, in Congress, tried to grapple with in the Superfund deliberations in 1984 and 1985 on victim compensation. There was a scheme developed in the Senate, the general concept of which was that it would try to use a federal funding source to pay certain compensation claims in somewhat of a last-resort type of option. If insurance was not there and the health claims could not otherwise be paid, under a fairly narrow set of restrictions, the claims could be made to a federal fund. Industry, by and large, uniformly opposed that provision. They did it for several reasons, one of which was because they were concerned that with the way the taxing mechanisms typically looked in Superfund, that tax would fall into a very narrow segment of industry as opposed to a broad-based type of segment.

In my view, however, the issue that may have most affected the members' judgments on whether or not they were going to pass that proposal in the Senate had much more to do with the previous experience that Congress had had with the Black Lung program. This program started out with high motivation and was originally expected to be a very small federal burden. In fact, it turned out to be a huge federal exposure, and the committees were constantly confronted with how to meet those federal commitments. Therefore, before they would tack on a new commitment that potentially meant an enormous future exposure, they were going to be very cautious about taking that step.

Another issue that was raised was the possibility of having cleanup contractors subject to a different set of liability standards, rather than the strict liability standards. That, in fact, was an issue in 1985 and 1986. It was an issue for which most people in Congress had a strong empathy. They did not pass Superfund with the idea that they were going to produce a result in which the people who were to clean up the site would wind up being potentially liable for all the results. That was not the intent. Nevertheless, the solution to it was much more difficult to arrive at because of a variety of reasons.

Our group talked about a lot of different elements that would be important issues to raise. The underlying theme at the end was that if we were to look where the next step of this type of process should go, that step would be to try to find a mechanism to get people to the table, to build their credibility in such a way that when the next legislative deliberations on Superfund or on hazardous waste laws take place, there is a trust that can be built up in the information that people want to bring into that discussion. One of the pivotal factors in making that type of trust exist is a belief by Congress that when we are talking, we are talking in a fair

A LEGISLATIVE PERSPECTIVE

329

manner; that we have treated our adversaries fairly and that what we are saying has a truth value to it that will be largely consistent with the other people who are to be involved in the debate. We have to come to the table with a quality of credibility that is not easy to obtain because of many of the reasons heard earlier today. Still, if we listen to what the papers suggested, we are going to have a broad-based solution. We are dealing with an adversarial process, so we must be persuasive. And we must also recognize that, in Congress, we are operating in a competitive world. It means that the people there, who are going to deal with us, are going to deal with us in such a fashion that they win as well. We must be prepared to give them information that will be acceptable and that will be publicly perceived as such. We will be frequently confronted with tough dilemmas in trying to achieve that kind of goal. Nevertheless, past experience suggests that failure to move in that direction will produce the same result as before: We do not become part of the final compromise when it is ultimately made.

IV TOWARDS AN INTEGRATED WASTE MANAGEMENT STRATEGY

13 A SYNTHESIS OF THE CONFERENCE

F. Henry Habicht II

Now I am paying the price for having avoided writing a paper in advance. Let me briefly set forth a series of observations for you. I think there are several themes and observations that have come through the discussion that are important for us to have clearly on the table as a foundation for future action. Then I want to suggest some ideas that have come to me both as a result of your discussions and through experience with another project in which I'm involved. These ideas will hopefully provide a basis for further action.

I must also pause for just a moment to commend Howard Kunreuther, the Wharton School, and Jack Mulroney and his support for this Center. I think we need to find a way to follow up on what we have discussed and to follow up through this Center as much as we can because it's a great beginning and an extremely sophisticated level of discussion — certainly more than I've heard at most of these conferences.

I will try to identify the threads that ran through all of our discussions in order to provide a foundation for further action. Let me begin with a series of first principles. I have often joked about having been on both sides of the hazardous waste problem, but one thing that is clear from these discussions is that there aren't two sides to the problem. It's a mul-

333

334 INSURANCE AND RISK MANAGEMENT FOR HAZARDOUS WASTE

tifaceted one, just as Howard Kunreuther pointed out at the beginning of the discussion.

There are many groups with distinct interests that have to be kept in mind. Obviously, the common thread is the public at large. What drives all the issues and problems and uncertainties and frustrations — and the source of the solution — is the public. We've heard this point often in these proceedings.

There are also a few other observations that came through in our discussions. First, we have to include the waste management sector as a separate and distinct group in our diagram of the various interested parties. That sector is growing, and deals with remediation, treatment, and recycling. This industry is looking to help provide the technologies to solve the problems inherent in waste generation, and I think that's very important.

Second, a principle that came through our discussions is that we have past and future dimensions to our waste problems. As Les Cheek said, the past dimension is casting a pall over our ability to address the future — sort of encumbering the surpluses. Obviously, we need to find a way to deal with the both the past and the future, but we can't let the past drag us down unduly in seeking solutions as we look ahead to future business activities.

Third, another common theme can be categorized as uncertainty and lack of information required for insurability. At some point, liability uncertainties can make the risk uninsurable, because joint and several liability bring other unforeseeable third parties into the liability mix and make it very difficult to calculate the risk of loss for a specific insured who might be liable for that third party's waste.

Fourth, to all of us there is frustration with *scientific* uncertainty. That uncertainty leads public institutions in the risk assessment process to assume the worst case. There is frustration on the part of the PRP community with the use of worst-case assumptions. On the other hand, we've heard Nick Ashford and others mention that even worst-case assumptions can understate risk in some cases because of synergy, chemical transformation, and that sort of thing. So uncertainty is bad for all of us.

Environmentalists don't have a stake in uncertainty, even though there is sometimes concern that the more uncertain things are, the more the laws will be pushed to the limits. I think everybody has a stake in certainty. Of course, Congress, which has been the butt of almost as many jokes here as lawyers and economists, obviously has a very difficult role. I think Jack Clough and Lee Fuller have described Congress's perspective extremely clearly. Congress often acts on far less than adequate data. We know that

A SYNTHESIS OF THE CONFERENCE 335

industry is in part at fault for not getting them the information policymakers need in the ways Lee described.

Finally, the courts foment uncertainty. Courts will always step in where the other institutions of government have failed to solve a problem, and the courts perceive that the public hasn't been satisfied with what the other institutions are doing. No matter what the Constitution says, no matter what conservative jurists assert, courts *will* step in to solve a problem, whether to compensate people who are perceived as having gotten a raw deal from a company or the system or to slap EPA on the wrist. The courts will intervene, and we have to keep that in mind as a source of uncertainty. We can't ignore that judges are inspired, as they say, by public concern. They read the newspapers too.

So this uncertainty is inherently destabilizing to what industry is trying to do, both by fueling public hysteria and by leading politicians to act on less than adequate information.

With the foregoing context in mind, I think that this diverse group is virtually unanimous in several views.

First of all, as we have said, the public's desires have to be understood; they have to be addressed somehow. The public is interested in zero risk, and there are two critical developments on this front that must be watched. One is the Community Right To Know program, which should be watched very closely because it was an outgrowth of distrust, a feeling that the public wasn't getting the facts. And as long as there is distrust, you are going to have regulatory deadlines and Community Right to Know programs. In addition, technological assistance grants and funds also ensure that the public is getting all the facts. Citizens may be allowed to hire their own separate layer of experts to look at what's happening at Superfund sites. These developments are very important in our search for solutions.

A second common view that all of us seem to share is a consuming interest in advancing technology. We have to create the conditions for promoting the rapid advance of innovation and technology. The unanimous agreement is that the structure of the current laws and regulatory system inhibit that advance and the introduction of new technology. Superfund promotes least-common-denominator approaches that often frustrate the introduction of new technology. Pressure to move toward permanent solutions, without inhibiting trying new things — political and otherwise — is very important.

A related concern that seemed to be unanimous was that, in addition to killing all the lawyers, we need to kill all the landfills. We need to get everything out of landfills that is at all possible. This seems to be a goal

336   INSURANCE AND RISK MANAGEMENT FOR HAZARDOUS WASTE

on which everyone agrees. I think Marcia Williams and others made a very important observation, which is that getting waste out of land as much as possible is a nice goal to have, but neither insurers nor PRPs or the public can lose sight of the fact that landfills are going to be with us for a long time. We need to work to educate everybody involved in this issue.

The term *hazardous waste* was used fairly loosely in the discussions, and we need to be able to differentiate in some meaningful way among wastes. Marcia suggested differentiating between raw waste and treated waste. There are residuals that have to go in the landfills no matter what we do, and we can't ignore that fact as we put together an integrated waste management strategy. Landfills are going to have to be part of that integrated strategy, so we've got to deal with the key distinctions among types of waste to ensure using needed landfills in the safest way. We must also find a strategy that will push us toward getting maximum waste out of landfills, again by promoting a technology that will do it. In short, there is a strong feeling that, as much as possible, society needs to get away from the old *pollute and cure* approach and move toward an *anticipate and prevent* approach. Everybody agrees on that, and I think that the environmentalists would certainly agree to that. Also, in any waste program, we need to have systems that will allow technology to catch up with public desires. I would like to develop the point that current law discourages new technologies, but time is limited.

There was also an interesting discussion about the differentiation between the big guys and the small guys; I haven't focused on this point in the same detail that emerged in this meeting. Often cited were the difficulties faced by smaller companies, in terms of getting insurance, self-insurance, insurability, and also being able to bring the best technology to bear on waste streams. In crafting solutions, I think it is important to distinguish that the big guys and the little guys are disparate in their positions and in their abilities to deal with issues. We need to see if the institutions in society can help to address that disparity.

Finally, as another view shared by the discussants, I think Martin Katzman had an important observation. As we look at the external forces that influence waste management behavior in our society, we of course focus on the legislative and regulatory regime. That's one set of external forces. There is also a second: the tort regime. The common law regime with its desire to compensate is, as we all know, a regulating agent as well. Of course all of these elements are fueled by the public in one way or another, so again the public's role is stipulated.

But there is a third external force, which is the marketplace. While insurance unavailability has had some salutary effects, as Jack Morrison

A SYNTHESIS OF THE CONFERENCE

337

has indicated with respect to pushing people toward more conscientious behavior, as an overall view I get the sense that most people agree that having insurance not available is a bad thing. In the American psyche, insurance is equated with security. Jack Mulroney was saying that this morning, and I think he is absolutely right. A marketplace without insurance is inherently destabilizing in terms of public hysteria. If insurance is the industry with the experts who assess and quantify risk — and if they are staying out of the market — you can imagine that it's going to be enormously difficult to reassure the public that things are going all right. So we have to find ways, without sort of skewing the world, to get insurance back in the marketplace.

We all have a stake in insurability, not only because of the financial security that it provides, but also because of the psychological security that it provides to the public. There is force to the assertion that if insurance isn't able deal with the risk, then the lay member of the public is not going to be able to do so either.

As we turn to solutions, I want to reemphasize that the public is the glue that must hold all elements of our integrated strategy together. The public is the key to a solution, but no solution will be easy. The locality seems to be the key element of the public that has come through all our discussions. Looking at the local community is our key first step, not looking at special interest groups and addressing each interest group one by one. The wisdom of Jefferson comes through here. Hazardous waste issues are certainly local in their characteristics. And because of the siting issues we've talked about, it is absolutely critical that we focus on the local community.

One of the discussions that we have had over and over again is that the public fears chemical risk as something near and immediate. But the true scope of the risk and what they are giving up for reducing that risk is something that is very remote and unclear to the average citizen. That is why people perceive risk the way they do. This remoteness or attenuation of the cost or burdens is something we simply must address, and the only way to do that is to deal closely with local communities. I'll have a suggestion on that as we move forward.

Let me make a few suggestions that I hope will keep up this momentum that has been building over the last 26 hours. Again, the underlying premises on which we build these suggestions are distinguishing the past from the future, the need for information and data, and the need to involve communities, particularly local communities, in this solution. We also must keep in mind the need to integrate approaches in a way that promotes new technology as much as possible, and we must keep in mind that insurance does have an important role. Getting the risk picture to the point that

338 INSURANCE AND RISK MANAGEMENT FOR HAZARDOUS WASTE

brings insurance back into the market seems to be a good objective. The list of observations I've just gone through and others that have come out during this conference are not exhaustive, but based upon the level of sophistication of this discussion and my familarity with the field, I think that we have an excellent list on which to base solutions.

In order to have a chance of succeeding, we have to have a strategy that addresses every one of those elements and any others that you think are important. Accordingly, the action items that we need to think about pursuing are, first, to deal with the past. It is dragging us backward. Past contamination is related to the future and, in a sense, we are solving the problems of the past as a way of helping to build the technology and ideas for dealing with waste in the future. To deal with the past, we have heard that we've got to find ways of bringing insurers and PRPs and other interested parties together in a negotiating forum, rather than in court. This is absolutely critical, and bringing these parties together is an idea that Les Cheek, I, and others feel the Wharton Center could constructively explore as a facilitator. At least Howard Kunreuther could convene a group of insurers and PRPs to talk about these issues and see that there are ways of reaching a rapprochement on some of these problems.

And we have also discussed the need to settle Superfund cases. We all have an interest in Superfund working efficiently and quickly and also working in a way that allows technology to catch up with the problem. We have to craft solutions to these sites and of course to have settlements that minimize litigation; in short, we have to deal with the past because it does relate to the future.

Second, with regard to the future, we heard some fascinating discussions about new approaches to insurance and how to get insurance back into the market. We have to bring back insurance. We obviously are in a transitional phase; people are first in shock and paralysis, and then they move toward coming up with innovative solutions. We've heard a helpful perspective on these things from our friends at Swiss Reinsurance. We heard some very good ideas for trying to get insurance back into the market. Dennis Connolly and others have talked about ways of restructuring the contractual relationship. The Wharton people have put together ideas about *layers of coverage* of losses in this area. All of these ideas are excellent. We only spent a couple of hours on them, and I think it is critical that at Wharton or somewhere else we put together a conference with sophisticated people like these to do nothing but brainstorm about new joint approaches to waste management by insureds and insurers, assessing risk, and bringing back insurance so that more companies can access the EIL market again, sooner rather than later.

A SYNTHESIS OF THE CONFERENCE 339

Again we've talked a lot about the liability issues. Maybe we have to leave the situation where it is now until Superfund's reauthorization but we need to look at whether or not joint and several liability could be modified, without being scrapped, in the ways that the insurance group talked about. In this way, insurance can build a relationship with PRPs in which they get the information they need on an ongoing basis and there isn't this cloud of uncontrolled liability that is too unpredictable for the insurers.

Next, the exploration of administrative tort claim or administrative claims for true cases of environmental illness is something that is worthy of a separate conference or separate effort or study.

We also need a conference focusing on promoting technologies, with the right people together, particularly from the treatment and service industry and the manufacturing sectors, as well as government, including career people. I found in my work in government that the career people who work in these programs have wonderful ideas for technology, and often are privately frustrated that the program that they have to administer frustrates the promotion of new technology.

We need to get these people together and talk about how programs can be fine-tuned. I know the NRDC is sitting down with individual companies, looking for ways to minimize waste on a voluntary basis. Companies under reasonable circumstances are agreeing to do this. They are looking both at revising government programs and at nongovernment voluntary agreement for dealing with waste minimization and advancing technology. We all have an interest in allowing technology to flower and to bloom. This is critically important.

There are a number of other ideas, but let me close with the two biggest issues and suggested approaches to them. The first one is uncertainty and lack of information. The second is the level of public emotion about hazardous waste. With regard to the lack of data, the Coalition on Superfund is designed to support good objective research. Industry can change its image. By supporting good objective research at the best institutions, nothing will be hidden from the public. In fact, the public will be involved in this gathering of data. Essentially, we hope the data will tell a story without any biases. We are making every effort to ensure that the public is involved in commenting every step of the way. In this effort, industry can create a base of information that will serve as an input for making decisions.

Sophisticated arguments alone will not convince anybody to say that a given level of risk is acceptable. One has to develop information and data. If the informed public still believes that society should move toward zero risk, this information will allow us to find the best way to get there. The

340   INSURANCE AND RISK MANAGEMENT FOR HAZARDOUS WASTE

data will allow us to look at the programs and say, for example, "Okay, if the public prefers not to abandon the goal of zero risk, we must find better ways to get there." Maybe we don't need to reach it tomorrow. If we wait a year or two years, which is the theory behind the Clean Air Act debate, we may find more efficient ways to do it that won't be disruptive to the marketplace and won't unduly disrupt the economy.

We're at the point where the polarization is still too great even to have rational debates among all interested parties about how to get from here to there. Such debates need facts. The coalition on Superfund is just one effort along these lines if it goes forward and achieves reasonable acceptance. That's a fairly long-term effort — at least a couple of years — but it is important to get started on it with regard to all these problems we have identified.

My final area of suggestions concerns the level of public emotion. We heard the phrase "Educate the public" very often. "Educate the public" is a loaded phrase. I think the better phrase is one that Marcia mentioned last night: "Involve the public." It would be very useful for an organization like Wharton to pursue and find a way to focus on local communities in the siting area. We are trying to do that with Bill Ruckelshaus in the State of Washington. There we seek to create an open process where the community feels it has an absolutely free choice as to the way cost and benefits are shared. There are incentives for the community to site facilities if they are informed and feel they have a choice.

It's truly the grass roots that bubble this emotional hysteria up to the members of Congress, so we have to focus upon the grass roots. The trade associations need to move out to the communities. Also, we must try to understand what makes local communities tick and to address the siting issue directly with them.

Issues like landfills are going to be with us for some time, and by going back to the grass roots, one may be able to find source of agreement between industry and local communities. I think these local agreements are a very good idea. Those agreements are enforceable in court but are negotiated to the extent feasible between the community and the businesses, rather than being arranged through a more cumbersome regulatory system.

Let me stop here. I think this Wharton Conference is a noble effort. We need ways to keep in touch with each other and to keep moving this process forward, rather than just going home and saying: "Boy, that was a great conference; I wonder what happens next." Following up with actions such as those I have briefly suggested is critically important, because this is the start of a major transitional phase. Let's not wait for the sky to

A SYNTHESIS OF THE CONFERENCE

fall before we make some important changes. We should begin today by committing ourselves to an open process of breaking down the hazardous waste problem into its components and finding technologies and processes that will inspire public confidence. Focused studies and conferences at places like Wharton can move this effort forward. I wish us all well as we proceed.

14 CONCLUSION

Howard Kunreuther and Rajeev Gowda

14.1 The Search For Solutions

The Wharton conference succeeded in bringing together representatives from the major interested parties — manufacturing, insurance, government, law, the environmental movement, and academe — which have a stake in the waste management process.

The papers presented at the conference raised key issues and pointed out different stances. The conference provided an opportunity for developing elements of an integrated waste management strategy that promises to clarify the costs and benefits of various actions, thus facilitating some compromises among the interested parties.

Integrated waste management means coordinating the production, transport, treatment, and disposal of hazardous waste across all media (air, soil, water) — dealing with waste from cradle to grave. Such a comprehensive concept links the system or systems of waste management with the objectives and agendas of the various parties. It challenges the old mind-set of *pollute and cure*, which has been partially responsible for the current anxiety about the quality of the environment, and advocates the approach of *anticipate and prevent*. It implies both control of the waste problem and a collective pulling together of all parties.

343

344 INSURANCE AND RISK MANAGEMENT FOR HAZARDOUS WASTE

Each of the interested parties has its own special concerns, but for institutional reasons, lack of interest or funds, or the sheer complexity of private- and public-sector systems, some voices may be excluded. The conference emphasized the continuing need to assure everyone a voice.

Crucial to the success of the conference were the small group discussions. The small groups were composed of participants representing a cross section of the interested parties. Their discussions revolved around papers presented on each of the viewpoints that characterize the environmental policy debate. The clash of perspectives at the small group sessions ultimately coalesced into a set of challenges and plans. These form the core of our recommendations in these concluding remarks.

The theme pervading the search for solutions is that collaborative action, as opposed to the adversarial attitude that has characterized the environmental policy arena in the past, is the key to a successful resolution of the colossal problems that still confront us.

Time magazine captured the magnitude of environmental problems by diverting from naming a usual man or woman of the year to declare "The Endangered Earth" the "planet" of the year. *Time* pointed out the tremendous problems both in locating and cleaning up waste sites and in simultaneously formulating strategies to control the volume of waste produced. And the magazine noted the worldwide impact of the waste crisis — "Few developing countries have regulations to control the output of hazardous waste, and even fewer have the technology or the trained personnel to dispose of it." To remedy the appalling lack of knowledge about the toxic effects of waste, *Time* called for increased funding for research.

The solutions proposed and discussed at the Wharton workshop — source reduction and recycling — are reflected in *Time's* article. As *Time* stated, "Landfills and incinerators, however harmful their emissions, will be needed as part of well-managed waste disposal systems for the foreseeable future."

Working diligently, workshop participants came up with a set of challenges and potential problems inherent in developing an integrated waste management strategy. We shall organize our recommendations in the same framework that guided the conference.

CONCLUSION 345

14. Towards Integrated Waste Management

14.2.1 Risk Assessment

Challenge: Need for state-of-the-art assessments
We need a scientific understanding of the risks posed by hazardous waste to human health and to the environment. Risk assessments must be able to evaluate the synergistic effects of hazardous wastes, in which combinations of wastes, typical of Superfund sites, could cause more damage than the individual chemicals that make up the wastes.
Potential problems
The fear of misusing less-than-conclusive research findings by parties with their own agendas makes other parties wary about adopting risk assessment in policymaking. Environmentalists, for example, fear that risk assessment could become a tool of those preferring to scale down the national hazardous waste management effort.
Challenge: Risk assessment for prioritizing cleanups
Risk assessment, in its present state helps regulatory agencies prioritize cleanup actions by providing data for rank-ordering sites.
Potential problems
Opponents of risk assessment in environmental policy decisions cite the uncertainties in the estimates and are reluctant to engage in tradeoffs implied by benefit–cost analyses.
Challenge: Risk assessment and the public
People need to be provided with data on the benefits as well as on the limitations of risk assessment. Attention must be paid to the underlying assumptions guiding any analysis.
Potential problems
Because people have difficulty comprehending probabilities, they are likely to understand only imperfectly the elements of uncertainty and ambiguity in risk calculations.
Challenge: Assessing risks associated with incineration and alternative methods of waste treatment and disposal
The risks posed by alternative waste disposal methods must be assessed before deployment. For example, although incineration is a promising alternative to land disposal, the threat of increased air pollution from incinerators and problems of disposing of ash are reasons for caution.
Potential problems
Short-cut solutions to the waste management problem, like transferring hazardous waste from one site or medium to another, must be avoided.

346   INSURANCE AND RISK MANAGEMENT FOR HAZARDOUS WASTE

Long-term solutions that actually get rid of the waste must be the top priority.

14.2.2 Insurability

Challenge: Cooperative insurer–insured problem solving
Cooperative efforts are vital to reduce transactions costs and speed the pace of waste management.
Potential problems
Insurers are reluctant to take a more direct role in risk management by conducting more intensive risk assessments or by monitoring clients' activities. Both industry and their insurers have incentives to pass the economic burden to the other.
Challenge: New Risks: Insurer risk assessments and monitoring
Insurers need to refine their engineering risk assessments and actuarial techniques in order to tackle the unique challenges posed by hazardous waste risks.
Potential problems
The insurance industry must overcome its reluctance to be an active and direct player in conducting risk assessments and managing risks.
Challenge: New Insurance Solutions
New alternative techniques to ensure financial responsibility are vital to the conduct of industrial activities. Alternatives could include insurance pools and/or governmental involvement through reinsurance.
Potential problems
There are difficulties in organizing cooperative pools between low- and high-risk parties. Further, governmental involvement in insurance markets may be viewed by the insurance industry as interference with the functioning of private competitive markets.

14.2.3 Risk management

Challenge: Regulating waste management across all sources
Regulation, as part of an integrated waste management program, needs to focus on *all* sources of pollution and hazardous wastes. The scope of regulatory oversight should go beyond merely industrial generators of wastes to other, often overlooked sources; for instance, agricultural chemical leachates, municipal-wastes, septic tanks, road de-icing salts, urban

CONCLUSION 347

runoffs, and other, perhaps more pervasive and serious threats to groundwater.

Potential problems

The complexity and far-reaching nature of the problem may require a revolution in methods of production and environmental management.

Challenge: Source reduction — product redesign and recycling

A product's lifecycle runs from conception, design, manufacture, use, deployment, through disposal. Pollution controls must be instituted as early as the design phase, when potential hazards can be engineered out. Recycling must also be made a viable option.

Potential problems

The structuring of incentives for source reduction can be a very complex task. Counterproductive overregulation must be avoided. Feasible recycling faces many operational hurdles: it requires networks for waste collection and renewal and the disciplined cooperation of people even before they have economic incentives to recycle.

Challenge: State-of-the-art waste treatment facilities

State-of-the-art facilities for treatment, storage, and disposal of wastes must be sited to handle the tremendous amount of waste that will continue to be generated and for which the available capacity is rapidly diminishing. Some type of benefit-sharing or compensation to the host community may be needed to facilitate local cooperation in the siting of these facilities.

Potential problems

Public opposition to the siting of waste facilities is based on fear of the potential damage to the neighborhood from hazardous waste and outrage at what the public perceives as unfairness in the site-selection process. The public must be convinced that these facilities are safe and will be closely monitored and controlled.

Legal and legislative arrangements for waste management will affect these approaches. Further, the perspectives of key interested parties like industry, the insurance sector, and environmental groups will have a significant impact on the type of waste management programs that evolve in the coming years. Finally, improved waste management across society requires an informed citizenry; risk communication has a major role to play in this regard.

14.2.4 Industry

Challenge: Encouraging cleanup by potentially responsible parties (PRPs)

Self-initiated cleanup by PRPs, which adhere to the required standards,

348 INSURANCE AND RISK MANAGEMENT FOR HAZARDOUS WASTE

are an ideal way of achieving the original goals of Superfund since they ease the burden on Superfund and increase the pace of cleanup.

Potential problems

Adversarial attitudes, rather than cooperative problem-solving, are encouraged by the existing waste management framework, which relies largely on new toxic torts to settle disputes. Thus, problems of determining financial responsibility for cleanup expenses seem resolvable only through the time-consuming and costly process of the court system. Strict liability for cleanup contractors also hinders voluntary cleanup.

Challenge: Information and technology exchange across companies

Small and large companies must freely exchange information, research results, and technological expertise through data bases and industry trade associations.

Potential problems

Companies are reluctant to part with trade secrets and research results that give them a competitive advantage.

14.2.5 Legislative

Challenge: Clear liability guarantees

Companies that adhere to regulatory standards must have their liability curtailed. Retroactively imposed liability that penalizes even companies that abided by regulatory standards destabilizes the environment within which industries can function.

Potential problems

Prevailing uncertainties in hazardous waste management, combined with the polluter pays principle dictate against limiting potential liability even for companies that follow the law.

Challenge: Fine tuning economic incentives for waste management

The liability regimes set up by RCRA and CERCLA succeeded in creating a market for wastes by making companies aware of the true social costs of the wastes they generate. Some firms have been encouraged to change their production processes in order to reduce the amount of hazardous waste they generate. Such economic incentives must be fine-tuned to realize specific public policy goals — for instance, ensuring effective enforcement.

Potential problems

The liability regime instead of creating a market for liability insurance could also drive insurers out of the market. Rules such as joint and several liability are crippling the critical financial responsibility component of the waste management effort.

CONCLUSION 349

Challenge: Legislative consistency

The various environmental statutes must have consistent standards even if they separately address different media like air, water, and land.

Potential problems

Legislation on a single topic is passed piecemeal, attacking different problems as they demand attention. Hence inconsistencies in standards could arise perhaps allowing greater pollution in some media than in others. Waste generators might be falsely encouraged to take short cuts and dispose of their waste in another medium instead of developing less harmful waste disposal technologies.

Challenge: More streamlined and flexible regulation

A cumbersome regulatory system will bog down the nation's waste management effort. The permit process, in many cases, could be streamlined if rules were laid down for individual proposals that follow a generic pattern. Proposals should be approved by default if they are not processed by regulatory authorities within prespecified time limits. Only those projects that deviate from an established framework or a common type need special attention.

Potential problems:

Balancing discretion and rules is not easy. Flexibility can be misused and lead to the slackening of enforcement.

Challenge: Alternative dispute resolution techniques

Alternative dispute resolution mechanisms such as mediation and arbitration can avoid the problems of the tort regime.

Potential problems

Such mechanisms that are enforceable and that satisfy all involved parties as unbiased are difficult to establish.

Challenge: Tackling small company problems

Small companies should be encouraged to perform risk assessments and develop risk management programs. Methods must be devised to bring these companies into an integrated waste management effort.

Potential problems

Adequate risk assessments are beyond the economic means of many small companies. It is also difficult for them to meet financial responsibility requirements.

14.2.6 Risk Communication

Challenge: The need for unbiased sources of information

Efforts must be made to establish credible, unbiased sources of infor-

350   INSURANCE AND RISK MANAGEMENT FOR HAZARDOUS WASTE

mation that can be relied on by the public and decision makers on all sides of the hazardous waste spectrum.

Potential problems

The uncertainties and lack of scientific expertise that characterize hazardous waste management create conflicts in the information presented by different interested parties.

Challenge: Improved public understanding of risks

For the public to make decisions on the tradeoffs between risks and benefits, it must understand risk better. SARA Title III, the Community Right to Know law, must be carefully implemented in letter and spirit.

Potential problems

The tendency for the media to highlight salient events could lead to public misperceptions of the nature and extent of the potential hazards from waste.

INDEX

2-AAF (carcinogen), 39
ACE program, 292
Acid rain, 319
Action Memo, 90
Adverse selection, 295
Agency for Toxic Substances and Disease
    Registry (ATSDR), 89, 90
AIA. *See* American Insurance Association
Aircraft industry, safety efforts over the
    life cycle, 110–111, 113, 123
Air guideline, ambient, 250, 267
Air transport dispersion model, 21
Alternative dispute resolution
    settlements, 301, 302
    techniques, 80, 81
American Corporate Counsel Association,
    55–56
American Home court, 48–49
American Insurance Association (AIA)
    insurance policies and waste generator
        reimbursement to EPA, 45
    proposal to overhaul CERCLA, 60
American International Group, 274

American States Insurance Co. v. *Mary-
    land Casualty Co.,* 49
Amoco-Cadiz oil spill, stigmatization, 204
Animal bioassay studies, 24
Antilitter movement, 220
Applicable or relevant and appropriate re-
    quirements (ARARs), 24, 94
Aquifers, 26, 27, 28
ARARs. *See* Applicable or relevant and
    appropriate requirements
Arbitration, 301, 302
Armco, Inc. v. *Maryland Casualty Co.,* 45,
    51
Arsenic ingestion potency, 19, 24
ASARCO smelter, risk assessment con-
    ducted by EPA, 21
Asbestos Claims Facility, 65, 66, 68, 72
Asbestos disease, 108, 109; *see also* Cancer
    early warnings, 111–113
    insurance coverage cases, 45, 57
    Mississippi law decision, probability of
        contraction, 62
    profile of claimants, 65

351

352INDEX

Ashland Oil/Pittsburgh spill, 258
Assets and liability evaluations, 177
ATSDR. *See* Agency for Toxic Substances and Disease Registry
Auditing programs
  for compliance with regulations and laws, 259, 268
  for waste monitoring, 306
  to assure that corporate standards are upheld, 248
Austrian industrial liability policies, 145
Ayers court, 48–49

Bankruptcy, cleanup costs taking first claim on the assets, 275
Barmet of Indiana, Inc. v. *Security Insurance Group,* 49
Basic Error Rates (BER), 114
BDAT. *See* Best demonstrated available technology
Benzene, 20, 234
Benzo(a)pyrene, 20, 39
BER. *See* Basic Error Rates
Best-available technology standards (BAT)
  determined medium by medium, 233
  and land-disposal ban, 231
  set by EPA, 260
Best demonstrated, available technology (BDAT), 228
Bhopal incident (India), 106, 107, 109, 110
  "hazardous clouds" accident, 258
  risk perception, 197, 201
Binding arbitration, 80, 81
Binghamton fire (New York), 109, 110
Bioavailability, 22
Biodegradation, 28–29
Biotransformation, 29
Biotreatment of waste, on-site, 306
Black Lung program, Congressional legislation, 328
Bottle bills, 222
Bravo (offshore oil platform), crude oil pollution of North Sea, 108, 109

CAER. *See* Community Awareness and Emergency Response

Cancer; *see also* Asbestos disease; Carcinogens
  caused by two fundamentally different processes in animals, 38
  and exposure to waste burial site, 5
  hazard estimation, 39
  lung, 19, 328
  skin, 19
  trichloroethylene as a carcinogen, 29
  vinyl chloride as a carcinogen, 29
Carcinogens, risk assessment guidelines adopted by EPA, 15–16
Carcinogen Assessment Group, 18; *see also* Environmental Protection Agency
CERCLA. *See* Comprehensive Environmental Response, Compensation, and Liability Act
CERCLIS. *See* Comprehensive Environmental Response, Compensation, and Liability Information System
CGL. *See* Comprehensive general liability insurance policies
Challenger accident, risk perception, 197, 201
Chem-Dyne Corporation, 296n.
Chemical conversions, 23
Chemical Manufacturers Association (CMA), 46
Chemicals
  concentration determination, 24
  evaluation for potential toxicity, 13
  health hazard assessment of commercial marketed ones, 1
  potency, 20
  risk assessment guidelines, 16
  toxic, exposure to, 19
Chernobyl, risk perception, 197, 201
Chlordane, 20
Chlorofluorocarbon CFC-12, 106
Chromosome damage, 20
City beautification, 218–219
Claims-made policy, 285–286, 288
Clean Air Act, 97–98, 172
  actual dates set for the elimination of pollution, 231
  finding of better ways to reach zero risk, 340

INDEX

353

section 112 and large industry shutdowns, 225
Clean Sites, Inc., 68
Cleanup contractors, strict liability standards, 301, 328, 348
Cleanup costs, first claim taken on bankrupt company, 275
Clean Water Act, 97–98
  Best Available Technique (BAT) standards for toxic pollutant discharges, 228
  compliance seen as costly, 240
  extended to cover spills of hazardous chemicals on or near navigable water, 226
  jurisdiction of EPA extended to onshore spills, 226
  oil-spill cleanup program by EPA, 225–226
  Section 208, 221, 223
  wastewater stream regulation by EPA, 262
Clean Water Act Amendments (1972), Senate Report on, 221–222
Clinical ecologists, 62–63
Clioquinol, 109
CMA. See Chemical Manufacturers Association
Coalition on Superfund, 339
Coast Guard, oil and chemical spills from vessels treated, 226
Coinsurance, 291, 295
Combination airshed effects, 270
Community Awareness and Emergency Response (CAER), 259
Compliance, with laws and regulations by an industry, 247–248, 250
Comprehensive Environmental Response, Compensation, and Liability Act (CERCLA, Superfund) (1980), 5–7, 44–45, 47, 86–87, 98; see also Environmental Protection Agency
adverse selection as a result of retroactive imposition of liability, 279–280
amendments requiring ARARs, 94
cleanup costs modified by SARA, 52–53, 55, 58
cleanup and management of hazardous waste facilities, 13, 141

cleanups financed by insurance pool, 128, 129
Congressional consideration of reauthorization in 1984, 63
Congress not admitting its adoption was wrong, 78
current average site cleanup cost, 44
depriving insurer of predictability needed to offer coverage, 14
emphasis on cleanup of existing sites, 273–274
exclusion of onshore oil spills, 226–227
government to handle large liability losses, 292
groundwater quality standards as the goal of cleanup, 230
insurer involvement, 281
joint and several liability interpretation, 277, 278, 279, 294
joint, strict, and several liability of corporations, 178–179
liability, 81
liability standards leading to corporate auditing system, 177
major effort to overhaul by Congress, 60
parens patriae actions, 64
potential sites, 10–11
prevention of an event from occurring, 69–70
reauthorization in 1990–1991, 61, 78, 293
and revision of Hazard Ranking System, 92
risk assessment approaches, 16
settlements needed to minimize litigation, 338
several liability standard incorporated, 177
site liability, 44
tax structure, 64
tort-law-based liability scheme, 59–60
worst-case scenarios and environmental liability, 166
worst-case scenarios and potential sites, 166
underground storage tank cleanup, 227
victim compensation issue, 328

354 INDEX

Comprehensive Environmental Response, Compensation, and Liability Information System (CERCLIS), 41–42; *see also* National Priorities List
Comprehensive general liability (CGL) insurance policies, 47, 51, 81
 no coverage for waste generators reimbursing EPA, 45
 pollution exclusion, 48
 waste site coverage issue, 79–80
Computer air-dispersion models, 250
Computerization, 322
Conservation Chemical Company, cost-recovery cases, 55
Conservation movement, 219–220
Continued associations, method of, 204–206
Control efforts, vs. emissions, 106–110
Corporate conscience, 177, 180
Cost-effectiveness of cleanups, and benefit justification, 100–101
Critical mass, 72
Cross-media waste-minimization program, 257
Crude oil pollution, 108, 109

Danish industrial liability policies, 145
DDT, presence in food, 107, 108
Deductibles, 295
Delaney Clause, 16
De minimis standards, 192, 234
Dermal exposure, 25–26, 34
DES, 108, 109
Destruction of waste, on-site, 306
Detoxification, 306
Dibenzofuran, contamination during Binghamton, N.Y. fire, 109, 110
1,1-Dichloroethylene, inhalation exposure toxicity, 26
Dilution, as solution to pollution, 107, 108
Dimethyl sulfoxide, dermal exposure toxicity, 26
Dioxin, 2, 5, 20, 108, 109
 bioavailability in soil and fly ash, 22
 contamination during Binghamton, N.Y. fire, 109, 110
 early warnings, 111–113
 and incineration, 185

not listed as hazardous in RCRA regulations, 172
potency factors, 19, 24
Seveso incident stigmatization, 204
toxicity effect on humans not fatal, 35
toxic tort case involving a tank-car derailment, 62
Dispersion models, 21
Disposable packaging, 220
DNA adduct formation, 20
Dose, 18
Dose-response
 curves, 17, 21
 modeling, 17, 19
 relationship, 16
Drinking water exposure, 23
Drum storage pads, permits for, 248
DuPont Company, waste minimization, 128
Dutch insurance pool, 131–132, 151–152, 162
Dutch Public Liability Policy, 151

EDF. *See* Environmental Defense Fund
Effective dose, 21
EIL insurance. *See* Environmental impairment liability insurance
Eligibility of waste disposal sites, 154–155
Emergency Planning and Community Right to Know Act (EPCRA) (1986), 86, 304, 309, 313, 335, 350
Emergency Preparedness Law, 86
Emissions control
 end-of-pipe control installed, 305
 Rohm and Haas program, 250–251, 266
 standards, 145–146
Empirical state-of-the-art approach, 258
End-of-pipe control, toxic air emissions, 305
Environmental Cleanup and Responsibility Act (ECRA) (New Jersey law), 179
Environmental Compliance Services, 274
Environmental Defense Fund (EDF), 229
Environmental impairment
 European industrial liability policies, 150, 151
 and insurability, 147–148
 and liability life policy, 157

INDEX

Environmental impairment liability (EIL) insurance, 272–273, 285–289, 298
  contract design changes, 285–286
  enforceability of insurance contracts, 286–287
  integration of a scheduled compensation system with the tort system, 288–289
  market needing reassessment, 338
  organizational form changes, 287–288
Environmental impairment liability (EIL) insurance covers in Sweden, 131–132, 152–153, 162–163
Environmental pollution, nature of the problem, 3–8
Environmental Protection Agency (EPA); *see also* Carcinogen Assessment Group; Comprehensive Environmental Response, Compensation, and Liability Act
  air-transport models, 23
  apportionment of cleanup liability, 46
  cleaning up of a landfill, 8
  difficulties in implementing RCRA, 223–224
  enactment of regulatory controls, 308
  Exposure Assessment Guidelines, make explicit what is now implicit, 38
  extent of polluters' responsibility, 5–6
  financial responsibility requirements mandated, 5–6
  first-come, first-served system of control, 232
  first series of hazardous waste regulations issued, 183
  groundwater strategy, 27
  guidelines for assessing the risk of carcinogens, 15
  hazard ranking system utilized, 14
  improvement of risk management, 86
  jurisdiction asserted over groundwater, 221
  performance standards for disposal facilities, 223, 224
  permit process challenges, 349
  permit system, 223
  potency factor reconsideration of hazardous chemicals, 24
  preliminary cleanup assessments for sites on the CERCLIS list, 42–43, 44
  preparation of the Record of Decision, 93
  regulation of goitrogens, 38–39
  regulatory powers partially passed to the insurance industry, 274
  rigid schedules for carrying out land-disposal program, 227–228
  risk assessment guidelines updated in 1986, 17
  risk assessment of ASARCO smelter, 21
  risk perceptions compared to that of American public, 197
  Science Advisory Board, 19
  small-generator exclusion preserved, 227
  spending too much CERCLA tax revenue on remedial actions, 61
  strict accounting for rapid progress by SARA, 229
  Superfund Manual, 23
EPA. *See* Environmental Protection Agency
EPCRA. *See* Emergency Planning and Community Right to Know Act
Epidemiology, 24, 39–40, 304
Ergonomics, 116
Ethylene, 258
Ethylene oxide, 258
Ethylenethiourea (ETU), 18
ETU *See* Ethylenethiourea
European insurance, 131–132, 149–153, 162–163
  contract design changes, 299
  words for industrial liability policies, 145–146
Exclusion clause, 49–50, 69, 80
Exogenous factors, 4–7
Exposure assessment, 16, 17, 18
Exposure pathway, 25–27
  elements of, 24
Externality, 3

Fault-tree analysis, 258
Federal Emergency Planning and Community Right to Know Act (1986), 304
Federal Water Pollution Control Act (FWPCA), 64, 220
Fixation, 306
Flixboro, "explosive cloud" accident, 258

FMEA, 113
FMECA, 113
Food, Drug and Cosmetic Act, 16
Foreseeable misuse, 116
Foster and Chrostowski exposure model, 25–26
Free-market competition, 3, 4
French insurance pool, 131–132, 149–150, 162
French Public Liability Policy, 149
FWPCA. *See* Federal Water Pollution Control Act

GAO. *See* General Accounting Office
GARPOL in France, 131–132, 149–150, 162
Gasoline, lead additives, 232
General Accounting Office (GAO)
castigation of EPA, 102
report on disparity between EPA's and GAO's inventory of hazardous waste sites, 41–42, 43–44
Generalized dispersion models, surface and groundwater, 21–22
Genotoxic animal carcinogens, 39
German industrial liability policies, 145–146
Goiters, 38–39
Goitrogens, 38–39
Government changes wanted
bias towards action by PRP's, 262–263
more basic research in atmospheric chemistry, 262
more monitoring, 262
performance standards for entire country, 260
recycling permits less troublesome, 261–262
shared funding used more innovatively, 263
solutions to be decided by community, 260–261
Gradual pollution limitation
conflict avoided by European setups to cover pollution risks, 132
European industrial liability policies, 150–151, 152, 153, 154, 162–163
immaterial, 155

insurability criteria, 294, 296n.
insurance techniques for hazardous waste sites, 161
and Uncertainty criterion for insurability, 277
v. *sudden* and *accidental* limitation, 145–146
GRAS-levels, 115
Great Lakes Container Corporation v. *National Union Fire Insurance Company,* 49, 51
Groundwater contamination, 25–31, 35–36
jurisdiction controversy, 220–223
protection efforts, 99–100
protection guidelines used by Rohm and Haas, 252–254, 267
removal time-consuming and expensive, 64
Ground water modeling, 22

Harris poll, risk perception, 196
Hazard
assessment, 114, 123, 117–119
definition, 116
identification, 17, 24, 116–117, 123
Hazard catalog, 113, 116–117, 119
Hazard effect categories, 117–118, 119
and risk reduction, 121
Hazardous and Solid Waste Amendments of 1984 (HSWA), 6, 172–174, 190
Hazardous waste management, three sources of policy, 218
Hazardous wastes
generation annually in U.S. and in Switzerland, 106
legal and economic consequences, 303–305
Hazardous waste sites, human health injury probability, 39–40
Hazard Ranking System (HRS), 88, 91–92, 97
HAZOP method, 109, 113, 258
HEM. *See* Human exposure model
Heptachlor, 20
Hobson's dilemma, 58
Hoffman-LaRoche subsidiary, explosion, 5
Hooker Chemical Company, 225

INDEX

Hormonally mediated process, 38
H.R. 4813, federal cause-of-action for bodily injuries, 167
HRS. *See* Hazard Ranking System
HSWA. *See* Hazardous and Solid Waste Amendments (1984)
Human exposure model (HEM), 21
Hyatt Regency Hotel, collapse of walkway, 106, 107
Hydrocarbons
   emissions, 262
   volatile, 233
Hydrogeological techniques, 27
   used for Rohm and Haas groundwater protection guidelines, 253, 267
Hydrogeology models, 28, 29

IARC. *See* International Agency for Research on Cancer
IELA. *See* Insurance Environmental Litigation Association
Immission standards, 145–146
Incineration, 2, 10, 107, 175, 306
   commercial market required under TSCA regulation, 178
   community receptivity problem, 185
   and legislative regulation, 176
   preferred alternative to recycling, 256, 267
   toxic waste reduction by chemical companies, 285
Indemnification coverage for cleanup claims, 45–46
Industrial liability, prudent action program, 192
Ingestion, and potency value, 23
Ingestion exposure, 23, 25–26, 34
Inhalation exposure, 25–26, 34
Insurability; *see also* Insurance
   conceptual framework for analysis, 282–285
   cover limits, 137–138
   criteria, 134–136, 276–280
      environmental pollution scenario, 293–296
      identification of losses, 277–278
      low correlation between the individual exposures, 277

moral hazard and adverse selection, 279
      probability distribution of future losses to be measurable, 278
      uncertainty of events, 276–277
   decrease of, 135–136
   degree of, 135
   and environmental impairment, 147–148
   helpful exogenous factors for hazardous waste risk, 142
   impeded by endogenous components of hazardous waste risks, 136–139
   impeded by exogenous components of hazardous waste risks, 139–142
      compensation based on worry (moral hazard), 140
      difficulty to relate carcinogenicity to chemical waste, 141
      generous awards to personal victims, 139–140
      joint and several liability, 140–141
      litigation between a waste producer and his insurers, 141
      maximum possible loss unsatisfied if upper limit defined, 142
      misinterpretations of wordings of insurance contracts, 141
      sensitivity of population to environmental impairment (moral hazard), 141
   insurer requirements limited, 156
   preconditions for, 272–273
   primary parties' involvement, 156
   psychological security, 337
   public policy criterion and fairness, 137
Insurability grid, hazardous waste problems, 280–282
Insurance; *see also* Environmental-impairment liability insurance; Insurability
   availability, decrease of, 135–136
   claims-made forms, 282
   claims-made policy used, 285–286, 288
   deep pocket, 298
   definition, 271
   different layers of coverage provided against losses, 290–291
   economic role as a free-standing capital-

INDEX

Insurance (*cont'd*)
ist institution creating value, 275–276
equated with security, 337
European liability policies, 145–146
European setups to cover pollution risks, 131–132
for Superfund cleanup contractors on a case-by-case basis, 274
free-standing capitalist institutions that have to attract capital, 298
functions, 143, 271–272
incentives in risk management process, 284–285
insurers as guarantors of financial responsibility, 275
insurers as monitors of hazardous waste disposal activities, 274
insurers as regulators for generation, storage and disposal of hazardous wastes, 274
insurers viewed as having a deep pocket, 275
layers of coverage, 338
limited in providing coverage except for vertically integrated firms, 193
pollution risk, availability restricted, 133–134
proposal for hazardous waste, 289–293
as risk-spreading mechanism, 275, 276, 298
role in an integrated waste management strategy, 282–285
self-insurance, 299–300, 305
techniques used for hazardous waste sites, 161
three-tiered program, 193
transfer of risk, 143–148
underwriter conservatism requiring over-design of product for insurability, 129
underwriting program goal for pollution liability, 126–127
Insurance companies, solvency, 53–57
Insurance Environmental Litigation Association (IELA), 46
Insurance pooling, advantages, 288, 290–292
Integrated waste management concept (policy), 305–308, 311–313

communication challenges and potential problems, 349–350
definition, 343
definitions confusing, 314
economic pressure as an incentive, 313
industrial challenges and potential problems, 347–348
insurability challenges and potential problems, 346
legislative challenges and potential problems, 348–349
policy instruments for promotion of, 308–310
preventive counseling function, 307
risk assessment challenges and potential problems, 345
risk management challenges and potential problems, 346–347
solution involving maximum waste removal from landfills, 336
Interagency Regulatory Liaison Group (IRLG), 16
Interior, Department of, surface water pollution control, 220
International Agency for Research on Cancer (IARC), 17
In vitro studies, 24
In vivo studies, 24
IRLG. *See* Interagency Regulatory Liaison Group
Italian insurance pool, 131–132, 150–151, 162

Jackson Township case, 47–48, 51, 141
contract wording allowing awards to 300 claimants, 142
groundwater contamination by a landfill, 278
nonmonetary damages and insurability, 295
Joint and several liability doctrine, 7, 184, 189
imposed by judicial decisions, 274
judicial interpretation of CERCLA, 277, 278, 279, 296n.
personal injury cases, 304
possible modification proposed, 339
precondition to insurability, 301

INDEX

359

Journal of Risk Analysis, 19
Justice Department, imminent-hazard
suits, 226–227

KAB. *See* Keep America Beautiful, Inc.
Keene Corporation v. *INA,* 65–66
Keep America Beautiful, Inc. (KAB),
219, 220

Land disposal facilities. *See* Landfills
Landfills, 2, 10, 175
cleanup, 7–8
cleanup and coverage litigation possibilities, 47
EPA jurisdiction, 224–225
getting waste out of them, 335–336
leakage, 225–227
minimal use by Rohm and Haas, 254–
255
redundant controls, 98
subtitle D facilities of RCRA, 43–44
Lead additives in gasoline, 232
Legal dumping, 172–174
Legal perspective, hazardous waste management, 193
Legislative process
four assertions about how a group can
affect Congress, 318–321, 326–329
practical consequences, 321–326
Leukemia, 20
Liability, judicial rulings, 7
Liability life policies (LLP), 104, 132–133,
156–159, 160, 166
Liberty Mutual Fire Insurance Company,
47–48, 49
Life cycle, 116, 118, 123
Linear nonthreshold model, 20
Litter control, 220
LLP. *See* Liability life policies
Love Canal, 171–172
abandoned waste dumps as serious na-
tional problem, 5, 226, 230
risk perception, 197, 201
stigmatization, 204
Low-dose exposures, 124
Low-dose linearity, 38
Lump-sum payments, 157

Maintenance programs, 119, 123–124
Maintenance-type coverage, 70
Major-Accident Prevention Program
(MAPP), 258–259
Manufacturer, integrated into the analysis
team and operation of hazardous
waste facility, 124
MAPP. *See* Major-Accident Prevention
Program
Maryland Casualty Company, 45, 49
Marsh and McLennan, insurance pool, 292
Maryland Casualty Company, 49
MAS-Pool in the Netherlands, 131–132,
151–152, 162
Maximum Contaminant Level Goals
(MCLGs), 60, 61, 98
Maximum Contaminant Levels (MCLs),
60, 98
for drinking water, 234
MCLGs. *See* Maximum Contaminant
Level Goals
MCLs. *See* Maximum Contaminant Levels
MERITS, 86
Metal stabilization processes, 175
Methylene chloride, 20
Mobile treatment unit provisions, 186
Moral hazard, 70, 295
Multi-user, 280, 299

Named-perils concept, 132, 147, 150, 152,
154–155, 162–163
insurance techniques for hazardous
waste sites, 161
National Academy of Science, descriptive
terms for steps in the risk assessment process, 16–17
National Contingency Plan, 226, 230
National Priorities List (NPL), 41–44, 88,
91, 92, 97; *see also* Comprehensive
Environmental Response, Compensation, and Liability Information
System
approximately 1000 sites on list, 101
CERCLA liability and higher taxes, 60, 61
coverage litigation, 47
deletion, from the list, 95
published list of sites with a certain risk
level, 88

INDEX

National Union Fire Insurance Company, 49

Natural resource damage claims, 64

NEPACCO decision, 45, 51

New federalism, 308–309, 313

NIMBY syndrome (not in my back yard), 214

NOEL. *See* No-observed-effect levels

Nongenotoxic animal carcinogens, 39

Nonthreshold (carcinogenic) effects, 24

Nonthreshold process, 38

No-observed-effect levels (NOELs), 21

NPDES permit system for wastewater discharges, 261

NPL. *See* National Priorities List

Occurrence
  probability of, 118
  redefined by courts, 278

Occurrence policy, 285–286, 288

Octanol-water partition coefficients, 26

Office of Science and Technology Policy (OSTP), 17

Office of Technology Assessment (OTA), cleanup cost estimate for hazardous waste sites, 41–42

On-Scene Coordinators (OSCs), 89

Onshore oil spills, 226–227

Organization for Economic Cooperation and Development (OECD), 278

OSC. *See* On-Scene Coordinators

OSTP. *See* Office of Science and Technology Policy

OTA. *See* Office of Technology Assessment

Overall risk characterization, 17

Owned-property exclusion clause, 279

Oxidation, 306

Ozone, 262

Parens patriae claims for damage to natural resources, 64

Partisans, risk perception, 202

Pathways, 92
  and insurance provider's risk assessments, 126

PCBs, 109, 110

contamination of a dairy's cows (stigmatization), 203–204

presence in Japanese cooking oils, 107, 108

risk perception, 198

use eliminated gradually by Rohm and Haas, 258

Permit and manifest system, administered by EPA, 223, 224

Permit-by-rule program, 269

Personal injury statutes of limitation, 63

Pesticide Root Zone Model, 27

Pesticides, risk perception, 198

PHA, 113

Plume gerrymandering, 172

Pollutants, definition, 221

Polluter pays principle, 7

Polychlorinated biphenyls, 178

Polycyclic aromatic hydrocarbons, potency factors, 24

Polycyclic organic compounds, 20

Pooling arrangements, 104; *see also* European insurance

Pool Inquinamento in Italy, 131–132, 150–151, 162

Population growth, 322–323

Post facto changes, hazardous waste disposal, 245–246

Potentially responsible parties (PRPs), 14, 43–47, 51–52, 55, 58–61, 66, 262–263
  CERCLA liability attached to status, 71
  challenge of encouraging cleanup in an integrated waste management policy, 347–348
  cooperative efforts common, 66, 68
  to meet with insurers in a negotiating forum, 338, 339

Prevention costs, 106, 107

Probability-severity grid, 103, 114

Process hazards plan, 250

Product stewardship, 305

Property-casualty insurance industry, American, 51–52, 53–57

Property rights, 3

Propylene, 258

Protection level, 114, 119, 120–121, 123
  determined by the "Zurich" Hazard Analysis, 121

INDEX

risk reduction established by "Zurich" Hazard Analysis, 122

PRPs. *See* Potentially responsible parties

Prudent action, 246–248, 263, 265–266
guidelines, 264

Psychometric paradigm, 198

Psychometric studies, 198

Quinoform-type drugs, 109

Radiation exposure, 15, 234

RCRA. *See* Resource Conservation and Recovery Act

Reagan administration, political support for stringent regulation of landfills, 225

Record of Decision, 93

Recovery, 177, 306

Recycling, 186, 219
labeling of recyclable products for consumers, 242
made more accessible by waste disposal laws, 222
off-site, 306
on-site, 306
permits confusing, 261–262
and pollution, 127–128, 172
as raw material in another company, 261–262
of solid waste in Rohm and Haas environmental program, 256
and source reduction, 347

Remedial cleanup levels, 93–95

Remedial costs, 2

Remedial investigations/feasibility studies (RI/FS), 42–43

Remedial prioritization, 91–93

Remedial programs, 87–89, 91
preliminary assessment, 87
site investigation, 87

Removal cleanup levels, 90–91

Removal prioritization, 89–90

Removal response, 86–87

Reservation-of-right device, 67

Resource Conservation and Recovery Act (RCRA) (1976), 5, 6, 42, 98, 223

cradle-to-grave approach to the regulation of hazardous waste, 273
drum storage pad permits, 248
eliminating environmentally unacceptable practices, 179
emphasis on preventive hazardous waste management, 169
four functions, 189
government to handle large liability losses, 292
hazardous chemicals, 86
imminent hazard provisions, 226
insurance and financial responsibility requirements, 282
minimum performance standards for wastes, facilities, and generators, 181
1984 amendments setting rigid schedules for EPA to carry out land-disposal program, 227–228
1984 reauthorization, 172–174
not read by any Congressman before enactment, 222–223
permit system for solid waste disposal, 256
regulations leading to corporate auditing systems, 177
role in establishing market for preventive hazardous waste management, 170–172
stronger federal enforcement authority than from Clean Water Act, 222
subtitle D facilities, 43
and Toxic Substances Control Act related, 188
waste treatment preferred to land disposal, 184
wastewater stream regulation by EPA, 262

Responsibility, and insurance, 95–96

Retroactive strict liability, 7

Rework, 249

RI/FS. *See* Remedial investigations/feasibility studies

Right-to-know laws. *See* Emergency Planning and Community Right to Know Act

Risk
definition, 83, 105–106

INDEX

Risk *(cont'd)*
significant, 239
unacceptable, 246, 247, 248, 263, 266
definition, 248
guidelines, 265, 266
Risk analysis
higher-order impacts, 199, 200, 201
methods, 113
social amplification, 199–202
Risk assessment, 11, 13
at site-remediation level, 98
cost-benefit analysis, 218
definition, 23, 83, 85–86
designed for the users' purposes, 125
need for restraining, 183
pollution loss potentials, documentation
of, 125
process establishing national minimum
performance. standards, 181
process limitations, 170
quantitative, 16
risk comparison over time, 31
site-specific factors, 181
steps in process, 15
to delineate differences between firms
considering insurance pooling,
291
to improve the effectiveness and functioning of hazardous waste programs, 241–243
underwriting program goal, 126–127
"Risk Assessment and Risk Management
Strategies for Hazardous Waste
Storage and Disposal Problems,"
xv, 3
Risk communication, 207, 208–210, 212–
213, 215–216
definition, 84
eight rules, 209–210, 215
objective for, 214
role, 9
Risk comparisons, 207–210
and chemical risks, 208
Risk engineering, 144–145
Risk management, 11, 161
cost included, 85
definition, 83–84
factors, 85
major elements, 85

regulations and joint and several liability
standard, 177
societal factors, 84
Risk perception, 195–199, 200
related to risk communication, 207,
212–213, 216
Risk premiums
cause-effect connection uncertain, 137
reduction steps, 138–139
safety loading or *fluctuation loading* contained, 136–137
Risk profile, 114, 121
grid, 119, 120
Risk reduction, 119, 120–123, 268
and corporate risk management, 305
with reference to the protection level by
"Zurich" Hazard Analysis, 122
Risk transfer, 143–148
Rohm and Haas Company
environmental control program, 245–259
management system and quality programs rather than remediation, 249

Safe Drinking Water Act (1974), 60, 98,
221
hazards of abandoned waste dumps
brought to suit, 226
1986 amendments increasing number of
water quality standards, 230
Safety, Health and Environmental Policy
Statement, 263–265
Safety organization, and risk reduction,
123
Sandoz/Swiss spill, 258
SARA. *See* Superfund Amendments and
Reauthorization Act
Schweizerhalle fire in 1986, 106, 107,
109,110
Scope definition, 114–115, 116
Scrap processors, and pollution, 127–128
Security Insurance Group, 49
Senate Committee on Environment and
Public Works, 60
Senate Environment and Public Works
Subcommittee on Superfund and
Environmental Oversight, 60, 61
Severity, 19
Seveso explosion (Italy), 5, 106, 107, 109

INDEX 363

dioxin contamination, 204
safe destruction of waste done by local
government, 119
Sewage treatment plants as source of toxic
air pollution, 233
Shell Oil Company
coverage costs at waste sites, 46
waste cleanup coverage case, 45
Shower exposure, 23
Sierra Club, 219
Silent Spring, 4
Site-specific parameters, 21
Solid Waste Disposal Act, 223
Solid Waste Disposal Act Amendments,
222
Solid waste disposal program
Rohm and Haas Company, 254–257, 267
destruction or biotreatment, 256–257,
267
elimination and reduction, 255–256
landfilling, 257, 267
recycling of solid waste, 256, 267
Solvent, 306
Source reduction of waste, 6, 144, 184,
188, 305–310, 347
incentives, 309–310
Spillover. *See* Externality
States
waste disposal regulation laws, 222
water pollution regulation, 221
Stigma
environmental, 202–207
social, key dimensions, 203
Strict liability, 177
Subtitle D regulation, 43–44, 99
Sudden and accidental limitation, 80
CGL pollution exclusion, 48, 49–50, 69,
71
conflict avoided by European setups to
cover pollution risks, 132
European industrial liability policies,
149, 150–151, 152, 153, 154, 162–
163
immaterial, 155
insurability criteria, 294, 296n.
insurance techniques for hazardous
waste sites, 161
interpretation in the Jackson Township
case, 141

and uncertainty criteria for insurability,
277
v. *gradual* limitation, 145–146
Summit Associates, Inc. v. *Liberty Mutual
Fire Insurance Company,* 47–48,
49–50, 51, 287
Superfund. *See* Comprehensive Environmental Response, Compensation,
and Liability Act
Superfund Amendments and Reauthorization Act (SARA) (1986), 6–7, 42,
44, 52, 63
authorized adequate insurance in the
market, 274
average cleanup cost per hazardous site
on NPL, 101–102
capacity planning required through its
new requirements, 187
cleanup program authorization expanded, 229–230
Congress won't admit its adoption was
wrong, 78
coverage issues not resolved through
amendment, 60, 61
emission and effluent data risk perception, 215
EPA held to a strict accounting for
rapid progress, 229
Title III and systematic risk and consequence analysis, 259
Title III implementation, 350
worst-case scenarios and potential sites,
166
Survival, 19
Swedish insurance pool, 131–132, 152–
153, 162–163
Swiss industrial liability policies, 145
System safety, 110–113

TCE. *See* Trichloroethylene
Technology controls system, 267
Technology transfer
horizontal, 312
vertical, 312
Tetrachloroethylene, 29
Thalidomide, 107, 108
3M Corporation, waste minimization,
129

INDEX

Three Mile Island (TMI), risk perception, 197, 201
Threshold (noncarcinogenic) effects, 24
Threshold pollutants, 21
Times Beach (Missouri), contamination and stigmatization, 201, 204
Toxic tort
  cases, 62–64
  claims, emotional distress and increased risk of disease basis, 304
  law, 9, 141, 292, 293, 304, 336
  system, 128, 288–289, 301, 302
Toxic Substance Control Act, 86, 188
Trade associations
  for building bridges between large and small companies, 269
  for distributing knowledge to lesser or smaller users, 314
Transaction costs, reduction of, 79
Trash regulation, 218–219, 220
Treatability, 19
Treatment industry, 176
Treatment residuals, 186
Treatment, storage and disposal facilities (TSDFs), 6
  RCRA third-party liability requirements, 296n.
Trichloroethylene (TCE), 23
  contaminant of water, 28, 29–30, 35
Triple trigger theory, 65–66
TSDFs. *See* Treatment, storage, and disposal facilities
Tumor dose-response data, 20

Uncertainty as a hindrance, 334–335
Underground tanks, 227, 254
Union Carbide incident, Bhopal, 106, 107, 109, 197, 201, 258
U.S. v. *Chem-Dyne,* 296n.
United States v. *Waste Industries Inc.,* 141–142
United Technologies Corporation (UTC), 46
Unlawful vs. lawful behavior, 146–147
  European liability, 152, 153–154, 162–163
  insurance techniques for hazardous waste sites, 161

and Swedish liability, 152, 153
Unpredictability, 136
Upper bound, 39
UST regulation, 99

Vertical horizontal spread (VHS) model, 21–22, 28
Vertically integrated firms, 280, 299, 300, 305
VHS. *See* Vertical horizontal spread model
Vinyl chloride, 23, 29–30, 35, 234
Voluntary cleanup, 186

Waste; *see also* Recycling; Source reduction of waste
  avoidance, 144, 161
  constructional and organizational approaches to sites, 144–145
  hazardous, kept out of ordinary landfills by RCRA mandate, 225
  hazardous to human health and environment, 1
  improved communication about associated risks, 192
  legal perspective on management, 193
  management
    Congressional Budget Office study, 2
    decision making done by Board of Directors, 180
    sector, 334
  minimization, 128, 177, 180, 184, 305
    and integrated waste management concept, 306–307
    program at Rohm and Haas Company, 250, 257
    voluntary company agreements, 339
  old vs. new, treatment and disposal, 187–188
  production
    annual rate in U.S., 1
    annual rate per capita in U.S., 1
  reduction, generated at the source, 306
  as resource, 179
  sites
    eligibility, 154–155
    incentives for operation, 159–160

INDEX

legislative regulation of, 174–177
site-specific adjustment of standards
preferred, 190
as social cost, 179
suppression, 144, 184
Waste Industries Inc., 141–142
Waste treatment centers, hazardous release and financial responsibility, 8
Wastewater treatment plants, construction, 107
Water guidelines program, Rohm and Haas Company, 251–252, 266–267
Weight-of-evidence, 17, 18
Westinghouse Electric Corporation, coverage of cleanup costs, 46
WHO. *See* World Health Organization
World Health Organization (WHO), 109
asbestos found to be carcinogenic, 112

Zero bonds, institution of, 159
Zero contamination, 322
Zero discharge, 98
Zero-risk level, 16, 119–120, 233–235, 241, 261
of environmental laws, 322, 323, 325, 326
information needed to help us move toward it, 339–340
"Zurich" Hazard Analysis, 113–114, 116, 117, 118, 119, 123
design phase sensitive to long-run impacts, 129–130
predictability, 168
protection level, 121
risk profile grid, 120
risk reduction with reference to protection level, 122